# Global Forensic Cultures

# Global Forensic Cultures

Making Fact and Justice in the Modern Era

Edited by
IAN BURNEY and CHRISTOPHER HAMLIN

Johns Hopkins University Press
*Baltimore*

Johns Hopkins University Press
2715 North Charles Street
Baltimore, Maryland 21218-4363
www.press.jhu.edu

Library of Congress Cataloging-in-Publication Data

Names: Burney, Ian A., editor. | Hamlin, Christopher, 1951– editor.
Title: Global forensic cultures : making fact and justice in the modern era /
    edited by Ian Burney and Christopher Hamlin.
Description: Baltimore, Maryland : Johns Hopkins University Press, 2019. |
    Includes bibliographical references and index.
Identifiers: LCCN 2018032699 | ISBN 9781421427492 (hardcover : alk. paper) |
    ISBN 9781421427508 (electronic) | ISBN 1421427494 (hardcover : alk.
    paper) | ISBN 1421427508 (electronic)
Subjects: LCSH: Forensic sciences—History.
Classification: LCC HV8073 .L724 2019 | DDC 363.2509—dc23
LC record available at https://lccn.loc.gov/2018032699

A catalog record for this book is available from the British Library.

*Special discounts are available for bulk purchases of this book.*
*For more information, please contact Special Sales at 410-516-6936*
*or specialsales@press.jhu.edu.*

Johns Hopkins University Press uses environmentally friendly book materials,
including recycled text paper that is composed of at least 30 percent post-
consumer waste, whenever possible.

# CONTENTS

# ACKNOWLEDGMENTS

This book began in presentations and discussions at the Locating Forensic Cultures Symposium, held in July 2015 at the University of Notre Dame London Gateway. Thanks are due to its sponsors: the Wellcome Trust and the Centre for the History of Science, Technology and Medicine at Manchester University as well as the Department of History, Center for Scholarship in the Liberal Arts, and the Reilly Center for Science, Technology, and Values at the University of Notre Dame. We thank Elizabeth Kuhn for coordinating the symposium, and Kat Hamlin for editing and indexing. At Johns Hopkins University Press, we thank Matt McAdam, commissioning editor, and Ashleigh McKown, freelance copyeditor, for their excellent work, as well as other members of the staff for their encouragement, professionalism, and hard work.

# Global Forensic Cultures

# Forensic Facts, the Guts of Rights

CHRISTOPHER HAMLIN

What's a forensic culture? You live in one.

Consider the phrase "Just the facts, ma'am." To Americans of a certain age and to partakers of popular culture, it recalls the fictional Sgt. Joe Friday of the Los Angeles Police Department, a crusty detective in the many versions of the radio and later television program *Dragnet*. The program began in 1949, long before the days of *CSI*, and has spawned remakes and parodies into the new century. That Friday, long played by its creator Jack Webb, never says these exact words makes their iconic status the more remarkable.[1]

Joe Friday is a paradigm and in some ways a progenitor of a familiar theme of contemporary popular culture, a faith that the best security is forensic security. Persons may lie, but the crime scene investigation (CSI) experts will make the physical facts—the "silent witness" of DNA and other traces—speak the truth.[2] While it is easy to associate the rise of the forensic fetish with the O.J. Simpson trial in 1995 or the gradual emergence of rapid DNA-sequencing techniques, it is worth considering it also as a post–Cold War phenomenon.[3]

In Friday's procedural view of detecting, there is no room for "culture"— exploration of the meaning worlds in which others live is distraction. Unlike his likeable and well-rounded partner, family man Bill Gannon, Friday, with his unsmiling gravelly monotone, seems gratuitously misanthropic. His protestation, often to victims, that he wants only whatever facts they can supply and not their stories, interpretations, or speculations, sets up expectations of forensic deliverance. That orientation is evident in the title itself. The criminal, like the fish the medieval Scandinavians (who coined the term "dragnet") caught in a net covering the breadth of a channel and dragged along it, cannot withstand technology—Friday and his colleagues will sweep up all, to expose truth—and guilty persons, too.[4]

While Friday is not a forensic specialist, forensic techniques do often play a role in *Dragnet* plots, and it is made clear that the routine investigations he undertakes will be reliable because they are so narrowly focused, skeptical, and objective—precisely the characteristics being ascribed to science in the early Cold War. And yet this facticity is to have a cultural clout—for, having discounted it in others, Friday does claim authority to interpret and moralize. As in the famous quip, his paternalism is sometimes directed at women but may be applied to whomever is challenging the fact-based social order, whether by speculations, embellished accounts, or naïve expectations. Security, he suggests, will come from properly positivistic and unemotional (white) men who, through their very disregard of all that is "cultural," will keep their Supermanian commitment to "truth, justice, and the American way." How very different from an earlier Los Angeles detective, similarly laconic and alienated. Sam Spade likes facts but investigates in no systematic way. It is understood that he will be part of his stories, not outside them.

References to "deliverance" or to faith invested in science are not facile. Nor is comparing the contemporary forensic fetish with the profound yet ambivalent status of the physical sciences in the early Cold War. Both guaranteed security through connecting a *cosmos*, a nature, with a *nomos*, the social ordering through which we respond to it. Those connections make clear that forensic investigation is not the alternative to culture but cultural in the strongest sense, that of cosmology.[5]

An exploration of the differences between CSI and Cold War science is no less illuminating. As well as enjoying the mystique of their profession, many Cold War physicists reveled in their roles as "defense intellectuals." With notable exceptions, the contemporary cult of forensics appeals to anonymity, universality, and routine. In the Cold War, the concern was with *national* security; in the cult of forensics, it is with the constant and worldwide security of each. The strangeness of places and persons need not threaten if we trust that the most authoritative elements of justice are in place everywhere in the world. Cold War security concerns united us as potential victims of thermonuclear bombs. By contrast, contemporary forensics individualizes us. And it offers each of us security from (and relies on our fear of) opposing Hitchcockian nightmares: either we are victim of a criminal so devious as to elude ordinary means of discovery or we are falsely accused, caught in suggestive circumstances by mere plodding humans, and needing superhuman means of extrication. A single solution dispels both. Peace of mind, along with a millennium of liberty and fairness, will come through techniques of

"God's eye" transparency, which reveal all that has happened. Increasingly, as these technologies supplant prejudiced or corrupt persons, justice will become ever more perfect; the worrisome randomness that comes from clever advocates or whimsical judges and juries, as well as from simple ignorance of the facts, will end.

The exaggerated attention modern publics have given to it does not mean that forensics itself—a composite of techniques and of professionals who use them within a variety of institutions of policing, prosecution, and jurisprudence—is cultural in any important way. Yet contemporary authorities worry that the faith invested is not warranted. "If all you knew about forensic science was what you saw on television, you might . . . [believe] that only the most sophisticated and well-researched scientific evidence is used to solve and prove crimes," writes Erin Murphy of the New York University School of Law: "reality is different."[6] Murphy's comments follow a series of critical public inquiries into the quality of forensic protocols and practices, of which the 2009 NRC Report (by the National Research Council / National Academy of Sciences) *Strengthening Forensic Science in the United States* was the most comprehensive.[7] Among its concerns was that a blind faith in forensic deliverance was changing jurisprudence in disturbing ways. The so-called CSI effect referred to expectations by judges, juries, and the public that a case would be presented mainly as forensic evidence and that such evidence would be authoritative. If so, popular culture might be shifting jurisprudential practices through a reciprocal process in which faith in forensics begat more forensics, with techniques, professions, and institutions being then both product and producer of culture. At worst, faith in forensic deliverance might warrant a cynical practice in which gestures of forensic authority became simply the smoke and mirrors used to obtain convictions, a perspective recently endorsed, in the view of some critics, by Attorney General Jeff Sessions.[8] "Juries want 'science'? Give them science"—precisely the strategy commonly attributed to witchfinders.

The chapters in this book recognize the reciprocity of forensic cultures and practices as norm rather than aberration. For it has made a great deal of difference where—in time, space, society, and culture—one interacts with forensic institutions (as well as who one is). When we "locate" forensic practices—the title of the project from which this book arose—we find them everywhere rooted in what is unambiguously cultural: with ideas of what kinds of persons the world includes; of what dangers it and they pose; and, at an even deeper level, of why things happen.

But the book recognizes, too, that much of the contestation in the present has to do with a presumed "scientificity." For one may argue that the public's recent fascination with forensic science reflects a sea change. We now reap the benefits of real achievements in chemistry, physiology, and anatomy, and in new instruments for the recording and analyzing events. Perhaps the most profound forensic instrument has been the camera itself. It stands for all those forms of "lenses" that make events visible, revisitable, and amplifiable. But while the witchfinders analogy certainly misleads, most of us have far too little appreciation of the ways forensic expertise is created and used.

I write as a relic of the *Dragnet* era and a survivor of the Cold War. Early in my career as a historian of science, I was struck by the enormity of the impact of forensic contexts on science. Water analysis in nineteenth-century Britain defied common understandings of applied science. There, science did not authorize; rather, the need for authority shaped the science, something that struck me as utterly odd.[9] But not only was there no conception of forensic cultures four decades ago, forensic practices were largely invisible. In the popular culture of the crime novel, the chief forensic reagents were Miss Marple's tea and sympathy and up-to-date British Railways timetables.[10] Among historians of science and medicine, there was still little sense that the contexts in which epistemic determinations were made differed in interesting ways across time and space, much less that the system of adversarial jurisprudence might be an important site of that variability, particularly where those determinations concerned matters of justice, public policy, and accountability.

Only some decades later do we have some sense of the profound implications of forensic cultures, thanks to the initiatives of such scholars as Ian Burney, Simon Cole, Tal Golan, Sheila Jasanoff, Michael Lynch, and many others. And yet there remains too little appreciation of their evolution, variation in different settings, and role in governing as well as in detecting. Any cruise through a law library or a science library will probably disappoint; there has been little concerted attention to historical and interdisciplinary assessment of forensics. The focus instead is the practices. In present mass culture, forensics remains a narcotic anodyne, antithesis of ideology, and final frontier of fairness.[11]

This book is not a comprehensive critical cultural history of forensic practices. Each author claims expertise over only a small fraction of the ideal territory of such an enterprise—which might lie at the intersection of multiple strands of history and anthropology, as well as of law, ethics, philosophy, and sociology. The chapters are exemplars. They immerse us in the details of unique forensic cultures—we follow experts going about their business of detection, making their careers, and promulgating their methods to the state, the courts, the public, and to their colleagues and successors. But the cases do allow us to raise critical and comparative questions about these institutions that have so largely escaped critical scrutiny. In the next section, I review some of the most general. The fact-fixation of contemporary American culture suggests one cultural dimension; below I outline four others: the philosophy of forensic science, its accreditation as testimony, its relation to institutions of jurisprudence, and its broader role in government.[12] Together these indicate how far rights have been translated into techniques. Legal systems have varied across time and space; in what follows, I use American (and more broadly Anglo-American) institutions as the departure point, using other traditions for apposite comparisons.

## Finding "Silent Witnesses": Facts Become Traces

A good start is the evolution of the term *forensic* itself. Initially, it referred to matters of the forum. This is reflected in both its major uses—the arts of debate and the application of specialized knowledge to public matters. Broadly construed, what from 1899 has been called "forensic science" includes and extends much of what was earlier known as "legal medicine" or "medical jurisprudence," recognizable as a domain of expertise from the sixteenth century. Its realm is not limited to matters of criminal investigation: terms like "crime scene investigation" or "criminalistics" capture only parts of it.[13]

It is conspicuous, however, in answering what have been called questions of "silent witness," questions of what happened when no one (at least no neutral party) was there to see.[14] In his 1953 *Crime Investigation*, the pioneering practitioner/philosopher of American forensic science Paul Leland Kirk (1902–70) (whom we will meet later in another context) extolled its virtues: "This is evidence that does not forget. It is not confused by the excitement of the moment. It is not absent because human witnesses are. *It is factual evidence.* Physical evidence cannot be wrong; it cannot perjure itself; it cannot be wholly absent." To recover an event through its residues was a soluble problem: if "all the resources of science and human understanding" were brought to bear, "the message . . . [would be] clear, complete, and unequivocal."[15]

Contemporary forensic "cultures" do indeed reflect "cultures" of science, and yet it is as a science, as Simon Cole and Paul Roberts point out, that forensics is unique, problematic, and contested.[16] For it is more than a sphere of application; it raises questions about how science is done, who counts as a scientist, and more generally about what we take science to be.

The problems of silent witness are not new. Often they will have been dealt with informally, rather than by reliance on some designated institution, profession, or set of procedures. Though it would overstate to call them forensic scientists, some persons in a community would have had specialized knowledge, gained from long memories of its characters and conflicts or local knowledge of the hazards of a place and of ways of living.[17] A coroner's jury was, in its early days, no random sample of the public but a panel of local experts who would know what was plausible and who was credible.[18] We should not romanticize—ignorance and venality exist in face-to-face communities.

By contrast with such local and lay expertise, contemporary forensics, as evidenced in the aforementioned "CSI effect," represents a faith in the principles of ultimate knowability and universal means of knowing. The approach is especially attractive in mass societies—where dangerous persons are no longer known neighbors but anonymous beings who flit from place to place. Findings are to be considered within well-developed theoretical frameworks linked to expansive and well-organized data sets. Conclusions will be representable either as necessary consequences or as reliable products of appropriate analytical or statistical protocols. The composite conclusions will require integration of specialized subdisciplines. Though someone will have to put all the facts together, most professionals will be concerned with the extraction/examination of particular phenomena.[19] Yet the assumption of transparency prevails. An expert's objective in testifying, notes Kirk, will be to help the juror "*see* the basis of the conclusion as clearly as the investigator saw it." Achieve this and "argument" will be unnecessary.[20] One senses Sgt. Friday's impatience: facts, seemingly untouched by human hands, obviate the need for talkative time-wasting witnesses with their complicated tales.

What is displacing the tangled skein of words, however, is not merely facts but traces. Consider the central assumption in modern forensics: the "exchange principle." Over the course of the last century, the casual suggestion of the early twentieth-century French forensic scientist Edmond Locard that at every crime site the criminal takes something away and leaves something behind has evolved from supposition into generalization, maxim, and axiom. It is sometimes presented as the most general guarantor of the scientificity of forensic sciences, a functional equivalent (translated from force to matter) of Newton's third law that every action produces an equal and opposite reaction, and, as axiom, will not be subject to empirical falsification.[21] Exchanges will have occurred. Failure to find them is a failure merely of practice; what facts we do find are the presumptive traces of that exchange.

The status of Locard's principle exemplifies much of what is contentious about the forensic sciences. It seems commonsensical. But does the principle itself do any actual work? Unlike theory-centric physics, concerned with producing ever more inclusive general statements, forensics is concerned with practices useful in reconstructing sequences of contingent events. In that setting, the exchange principle arguably does no work, or no good work. Whether or not a crime scene investigator endorses Locard's principle does not affect what he or she does or does not find, which may or may not be traces of a crime. But if it does any work, Locard's principle will bias interpretation. Since evidence *must* exist, what is found will be evidence. While Kirk represents exchange as did Locard, simply as a likely consequence of criminal activity, he nevertheless assumes that investigators can unproblematically separate the significant traces of the crime from the mass of "accidental" circumstances that have nothing to do with it.[22] Locard's principle does much dubious work in popular culture as well. Overwhelmingly, the plots of the CSI dramas are expositions of Locardian axiomatics. There its status is not epistemic but exhortatory. If we have failed thus far to link criminal to crime, look again. Use new instruments. Reinterpret the evidence. But never doubt that there is a recoverable signal among the noise. We must find it, for psychopaths walk among us.

At the heart of the question of the scientificity both of the individual forensic sciences and of forensic science (for, as I suggest below, different problems are at stake) is the view that there is one proper way of being scientific, a view nurtured by the myth of a single "scientific method." The 2009 NRC Report, a response to multiple crises about the foundations of several well-accepted forensic techniques, represented forensic science properly so-called as a theory-testing endeavor, and in doing so sought to highlight commonalities with exemplary forms of science and with a canonical scientific method.[23]

In many respects, however, the trajectory of modern forensic science (here in its unitary sense, distinct from the several individual forensic techniques) is anomalous and even opposed to the trajectory of many other sciences. In two respects, it resembles practices of reasoning and concerns common to the premodern era that were jettisoned from science in the Enlightenment. First, it was concerned with moral accountability, and second, with seeing events individually and as instances of intent: thus its similarity to that premodern world in which the lightning strike came from God's finger and the cow's milk failed from the witch's spell. By contrast, post-Enlightenment science would represent physical or social events as accidents, composite consequences of general laws. However impactful on one's life an event might be, it could be understood in actuarial terms by means of the law of normal distribution. Ironically, even the lightning strike, though still an "act of God" in legal settings, is seen in precisely these terms—lighting must strike somewhere; divine intent is moot.[24]

The law, however, remains the last bastion not only of culpability but also of uniqueness. A decade after *Crime Investigation*, Kirk, struggling to characterize the uniqueness of a science of the unique, declared forensic science to be a *science of individualization*, a phrase that was both more general and more accurate than *identity*, since demonstrating uniqueness—that something happened in this way and not any other—was the criminalist's goal. His declaration came in a discussion of the "ontogeny of criminalistics," an exploration of the distinctiveness of the enterprise. Recognizing that forensic practice involved integrating multiple ways of knowing with regard to singular events, he acknowledged that it might alternatively be understood as a profession or even a skilled occupation as well as a science.[25]

Here it is important to articulate (and to resist the common conflation of) two kinds of forensic science. Kirk called them problems of people and problems of things, though we might call them integrative and analytical.[26] The latter is the chief concern of the NRC: the validity of particular forensic techniques and their competent use. Here the issues are those of variability within populations of similar objects. The determinants of that variability and the capacity of analytical techniques (and analysts) to distinguish same from different are problems common to any analytical science. The former is an archeological problem of reconstructing sequences of events through the reduction of all possible scenarios of what might have happened to what did happen.[27] It involves a broadly Bayesian problem as one seeks reasons to "shrink to smaller and smaller numbers" the population of possibilities, though probabilistic reasoning will always be asymptotic, Kirk admits. One cannot claim "absolute identity," he asserts, but one can claim that "for all practical purposes" the population of scenarios is reduced to one.[28] While integration will depend on analytics, it would be naive to think of it simply as the sum or product of the several forensic techniques. Not only may the variables, presumably many and interacting, not be fully known, but also the individuation claim—that it could not have happened otherwise—involves us in proving a negative.

## Letting Silent Witnesses Talk: Traces Become Testimonies

I surely know. Can I be an expert? How will I be received at the courtroom door? Inasmuch as finding facts is what forensic scientists do, it may seem odd that in legal contexts, "finder of fact" refers to juries (sometimes judges). Moreover, no matter what an "expert" may (claim to) know, what (if anything) that expert can say in a legal setting is another matter, as is the significance those statements will have.

That courts should hear so-called forensic experts at all is not obvious. They form a unique class of witnesses. In ordinary legal settings, truth making in law relies simply on the competence and honesty of witnesses with relevant testimony. Competence is common perceptual, cognitive, and linguistic functioning; honesty is secured by swearing before a God who watches out for truth. (The cosmic heavy-handedness was necessary—if perjury could be easily exposed, there was probably no need for the witness's testimony in the first place.)

But how can someone who did not witness become a witness? In charting the origins of expert witnessing in Anglo-American law, Tal Golan described the grudging accommodation in eighteenth-century English courts of a select few who had not seen the events in question, their testimony becoming admissible because they could deduce consequences of natural laws or owing to their great experience with similar matters.[29] One function of medical licensing had been to accredit such expertise: someone like a midwife, in command of a "science" whose application was too complicated to evaluate in a particular case, might be granted unique status as an "authority."

But while the inclusion of that person's expertise still (ironically) involves swearing-in, views and practices vary as to what can be sworn to. Behind these controversies lies Kirk's puzzlement about how forensic specialists should understand themselves: as scientists, members of a profession, or skilled technicians? While these may seem aspects of careers rather than alternative identities, they bring differing ethics and expectations. "Occupation" serves well for matters of fact. In the same manner as any other witness, a forensic technician can swear to having carried out particular operations on particular samples and to the results of those operations. How much to trust to those results and how to integrate them into a composite conclusion are essential characteristics of a professional's judgment, a matter of ineffable experience irreducible to a set of singular demonstrations. In either case, issues of how much certainty to claim will arise.[30] But even when they are shown to have occurred, errors of forensic judgment will rarely be seen as incompatible with good faith or as subject to perjury.

As for scientificity, Kirk confessed that there might be no single scientific framework that could integrate the several kinds of forensic evidence obtained: here forensics was fundamentally unlike physics. But even with regard to the individual techniques, problems have arisen. Sometimes they arise because the techniques are unique to forensic contexts; in other cases, the adaptation of common analytical techniques to the answering of forensic questions has strained the connection with the empirical and theoretical foundations of the more general sciences. That anatomists or developmental biologists have paid little attention to the uniqueness of hair or the generation of fingerprints—comparative domains of recent concern—is less the problem than that these forms of expertise then become subject to the critique of self-referentiality. While usually the integration of multiple forms of evidence will invite cross-calibration, problems may arise wherever highly technical skills rely on internal means of confirmation and credentialing.[31]

Finally, one great guarantor of scientific knowledge is tentativeness—that it is subject to criticism and revision. As an institution, science pursues (well-warranted) novelty; it is transgressive of itself. Institutions of jurisprudence put a premium on stability, manifested in procedures of due process. A century's use of fingerprinting is presented as proof of its validity rather than, as Cole points out, simply a record of its application in an institution that makes a virtue of inertia.[32] (In this light, Kirk's endorsement is striking: "the almost universal modern use of fingerprints for personal identification *testifies* to their reliability and utility").[33] But resistance to reopening cases on the grounds of evidence obtained through new forensic techniques is understandable if fairness is more important than the finding of fact in particular cases. For what is allowed in one case must be allowed in all. A science that opens new possibilities will be fertile with new forms of argument, but how often must the same cases be retried?

These concerns are evident in efforts of courts to reassert control over science—as in the controversial "Daubert" rules used in some US courts. In their emphasis on established procedures with known error rates, these rules have, in the views of some, opened the door to much-needed validation studies. Yet others complain that they block novelty. Many of the approaches described in subsequent chapters might have been subject to Daubert challenges: we invite readers to consider what the results should have been.[34]

## Trusting Silent Witnesses: Traces Become Truths

The Daubert guidelines concern trust in science. But what of the individual experts who translate for the silent witnesses in particular cases? The familiar clash of opposing experts on matters of fact and of interpretation suggests the trickiness of that translation.[35] Past decades have seen concern that delivery of forensic certainty is troubled, compromised even at high levels (e.g., the Federal Bureau of Investigation) by utilization of undervalidated methods and incompetent practice, but also by confusion about what kinds of claims an expert should make and how those claims should intersect with institutional structures. The recent spate of DNA-based exonerations may seem to confirm faith in forensic techniques, yet the convictions now overturned were often themselves secured by the prevailing forensic science of the day.[36]

The gravest structural problem concerns the creation of forensic evidence and its application in trial. In much of the world, forensic evidence is commissioned and presented by a united police-prosecutorial service. It may form the crux of its case. But how close should the relation be? Should it be conceived as part of a prosecution case, as it is the prosecution that will orchestrate its presentation? Or is it independent, only incidentally a component of a prosecution case? Exacerbating matters is the equation of legal truth making with winning the sentiments of a nonexpert jury. An attorney will decide what facts to present and how, and may regard some avenues of physical inquiry as superfluous, as complicating a simple picture. One finds, then, in popular culture and in practitioners' reflections, both an easy conflation of fact presentation with case making and a frustration with the abilities of clever debaters to obscure what should be the concern of the trial: reconstruction of what happened (a physical event manifest in physical facts).

That ambivalence is apparent in the creation of forensic evidence. As suggested above, forensic investigation should not be seen as naive empiricism—Sgt. Friday's "just the facts" begs the question of "which facts?" Hypothesis making and case building cannot be postponed until all the facts are in; rather, fact collection and evaluation will more often be iterative and entangled with case building. In such cases, a simple suspicion or an *idée fixe* may become a self-fulfilling prophecy. Defense advocates, rarely able to influence this process, especially in its earliest stages, may challenge both the findings themselves and the path dependence of inquiry—the hunches or hypotheses that concentrate attention on some possibilities to the exclusion of others. But even if finances allow, they will rarely be able to mount a full-scale countervailing inquiry simply because forensic traces rarely last long enough.[37]

To Kirk, a close and "reciprocal" relationship between forensic analysts and the police was a means to reach closure quickly and cheaply. To the NRC committee, however, inadequate insulation between analysis and case building was worrisome. It noted the insidious tendency toward "'anchoring,' . . . to rely too heavily on one piece of information" (often an early finding) and cited studies finding confirmation bias in feature comparison tests: analysts will unknowingly seek to support the organizations of which they are a part.[38] But ambiguity about the expert's role in the courtroom was deeply seated: Kirk urged experts preparing to testify to "be factual, not argumentative" but "to develop the evidence so that it *presents a case without a real loophole.*" Cross-examination of expert testimony did "not serve the ends of justice"; its sole purpose was to "cast doubt." He was little happier with direct examination; counsel was an unneeded filter in what was to be a relationship of transparency—"no argument is ever necessary if the juror can see the basis of the conclusion as clearly as the investigator saw it." Advocacy, he suggests, is properly subordinate to fact, necessary only where matters are uncertain. The facts are partisan; the unbiased silent witness *has seen.*[39]

What Kirk has represented is a hybrid system of jurisprudence that is at war with itself. Adversarial institutions that rely on persuasion of lay juries by skilled advocates are not ideal institutions for expert assessment of physical facts and determinations of the probability of various scenarios. That state of affairs is a product of the evolutionary history of legal institutions. The maladaptation is most acute in Anglo-American criminal procedure, yet, as partly an extralegal issue, its significance has been under-recognized until recently. It underlies concerns about conflicts of interest, being expressed by prominent forensic scientists like Max Houck, and the criticisms of Attorney General Jeff Sessions, a former prosecutor who has challenged the need to scrutinize the validity of forensic methods that have worked well in securing convictions. For, as Simon Cole points in the afterword to this volume, all this history is still very much with us.[40]

The problem is one of roles—of judge, jury, experts, and advocates. The sort of integrative criminologist Kirk imagines himself to be works with police and prosecutor to infer a best explanation from the composite of physical evidence and apply it to the case that is being created. But, notably, the best-known pioneers of this integrative criminalistics, Hans Gross and Edmond Locard, operated in continental, nonadversarial, or quasi-"inquisitorial" legal systems, which were the original context of the development of forensic medicine in early modern Europe.[41] In such a system, the judge's role was not merely to preside over a case but to investigate it, calling in whatever technical expertise was necessary (thus Gross was an assistant magistrate). No trial or resolving event would occur until the judge had determined the facts according to high and explicit standards of proof. Fact having been found, the purpose of the trial would be to decide what was to be done—any advocacy would focus on mitigation. Such systems recognize a disjunct between justice and the arts of partisanship: winning cases does not register, for finding fact is not a game.

In English jurisprudence, a recognizable system of adversary procedure, characterized by John H. Langbein as "lawyerization," only took shape in the eighteenth century, with the continental institution of medical jurisprudence—and, separately, expert witnessing—being grafted on only at its end.[42] Establishment of a general public prosecutorial service (in place of prosecution undertaken by an injured party) came even later, following reform acts beginning in 1879.[43] In America, where matters differed state by state, things were even more complicated.

Not surprisingly, different legal institutions demanded different forms of forensic reasoning.[44] In recent forensic science, singularity—Kirk's individualizing—is central. By contrast, early modern forensic texts put a premium on generality. Much of the expert's job was to categorize a person or event so that the right rule could be applied—proper classification allowing juridical uniformity. By contrast, adversarial systems, particularly common law systems relying significantly on case precedent rather than statute, encouraged ingenious and opportunistic advocates to exploit whatever variations on conventional knowledge they could extract from the rapidly changing research frontier. Especially in the areas of mental competence and toxicology, that openness led in mid-nineteenth-century England and America to a virtual limitlessness of possibility. "Experts" could be found who would assert on plausible grounds any claim or its opposite.[45]

With differences in jurisprudential institutions came differences in the ethical implications of experts' testimony. If one knows one's statements will be thoroughly challenged by advocates concerned only with persuading an inexpert and unpredictable jury, one may feel freer to engage in combative case building than if one knows that the judge will rely directly, perhaps solely, on one's professional judgment in fixing the fate of another.[46] In the early American republic, James Mohr has noted, the champions of forensic medicine had been concerned with protecting the innocent. Accusation was easy; science necessary to protect unpopular persons from mob justice. The standard for conviction—"beyond reasonable doubt" or its predecessor, "moral certainty"—was rightly high. By the mid-nineteenth century, that ideal had succumbed to Langbein's lawyerization, the era of advocatory anarchy that generated the familiar reference to experts being the most venal form of witness. The twentieth-century professionalization of policing and public prosecution (though the latter and sometimes the former remained political) would effect a reversal of forensic polarity.[47] Finding the guilty would be more conspicuous than protecting the innocent. And while the state might bear the burden of proof, in possessing forensic science, it also owned the means to meet that burden.

## Planting Silent Witnesses: From Tales to Rule

There is one more door to open. In keeping with the term's origins, "forensic"—that is, matters relating to the forum—need not be exclusively a matter of investigating unique crimes. The state—or whoever owns the forensic techniques—possesses whatever moral or epistemic power comes with those techniques. There is no obvious merit in relying on the chance that a silent witness will be present where needed; a responsible state will install them, both to discourage crime and to aid rapid response. Because the list of potential criminals includes everyone, we have surveillance cameras, but they are merely a continuation of the street lighting that eighteenth-century townsfolk established for precisely these reasons. Some of the technologies are so familiar that we rarely see them as forensic, for example, the licensing of automobiles and of drivers, and the many other forms of identification one is asked to supply: not only of personal identity, but of status (family, nationality, sex, age, race, criminal history, disease history). Governments have often surveilled foreign visitors (epidemic "intelligence" meshes readily with concerns about subversion); they photograph and follow financial transactions, phone calls, and Internet usage. States also use forensic techniques to define status as well as to determine fact. They are used to certify or deny competence to serve in the military, withstand torture, be tried for one's actions, dispose of property, and avoid institutionalization. They may be used in contexts of production and commerce—one may track and evaluate both the performance of work and its products. Forensic rule has even infected recreation: in the context of school sports, the old phrase "pay to play," referring to urban corruption, is transformed into "pee to play" (and, often, it is pee to work as well). And we should not forget that some of these powers—those that so easily record sound and sight—can be used to resist the forensic state.

While use of such techniques in jurisprudence often relies on their more general use in governance, it provides occasion, too, for distinguishing the rule of law from mere rule. A common feature of American forensic popular culture is the contentious constitutional issue of probable cause: independently of the validity of the techniques used, the bases on which one's silent witnesses have been singled out for interrogation by the authorities constitute a key procedural issue. But several chapters in this volume document what may be called forensic regimes in a literal sense, where "reasons of state" are sufficient grounds for routine surveillance. For overlapping reasons, silent witnessing loomed large in imperial governance. It was a means of mediating among cultures (including distinct legal cultures) and of dealing with civil, demographic, military, economic, and epidemiological forms of insecurity. The policing of vast and unfamiliar spaces, with unfamiliar dangers, natural and criminal, brought technical (forensic) responses as colonial powers looked to science to compensate for their limited abilities to understand the situatedness of crime or the cultural filters affecting their efforts to govern—what will people find objectionable, where do loyalties lie, what will they reveal, and in what ways, and to whom?[48]

Thus there are some takeaways:

- Forensic "cultures" are here, now, and always, not long ago and far away.
- The moral universe of CSI, often seen as a timeless and natural ordering of police, forensic investigators, and prosecutors who must overcome the skepticism of fiendish defense advocates to convince dubious judges or irrational juries, is recent and hardly universal. It is a product of a forensic culture whose power is evident in its invisibility and in the unthinkability of alternatives.
- Investment in forensics is more a matter of the intensity of demand than of the adequacy and quality of supply.
- Forensics is a language for telling stories about events, but mostly about persons. Our attention to those stories reflects prominent fears. Sometimes the making of those stories does, too.
- As a matter of fact if not of principle, the rights humans enjoy are products of forensic regimes.

## Forensic Fantasy, Forensic Invisibility

If forensics has had so profound an effect, why has it been historically so invisible? Partly for the same reasons, its practice—as distinct from its place in popular fantasy—still is. Either it belongs to the structures of the ordinary or, normally out of sight, it is out of mind. But it has also fallen between cracks in historians' division of the past into disciplinary categories. General and political historians (and political and legal thinkers) have paid more attention to constitutions and legislation than to implementation by police officers, medical examiners, or technicians. Studies of policing have focused more on class or race tensions than on routine investigation. Historians of science have generally overlooked obscure corners of interdisciplinary application. While acknowledging the inclusion of forensic instruction in medical curricula, historians of medicine have followed the prejudice of most doctors: medical jurisprudence is a distraction from practice, an onerous service the state may require (sometimes with little compensation). Histories of techniques rarely highlight forensic applications. Cinema and photography are art forms; our archives are not the endless nightly comings and goings in a convenience store. Blood typing is a boon to surgery, not a sneaky way of detecting identity.[49] Besides, these sifters of trash and pluckers of dust are abject and disgusting. Some may even follow our flushings.[50]

Scholarly exploration has been both scanty and scattered. The best general departure point, like this an edited volume (by Michael Clark and Catherine Crawford), appeared a generation ago.[51] There have been fine textbooks (Watson, Adam)[52] and excellent treatments of particular techniques, for example, on fingerprinting (Cole, Sengoopta),[53] anthropometry (Davie, Becker, and Wetzell), toxicology (Burney, Watson, Whorton, and Arnold),[54] lie detection (Alder, Bunn, and Littlefield),[55] criminalistics (Becker, Burney and Pemberton),[56] and DNA profiling (Aronson, Lynch et al.).[57] There have also been excellent studies of professional practices. Following in the wake of Charles Rosenberg's study of psychiatric testimony in assessing the guilt of Charles Guiteau, assassin of President James Garfield, Joel Eigen explored similar issues in Britain over nearly two centuries, while James Mohr considered medical testimony in nineteenth-century America more generally.[58] There has been sporadic interest, too, in the history of institutions—Ian Burney has studied the English system of inquests and Jeffrey Jentzen the longstanding tension in the investigation of deaths between the lay coroner (elected or appointed) and the expert medical examiner in the United

States.[59] The most general treatment of the recognition of expert testimony as a distinct mode of evidence is Tal Golan's work.[60] But behind all this work lies the legacy of Michel Foucault, explorer of the quotidian technologies of governance, though Foucault himself had little interest in particular forensic techniques nor much sympathy for the tasks to which they were applied.[61]

The chapters in this book explore forensic cultures in many places: Germany, France, and Spain, as well as America and Britain, the latter conceived mainly as a colonial power. South Asia receives much attention—owing to the political, cultural, and social complexity of the region as well as to the variety of ways the British exploited it—initially as agent manager for local rulers, later through the efforts of a single imperial government attempting to accommodate the multiple cultures in an immense area. They also explore Egypt, Palestine, and South Africa, where historical and cultural contexts brought unique administrative challenges, and Thailand, where sovereignty was compromised by the practice of extraterritorial jurisprudence—rules and procedures differed sharply according to citizenship, race, and ethnicity.[62]

The chapters focus on the period from 1800—a transformative era in anatomy and chemistry, in which photography and other forms of nontextual reproduction became available, and in which state bureaucracies had arisen for the creating and ordering of forensic information. Individually and collectively, these studies are rich, disturbing, and provocative. Each explores a single element of forensic practice, recognizing the entanglement of techniques, professions, and systems of jurisprudence/governmentality. There are common themes: forensic innovation has often been demand driven, dictated by local anxieties. The techniques are more often gropings of authorities for authority, not downloadings of applicable general knowledge matured in metropolitan centers of intellectual production. In this bootstrapping, groups of would-be professionals, arriving with little formal sanction, extend plausible approaches into essentials of defensible jurisprudence and comprehensive government. Not all succeed, but some, recalibrated or more carefully defined and validated over time, persist, flourishing with the professions that have tied authority to them. Or their fate may be affected by publics—outraged or anesthetized by the narcotic of forensic security.

Overall, the findings echo those of the 2009 NRC report: the domain of forensic science is a makeshift of plausibilities and poorly understood probabilities, a fabric of contingencies more than a well-ordered decision tree. That finding is indeed disturbing if we begin with a belief that forensic techniques can and should liberate. Even more disturbing in these chapters is a variable that repeatedly emerges as the differentiator of forensic practices: race or ethnicity. Early modern forensics was conspicuously concerned with gender; here, at least as often as a universalistic human science is represented as a means of transcending race, it is in fact used as a means of reifying or amplifying racial distinctions.

Thus the integrative agenda of this book: whatever may be the rules, goals, and contexts, the details of forensic fact making do matter. For each we may ask the following questions. "How has this method come to exist?" "Who uses it?" "What goals is it serving?" "Who will delight in, defer to, resist, or co-opt it"? "How does it contribute to something conceived as justice?" Whether or not such questions have been well asked, to recognize them, as these chapters do, remains important. These examples of the making and using of facts from the forensic past alert us to the profound and sometimes disturbing uses of forensic fact making in the present and future.

One could order these chapters in many ways. Here we have put them in three parts, ordering them chronologically within each. Part I, "Evidence and Epistemology," concerns the emergence and adaptation of forensic techniques. Part II, "Practices of Power and Policing," concerns their integration into institutions of jurisprudence and government. Part III, "Training and Transmitting," explores groups of forensic professionals who have acquired an authority as embodiers of what is tantamount to summary judgment, which they can then replicate as professionals training successors in their arcane art.

We begin with knowledge making in one of its most familiar forensic forms: someone has died, perhaps from something they ate. Provided that means of chemical analysis exist, this should be an easy matter of checking for substances that produce the fatal symptoms. While there may seem no obvious reason why we would ask the question of what is in something differently in a jurisprudential setting than in basic science, or in other arenas of application—analyzing foodstuffs or mineral samples, for example—the forensic context did alter chemists' practices, the philosopher of science Marcus Carrier demonstrates. Carrier focuses on toxicological practice in late nineteenth-century Germany, a context in which chemistry had become a well-ordered discipline and a well-recognized fount of authority. But in a legal setting, some criteria—particular questions and evidence-handling procedures—loomed larger than others, like precise quantification. Where juries made decisions, the communicability of procedures mattered: there was a premium on methods that could be easily explained to laypeople even if these were not the preferred methods in the laboratory. There was also a premium on nondestructive tests: should questions arise later, there might be need for retesting. While one may say that any investigator deploys an analogous set of epistemic criteria for drawing and presenting conclusions, Carrier's findings remain troubling. They echo a criticism of recent investigations of forensic practices, of a gulf between what suffices for courtroom performance and what is done in a research laboratory, with expediency and familiarity looming larger than comprehensiveness and exactitude. And one may ask a further question: Will epistemic criteria useful to a prosecution be equally appropriate for a defense?

As in chemical analysis, much of forensics operates on the pretense of universality. A fingerprint is not—or at least should not be—a different thing to different cultures. What of blood? Unambiguous? Monovalent in meaning? At the heart of Mitra Sharafi's chapter is a pun, the common if archaic English phrase "the blood will out." Perhaps, but what does it mean to read blood?

Sharafi, a legal historian and scholar of comparative legal institutions, explores an obscure functionary in early twentieth-century British India, the "imperial serologist." While the new science of serology allowed the distinguishing of human from animal blood, what warranted such an appointment (only in south Asia) was the use of displays of blood to deceive forensics-based rule. A public display of what was in fact animal blood was a way to frame an enemy. But through its imperial serologist, the Raj, buoyed perhaps by Lady Macbeth's recognition that "all the perfumes of Arabia" cannot "sweeten . . . the smell of blood still" (V, i), can respond. Yet serological policing merely brought an escalation: a display of human blood or, better, of a bloody human body, sacrificed by one group to implicate another. In such circumstances, the victim would, and in two distinct senses, be "lying in her own blood" (it was often a woman, Sharafi makes clear).

Sharafi's point goes beyond the mere subversion of imperial jurisprudence, however, to recognize the coexistence of distinct systems of justice. Official justice concerned individuals—perpetrators and victims. Unofficial justice concerned families or other groups. In a feud, to sacrifice (in suggestive circumstances) a member of one's group was not a bad tactic if one could thereby induce the officials to punish one's enemies and thus satisfy a *lex talionis*, a retaliation that would, at least in the actors' view, restore a proper balance. Sharafi's chapter suggests that where multiple systems of identity (and therefore morality and justice) coexist, ways will be found to make each serve the other. If we are tempted to see this example as idiosyncratically exotic, the familiar phrases of popular or vigilante justice suggest that the coexistence of concepts of justice is more norm than aberration. Consider cultures of lynching—"we're taking the law into our own hands" or "they're getting what's coming to them." Even "there's a higher justice."

Race and imperial anxiety loom large, too, in Projit Mukharji's study of "The Hardless Detective Dynasty." Mukharji takes us far from remote crime scenes to urban matters of commerce and property and to a different forensic skill, that of linking a handwritten document to its writer. He follows generations of a single family, that of Charles Hardless and his descendants, who effectively held an epistemic monopoly on graphology in Bengal for fifty years, beginning in the late nineteenth century.

Is this a matter of the crime scene? Although marginalized in the modern forensic cult, issues of property have often loomed larger than those of bodily harm, even murder. Its importance is evident in the antiquity of royal seals and the heavy reliance on notaries in many literate cultures. The English coroner was initially no mere advocate for the wrongfully dead, but a crown official representing crown interests: the heir who impatiently nudged his sire into the next world certainly ought not to inherit, but the state might happily do so. The legal problem of proxies for identity remains immense and immediate in this age of agonies about password security and "identity" theft.

From an epistemic standpoint, too, graphology exemplifies Locardianism in its axiomatic form: a document will bear the unique mark of its maker; just as surely as the "blood will out," so too the "hand will out." But the graphologist will often go beyond "same or different" to address the broader question of whether a document is what it claims to be. Assessing genuineness in turn involves considering both modes and motives of forgery; recognizing traces will thus be entangled with assigning meanings and making tales—as in other areas of forensics, it will be hard to avoid the hermeneutic loops in which assumed meanings dictate fact making.

While distrust will be the default state of colonial governance, reliance on the Hardlesses, as on the imperial serologist, reflects a Raj paralyzed by deception anxiety. Language and culture (and race) matter. The multiple south Asian cultures have a long experience of dealing with one another and with Europeans. Claiming to know whether any sort of person would write like that, the Hardlesses claim epistemic authority over documents from a range of cultures, but as Eurasians, they also claim neutrality. Not only will they not be subject to ethnic loyalties, but also their own hybridity may perhaps make them especially sensitive to attempted deceptions, manifest as crossings of class or culture.

Did the Hardlesses themselves deceive? Recognizing the vulnerability of their own authority, they took steps to translate their skills into European technocratic motifs by quantifying and even mechanizing them. Yet in doing so, they were protecting their monopoly, leveraging personal expertise into truth algorithms and truth-making instruments.

The arcane practices of the Hardlesses, like those of the imperial serologist and the German toxicologists, serve the state. They will be antidotes to popular prejudice. Yet the populace and its newspapers shape forensics, too, notes cultural historian Ian Burney—precisely the current worry about the CSI effect that has developed from contemporary forensic entertainment. Burney's chapter also begins as a tale of blood, blood in the Cleveland bedroom of Marilyn and Dr. Sam Sheppard in 1954. Initially, Clevelanders trust their police and its forensic investigators. Their blood-charting techniques secure Dr. Sheppard's conviction for Marilyn's murder. Enter Paul Kirk, the Berkeley biochemist and experimental criminologist we have met as author (a year earlier) of *Crime Investigation* and (a decade later) of "The Ontogeny of Criminalistics." Kirk, hired by the Sheppard family, reconstructs events differently using the new technique of examining blood spatter, which relies on the trajectories of flying droplets. He will offer "facts" not theories, yet Ohio law blocks reopening of the case until, a decade later, the US Supreme Court overturns the conviction.

At one level, Burney's chapter deals with vindication through forensic progress. At another, however, it concerns the multiplicity of forensic stories and their creation and consumption. Taking us deep into the voluminous trial transcript, he exposes not only the interplay of fact stating with story making (and unmaking), but the roles of the local and national press as amplifier and sounding board. Locally, righteousness in the conviction was initially coupled with contempt for Kirk as a "fact"-flaunting California interloper. Yet, within months, a countervailing righteousness had taken hold— remorse at a rush to judgment (already being recognized as a characteristic of Cold War red-baiting), with blood spatter techniques emerging (as DNA is now) as the last line of defense of human rights. There was fascination, too, with the prospect of vindication by polygraph, seen as the apotheosis of forensic practices. Its enforced honesty would usher in a world of transparency (in the process, rooting out covert Communists). With it would come acceptance of human difference, including acceptance of Dr. Sheppard's infidelities.

Yet for the flamboyant and crusading defense attorneys who took up Sheppard's case, it would be less the actual forensic evidence (which, they insisted, indicated a left-handed assailant bloodied in the attack) than *the idea* of forensic vindication (and its disturbing antithesis of forensic persecution) that would keep the case alive. The "reasonable doubt" that led to Dr. Sheppard's ultimate exoneration came in the form of a forensic imaginary.

If the Sheppard stories suggest the public's fickleness toward forensic deliverance, that fickleness reflected the contemporary ambivalence of the American public as civil rights concerns began to displace Cold War preoccupations. Desperate for security, it was becoming uneasy with policing. But underlying all was faith in the accountability of forensics as guarantor of the rule of law and the justness of the state. For in all these chapters, forensic practices and practitioners were seen to play merely an instrumental role. The forensic arts had been enlisted to meet real demands for resolution by authorities seeking the best fact-making technologies of the day. There were no vested agendas about what is at stake or who should prevail for the German chemists, the imperial serologist, the Hardlesses, the Cleveland police, or even the hired expert Kirk. While the experts might sometimes disagree, they were trusted as citizens of good will with no agendas other than happily applying their talents to the common good of truth finding. If they were sometimes unduly tied to a prosecutorial state, that was not their fault.

Yet it is a short distance from "state with efficient policing" to "police state," and forensic technologies may be the well-waxed skis that take us down that slippery slope. I have alluded already to the two chief sites where rule of law may diverge from mere rule: first, unequal accessibility to techniques that might vindicate a person; second, the routine use of forensic techniques, perhaps on certain populations, not just to track and convict, but to surveil and deter (or intimidate). It is naive to think of forensic technologies coming into play only after crimes—pattern-recognizing systems, whether fingerprint banks or surveillance cameras—require a database for comparison. Often they will allow rapid response; they will also deter, and undermine privacy both personal and political. Having one's child fingerprinted (or chipped) *may* foil abduction; it *will* strengthen state power.

One need not imagine any overt intent to concentrate power. Rather, what one sees in the cases in Part II is a reciprocity, with forensic practices becoming critical elements of governance in states cognizant, and deeply anxious of their own vulnerabilities. We begin with a forensic problem as familiar and as seminal as toxicology, the professional assessment of a "body found dead."

Jeffrey Jentzen, pathologist as well as historian of medicine, asks an ingenuously simple question. What happens to a single medicolegal institution, the English coroner's inquest, when it is exported to several colonial settings in the British Empire? The institution itself represents universal elements of forensic cultures: a technique (a medical inspection, perhaps including an autopsy); professional expertise (usually the modest expertise of a general practitioner or a military surgeon); and a simple jurisprudential procedure (an initial determination of culpability by coroner and jury). Sometimes, the coroner's inquest is depicted as an exemplary forensic institution, a union of intimate social knowledge with biomedical expertise. Yet Jentzen's survey reveals extraordinary variation.

The inquest is a communal institution, hence its extension must be made to map community. But race and socioeconomic or legal status baffle its appliers. In British North America, they puzzle over its applicability to chattel slaves or indigenous persons. They adapt standards. In India, endemic malaria among plantation laborers lowers the culpability of masters in beating deaths, which have surely occurred only because these persons have bloated their spleens with malarial plasmodia, doubtless to avoid work and defy normal forms of physical discipline.

More broadly, Jentzen shows how the several purposes and circumstances of colonization affected medicolegal institutions. Neo-European settlements, places of large-scale displacement or destruction of indigenous populations, raised different issues from those of military or commercial colonies. Sometimes, as in India, British colonists sought to co-opt or adapt to existing jurisprudential institutions; in others (e.g., South Africa), they inherited institutions from an earlier colonial power or, as in Egypt, developed a hybrid institution with another colonial power. And, necessarily, death investigation responded to accidents of geography, like sparsity of settlement, and to new hazards ranging from parasites to poisons.

Following Jentzen's broad sweep is José Ramón Bertomeu-Sánchez's charting of the subtle and disturbing chronology of the co-production of a national system of fingerprinting with the means of (ultimately dictatorial) governance in early twentieth-century Spain. Since Foucault, it has been easy to demonize anthropometric technologies. Paging through Lombroso's catalogs of portraits suggests how dossier creation could so easily convert any of us into some sad, sick, or criminal object; given the heritage of Franco, it is similarly easy to find menace in a Spanish fingerprint index. Bertomeu-Sánchez offers instead a story of anthropological curiosity and meliorist modernity.

In late nineteenth-century Spain, as elsewhere, modernizing ambitions brought to the fore new expert professions and growing expectations of expertise in public administration. Transparency and accountability (and traceability) came as components of administrative rationality, but also simply as science. It might be, hypothesized one of the newly professionalized physical anthropologists, that patterns of lines on fingers corresponded to other important differences among humans. Collection of these data, partly from prisons where they could be conveniently collected, led to other uses, and ultimately to networks among progressive scientists, lawyers, police, and prison officers, all wishing to respond effectively to the crises of Spain's disappearing empire, conflicting ideologies, and widespread disaffection. The "scientific policing" that resulted was distinctively academic. The fingerprinting arose within an infrastructure of standard setting, certification programs, and journal creation that evolved through mutual enlistment of professionals in Spain and abroad.

In this story of structures and networks, we see no hint of the coming menace of a total institution, but only social scientists and others—social workers (considered broadly) and state agents doing their professional work. Behind that banality is a modest benignity. Like Deborah Kerr in *The King and I*, they are "getting to know you, getting to know all about you / Getting to like you, . . . to hope you like me." But we are left with a more general and more troubling question: Is there some single point at which the power accrued in rational reformism suddenly turns sinister?

The odd trajectory of Spanish fingerprinting raises similar questions of how other innovations become complicit in state building. Binyamin Blum tracks the growing importance in British-ruled Palestine in the middle 1930s of a new and multivalent forensic technology, the police dog, in this case a Doberman, combining a tracking nose with an intimidating growl and powerful teeth. In Palestine, a region slipping into civil war with its police force split between the rival ethnicities, "political crimes" constituted some significant fraction of crimes. Remarkably, tracking dogs, like fingerprints, were to represent a demilitarized reformist alternative, part of a campaign to transform occupation into harmonious multiethnic community policing. With science, the authorities might transcend the doubts and conflicts that came through relying on testimony. Dogs, surely, were unprejudiced.

As a probative tool, the dogs were problematic: they could never be reliably calibrated nor cross-examined. Their effect was cultural. While Holmes— "Come Watson, the game is afoot"—might invoke England's hunt-and-hound cultural heritage, dog tracking had been unacceptable in English courts, abhorrent both to judges and to public sensibilities. Mainly the impact of their use was intimidatory, chiefly to the Arabs who were increasingly the targets of this tracking and among whom dogs (and, by association, their masters) were both feared and reviled. A common response was submission to forensic authority. A dog's approach had much the same effect as the uniformed officer's statement: "I clocked you at." A modest means of multiethnic state building had ended up exacerbating ethnic division.

In Thailand, the evolution of forensic or "truth" regimes into central components of political culture and state authority was subtler. Trais Pearson begins by dissecting a modern case, one based on DNA evidence. What we now look to as a way beyond culture and a means of universalizing justice was, in this instance, part of a legal culture in which ethnicity—here Burmese or Thai—was central. But that culture, in turn, was founded on Thailand's nineteenth-century heritage of extraterritoriality. Though Thailand was never a colony during the age of imperialism, it was not wholly independent. European-style forensic science arrived in the 1890s. If on the one hand it was a means of building a modern Thai state, on the other it was a response to the demands of the great European naval powers that their citizens, in Thailand to pursue commercial ventures, be accorded a justice similar to what existed in their own nations. Ironically, then, Thai sovereignty required capitulation to foreign forensic standards, particularly in the investigation of unnatural deaths, where foreign nationals were owed the possibility of forensic exoneration according to their own standards. What the modern DNA case reveals, then, is how far a political culture tied closely to the defense and articulation of nationality has crept into routine forensic practice. Relations with Myanmar are among the most fraught in contemporary Thailand; that these tensions would manifest in forensic practice is not surprising, but here, too, the situation is hardly unique. For, lest this episode seem a bizarre holdover from the age of opium war imperialism, multiple forms of compromised sovereignty persist, as do practices of forensic racism or nationalism. Americans are notorious but hardly unique in expecting their own legal institutions to apply elsewhere. If the FBI is not involved, surely it should be.

The chapters in Part III go further. They concern not the co-optation of forensic techniques but the creation of self-accrediting forensic authority. Here, too, we should not be surprised—the insistence that only fellow professionals can assess professional work is general. The authors here describe three different ways in which professionals escape state control, making themselves, in the terms of an actor-network theorist, mandatory sites of passage.[63] Thus we confront a problem outlined earlier: we may ask the expert to swear, but, as is not the case with ordinary perjury, we have no good way of auditing that oath. One can imagine, with President Eisenhower, justice held "captive to a scientific-technological elite," a dictatorship of crime scene inquirers with arcane methods and their own agenda, to which the dependent state would have to capitulate. Consider, for example, computer hacking. Anxieties about experts arise across the realm of forensics, and they should. On matters that matter, one is to entrust one's fate not only to putatively valid methods but also to the persons who claim to embody them. However much we bemoan expert disagreement, expert unanimity is more sinister.

One route to such transcendence is the iconoclastic individual. Devotees of CSI will know what a morgue is and what happens there. They will not know it as La Morgue, a building in Paris, nor will they know of Ambroise Tardieu, the mid-nineteenth-century Parisian pathologist principally responsible for the exacting and authoritative detail of the forensic postmortem. Tardieu's story is one of charisma and entrepreneurship. Bruno Bertherat charts his success in an intensely competitive professional world. In Paris, cadavers were the primary currency of medical epistemology (and authority) for the clinicians developing the new clinical-anatomicopathology. By cultivating a career in the obscure sideline of legal medicine, Tardieu obtained not only access to the bodies of those who died in dubious circumstances outside the hospitals, but also authority over forms of death that rarely occurred in hospitals. Infants comprised one large group of victims, and among these Tardieu frequently detected a feature, subpleural ecchymoses, still known as "Tardieu's spots." Bertherat explores Tardieu's insistence in the face of the skepticism of colleagues (and successors) that these proved suffocation and were grounds for infanticide convictions. That insistence reflected not only the consolidation of forensic authority, as Tardieu insisted that only he could speak on these contested matters, but also its creation as Tardieu, aloof toward the other centers of forensic expertise emerging both in France and in Germany, dictated his own truths on a range of forensic subjects through writing textbooks and editing encyclopedias. He would be the center.

Tardieu successfully forced his way into a position of command. Colonial states faced a different problem: what to do with existing forensic elites. For they relied on forensic regimes both to govern and to authorize their own authority—their administration could be just and efficient only if it were able to extract, integrate, and respond appropriately to information. But often that required relying on indigenous authorities, those expert in navigating the social and natural landscapes of conquered regions. Finding no alternative to trusting the ruled, colonial rulers enlisted the expertise of the guide, the interpreter, the midwife, or wise woman. The risks were justified (and celebrated) in Kiplingesque success stories, versions of the Sacajawea myth: pregnant Nez Perce teenagers like to help lost Jeffersonian explorers and have no agendas of their own.

Tracking exemplifies such "strange sciences," as Gagan Preet Singh makes clear: the term conflates a single practice (footprint recognition) with more general deductive skills (ability to find all manner of Locardian traces and to reason from them). The trackers to whom the colonial police appealed in late nineteenth-century Punjab were skilled distinguishers of footprints, specializing either in plains or in forests. The officials faced four successive and yet intersecting problems: first, could they trust trackers, both in particular cases and in general? Second, could tracking be disciplined by making trackers servants of the state? Third, how could their expertise be accredited in legal settings? Finally, could it be appropriated, that is, universalized and codified? All were problematic. The trackers' most general concern was their own independence. Their expertise was valued in civil as well as in criminal matters, and many preferred private practice, deciding what cases to take and negotiating their own, relatively high, fees.

Underlying the officials' unease was an anthropological question. Unlike the serologist's claim that the species from which blood had come could be determined from chemical changes occurring when some mysterious serum was added to a sample, the trackers' skill of distinguishing footprints on various surfaces seemed comprehensible and commonsensical even if unreplicable. But at issue was *who* was supplying the authority: confidence in the imperial serologist was as much endorsement of the English or Scottish families, schools, and networks from which colonial officials came as it was in the recondite science of Paul Ehrlich. To the British, the trackers exemplified the chronic anxiety of the Raj: rule by co-optation rather than open conflict was often a matter of mutual dissembling. Such anxieties had led the British to rely on tracking dogs in Palestine. Skills were one thing, loyalty quite another.

Yet one should not view the formation of relatively autonomous, self-policing communities of forensic experts merely as an expression of subaltern resistance. Rather, it is a general feature of professions. Complementing Singh's local focus is Heather Wolffram's study of informal networks of early twentieth-century British colonial policing. Wolffram is interested in how innovations in diverse colonial settings (ranging from Egypt to Ceylon) became, through the casual contacts of these mobile colonial officials with one another over a long period, a set of administrative and technical practices that would be replicated in new generations of officials. Moving from posting to posting and colony to colony, these midlevel officials, acting equally as mentors and mentees, brought experience garnered in one place to the problems of others, thus developing an adaptive "make it up as you go" forensics. Thus Palestinian dog tracking would be founded on experience from Kenya and, at one remove, from South Africa.

With new techniques came questions about validity and quality control—the training and accreditation of those who will use them. Who, one may wonder, organizes this work? Where are the centers of power? In fact, there was neither organizer nor center, but only adapting network. Wolffram's review of the correspondence of these world-wandering police administrators reveals not only the perpetuation and extension of their networks, but also their effective escape from the scrutiny of the metropole, with its expectations of rationality and routine. For here not merely forensic authority but also forensic legitimacy are reduced to traditions of practice.

These few studies should suggest how little we know of the guts of the rights we enjoy. Those rights rest on technologies. And while we know much about the history of both technology and legal institutions and principles, we know little about their intersection: hence "guts"—essential, hidden inside, and disturbing when examined too closely. The glimpses that this book offers are readily extendable: only multiply the variety of forensic techniques; their evolution over time; and the many social, geographic, and jurisprudential settings of their use, and one has a long agenda of important historical research. Each of these stories is the tip of the tip of an iceberg. They are provocative, disturbing, and relevant, as the sociologist and forensics policy scholar Simon Cole points out in his afterword to this volume. These historical problems remain with us in many forms. Who will calibrate the dogs (or the hair comparers)? How can the effects of forensic intimidation be kept separate from the valid results of forensic knowledge? Are we safer in the hands of dictators of forensic truth like Tardieu (or the FBI), or a group of cowboy colonials bringing their forensic rough justice to the wild east, such as Wolffram sketches? And how shall we stop innocent constellations of reformist technics, designed in the pursuit of rationality and accountability, from becoming the tool kit of fascist intimidation?

ACKNOWLEDGMENTS

The author thanks Ian Burney, Barbara Fick, and Barbara Walker—wonderful, informed friends and great sounding boards.

NOTES

1. "Dragnet—'Just the Facts,'" accessed April 20, 2017, http://www.snopes.com/radiotv/tv/dragnet.asp.

2. Representation of the physical traces of a crime as the "mute" or "silent witness" is often attributed to the pioneering French criminologist Emile Locard.

3. Michael Lynch, "The Discursive Production of Uncertainty: The OJ Simpson 'Dream Team' and the Sociology of Knowledge Machine," *Social Studies of Science* 28 (1998): 829–68; idem, "Science, Truth, and Forensic Cultures: The Exceptional Legal Status of DNA Evidence," *Studies in the History and Philosophy of Biological and Biomedical Sciences* 44 (2013): 60–70.

4. *Oxford English Dictionary*, 2nd. ed., s.v. "dragnet." The term's appropriation for systematic manhunts had occurred by the early twentieth century.

5. Peter L. Berger, *The Sacred Canopy; Elements of a Sociological Theory of Religion* (Garden City, NY: Doubleday, 1967).

6. Erin L. Murphy, "Sessions Is Wrong to Take Science Out of Forensic Science," *New York Times*, April 11, 2017; https://www.nytimes.com/2017/04/11/opinion/sessions-is-wrong-to-take-science-out-of-forensic-science.html.

7. National Research Council (NRC), Committee on Identifying the Needs of the Forensic Sciences Community, *Strengthening Forensic Science in the United States: A Path Forward*, Document No. 228091 (Washington, DC: NRC, 2009).

8. Stephen Hsu, "Sessions Orders Justice Department to End Forensic Science Commission," *Washington Post*, April 10, 2017, https://www.washingtonpost.com/local/public-safety/sessions-orders-justice-dept-to-end-forensic-science-commission-suspend-review-policy/2017/04/10/2dada0ca-1c96-11e7-9887-1a5314b56a08_story.html. See also opinion pieces by Radley Balko, "Jeff Sessions Wants to Keep Forensics in the Dark Ages," *Washington Post*, April 11, 2016, https://www.washingtonpost.com/news/the-watch/wp/2017/04/11/jeff-sessions-wants-to-keep-forensics-in-the-dark-ages, and Murphy, "Sessions Is Wrong." Cf. Simon A. Cole, "Forensic Culture as Epistemic Culture: The Sociology of Forensic Science," *Studies in History and Philosophy of Science Part C: Studies in History and Philosophy of Biological and Biomedical Sciences* 44 (2013): 36–46, at 42. As Cole points out, Sessions's point is familiar in law. See his "Who Speaks for Science? A Response to the National Academy of Sciences Report on Forensic Science," *Law, Probability & Risk* 9 (2010): 25–46.

9. C. Hamlin, "Edward Frankland's Early Career as London's Official Water Analyst, 1865–1876: The Context of 'Previous Sewage Contamination,'" *Bulletin of the History of Medicine* 56 (1982): 57–76.

10. Writing in 1953, the criminologist Paul Kirk assumed that both police and the public (represented by writers of detective fiction) were largely uninterested in physical evidence. Kirk, *Crime Investigation* (New York: Interscience, 1953), 3.

11. This generalization refers to matters at the intersection of philosophies of science and philosophies of law. The literature on applications of science to legal matters is enormous, though vastly underexplored. Much of the work we can do in the field was made possible by Jaroslav Nemec, a legal scholar and bibliographer at the National Library of Medicine. I draw heavily on three of his works: *Highlights in Medico-Legal Relations* (Bethesda, MD: National

Library of Medicine, 1976), *International Bibliography of Medicolegal Serials, 1736–1967* (Bethesda, MD: National Library of Medicine, 1969), and "International Bibliography of the History of Legal Medicine," Bethesda, MD: National Library of Medicine, 1974.

12.  Roughly speaking, this division corresponds to what I have elsewhere called technologies of witness, of testimony, and of judgment. See Christopher Hamlin, "Forensic Cultures in Historical Perspective: Technologies of Witness, Testimony, Judgment (and Justice?)," *Studies in History and Philosophy of the Biological and Biomedical Sciences Part C: Studies in History and Philosophy of Biological and Biomedical Sciences* 44 (2013): 4–15.

13.  *Oxford English Dictionary*, 2nd ed., s.v. "criminalistics." The term, first used in English in 1910 as the title of a journal, derives from the German *Kriminalistiks*. More generally, see Ian Burney and Neil Pemberton, *Murder and the Making of English CSI* (Baltimore: Johns Hopkins University Press, 2016).

14.  Frank Crispino et al., "Forensic Science—A True Science?," *Australian Journal of Forensic Sciences* 43, no. 2–3 (2011): 157–76. For the 1899 introduction, see the *Oxford English Dictionary*, 2nd ed., s.v. "forensic science."

15.  Kirk, *Crime Investigation*, 4.

16.  Cole, "Forensic Culture"; Paul Roberts, "Renegotiating Forensic Cultures: Between Law, Science and Criminal Justice," *Studies in History and Philosophy Science Part C: Studies in History and Philosophy of Biological and Biomedical Sciences* 44 (2013): 47–59. On questions stateable as science, see Alvin Weinberg, "Science and Trans-Science," *Minerva* 10, no. 2 (1972): 209–22.

17.  J. Andrew Mendelsohn and Annemarie Kinzelbach, "Common Knowledge: Bodies, Evidence, and Expertise in Early Modern Germany," *Isis* 108 (2017): 259–79.

18.  Katherine Watson, *Forensic Medicine in Western Society: A History* (London: Routledge, 2011).

19.  The tension between general and specialized competence in the several forensic sciences is central but underexplored. See Kirk, *Crime Investigation*, v–vi.

20.  Kirk, *Crime Investigation*, 521. Emphasis mine.

21.  For a defense of Locard as the foundation of forensic science, see Crispino et al., "Forensic Science."

22.  See Kirk, *Crime Investigation*, 4, 9. The multiple versions in Crispino et al., "Forensic Science," discloses an important ambiguity. Does the principle apply peculiarly to crimes, as Locard seems to suggest, or to any contact?

23.  NRC, *Strengthening Forensic Science*, 112–13.

24.  Alain Desrosières, *The Politics of Large Numbers: A History of Statistical Reasoning* (Cambridge, MA: Harvard University Press, 1998). For the continuing undercurrent of an individualistic interpretive approach to science, however, cf. Carlo Ginzburg, "Morelli, Freud and Sherlock Holmes: Clues and Scientific Method," *History Workshop*, no. 9 (Spring 1980): 5–36.

25.  Paul L. Kirk, "The Ontogeny of Criminalistics," *Journal of Criminal Law, Criminology, and Police Science* 54, no. 2 (1963): 235–38, https://doi.org/10.2307/1141173.

26.  Kirk, *Crime Investigation*, 3.

27.  This recognition, often overlooked, is well articulated in the textbooks of Max Houck. See Houck and Jay A. Siegel, *Fundamentals of Forensic Science*, 2nd ed. (Burlington, MA: Academic Press, 2010).

28.  Kirk, *Crime Investigation*, 13–15. For discussions of appropriate Bayesianism, see I. W. Evett, "Expert Evidence and Forensic Misconceptions of the Nature of Exact Science," *Science & Justice* 36, no. 2 (1996): 118–22, https://doi.org/10.1016/S1355-0306(96)72576-5; idem, "The Logical Foundations of Forensic Science: Towards Reliable Knowledge," *Philosophical Transactions of the Royal Society B: Biological Sciences* 370, no. 1674 (August 5, 2015):

20140263, https://doi.org/10.1098/rstb.2014.0263. As Cole points out in "Who Speaks for Science?," experts still assert identity in unqualified ways.

29. Tal Golan, *Laws of Men and Laws of Nature the History of Scientific Expert Testimony in England and America* (Cambridge, MA: Harvard University Press, 2004).

30. Simon A. Cole, "Splitting Hairs? Evaluating 'Split Testimony' as an Approach to the Problem of Forensic Expert Evidence (21st Century Challenges in Evidence Law)," *Sydney Law Review* 33, no. 3 (2011): 459–85; idem, "Who Speaks for Science?" Kirk, *Crime Investigation*, 16–17, recognized different kinds of expertise but struggled with the general question of how evaluate expert judgment.

31. This is the chief concern of what has been recognized as "golem" science. H. M. Collins, *The Golem: What Everyone Should Know about Science* (Cambridge: Cambridge University Press, 1993).

32. Cole, "Forensic Culture as Epistemic Culture," 39. NRC, *Strengthening Forensic Science*, 111–23, strongly represented science's tentativeness as a mark of its validity.

33. Kirk, *Crime Investigation*, 13. Emphasis mine.

34. David Michaels, "Scientific Evidence and Public Policy," *American Journal of Public Health* 95, Suppl. 1 (2005): S5–7; Gary Edmond, "Just Truth? Carefully Applying History, Philosophy and Sociology of Science to the Forensic Use of CCTV Images," *Studies in History and Philosophy of Science Part C: Studies in History and Philosophy of Biological and Biomedical Sciences* 44 (2013): 80–91.

35. Criminal cases rarely exhibit the freewheeling circus of countervailing expertise familiar in the deep-pockets litigation of mass torts or of some patent cases. Yet as Edmond, "Just Truth?," 81, 88, observes, standards of forensic proof are frequently higher in civil than in criminal cases. See also Edmond and David Mercer, "Litigation Life: Law-Science Knowledge Construction in (Benedictin) Mass Toxic Tort Litigation," *Social Studies of Science* 30 (2000): 265–316; C. Hamlin, "Scientific Method and Expert Witnessing: Victorian Perspectives on a Modern Problem," *Social Studies of Science* 16 (1986): 485–516.

36. Simon A. Cole and Troy Duster, "Microscopic Hair Comparison and the Sociology of Science," *Contexts* 15 (2016): 28–35.

37. William C. Thompson, "What Role Should Investigative Facts Play in the Evaluation of Scientific Evidence?," *Australian Journal of Forensic Sciences* 43, no. 2–3 (2011): 123–34, https://doi.org/10.1080/00450618.2010.541499.

38. Kirk, *Crime Investigation*, 9; NRC, *Strengthening Forensic Science*, 123–25; US President's Council of Advisors on Science and Technology, *Report to the President. Forensic Science in Criminal Courts: Ensuring Scientific Validity of Feature-Comparison Methods* (Washington, DC: President's Council of Advisors on Science and Technology, September 2016).

39. Kirk, *Crime Investigation*, 517–21. Many expert witnesses have reiterated Kirk's views. See Connie Fletcher, *Crime Scene: Inside the World of the Real CSIs* (New York: St. Martin's, 2006); Hamlin, "Forensic Cultures," 13–14. While Kirk takes the perspective of the prosecution, his best-known intervention would be as a defense expert, as we shall see in chap. 4.

40. Max Houck, "Intellectual Infrastructure: A Modest Critique of Forensic Science," *Science & Justice* 53 (2013): 1; Murphy, "Sessions Is Wrong"; Cole, "Forensic Culture as Epistemic Culture," 40; idem, "Afterword, "A Tale of Two Cities? Locating the History of Forensic Science and Medicine in Contemporary Forensic Reform Discourse," this volume.

41. Catherine Crawford, "Legalizing Medicine: Early Modern Legal Systems and the Growth of Medico-Legal Knowledge," in *Legal Medicine in History*, ed. Michael Clark and Catherine Crawford (Cambridge: Cambridge University Press, 1994), 89–116; Burney and Pemberton, *Murder and the Making of English CSI*, 9–11.

42. John H. Langbein, *The Origins of Adversary Criminal Trial* (Oxford: Oxford University Press, 2005). For the earlier period, see also his *Prosecuting Crime in the Renaissance:*

*England, Germany, France* (Cambridge, MA: Harvard University Press, 1974). For forensic involvement, see K. D. Watson, "Medical and Chemical Expertise in English Trials for Criminal Poisoning, 1750–1914," *Medical History* 50 (2006): 373–90; Tal Golan, "The History of Scientific Expert Testimony in the English Courtroom," *Science in Context* 12 (1999): 7–32.

43. *Encyclopedia Britannica*, 13th ed., s.v. "Criminal Law," 7:454–64. Some serious crimes had been offenses against the state much earlier.

44. Arguably, the ideal balance of roles arose, largely accidentally, in nineteenth-century Scotland, reflected in the distinct duties of procurators fiscal, sheriffs depute, police surgeons, and professors of forensic medicine. See Douglas Maclagan, "Address In Forensic Medicine, Delivered at the Forty-Sixth Annual Meeting of the British Medical Association, Bath, August 6th-9th, 1878," *BMJ* 2, no. 920 (August 17, 1878): 233–39; Brenda White, "Training Medical Policemen: Forensic Medicine and Public Health in Nineteenth Century Scotland," in Clark and Crawford, *Legal Medicine in History*, 145–63.

45. Charles Rosenberg, *The Trial of Assassin Guiteau: Psychiatry and Law in the Gilded Age* (Chicago: University of Chicago Press, 1968); James C. Mohr, *Doctors and the Law: Medical Jurisprudence in Nineteenth-Century America* (Baltimore, MD: Johns Hopkins University Press, 1993); Ian Burney, *Poison, Detection, and the Victorian Imagination* (Manchester: Manchester University Press, 2006).

46. Langbein, *Origins of Adversary Criminal Trial*, 334–35, makes this point with regard to the unwillingness of late eighteenth-century juries to convict for capital crimes. They did not trust adversary procedure.

47. Edmond, "Just Truth?," 80–81; Lynch, "Science, Truth, and Forensic Cultures," 62; Mohr, *Doctors and the Law*, 10.

48. Pioneering in this regard is David Arnold, *Toxic Histories: Poison and Pollution in Modern India* (Cambridge: Cambridge University Press, 2016).

49. Arguably, DNA is an exception. Its significance as a forensic technique may be on a par with its more general biological and medical significations.

50. "Bowled Over: Assessing the Contents of the Toilet Bowl in the Name of Crime Prevention," *Nature* 537 (September 15, 2016): 280.

51. Clark and Crawford, *Legal Medicine in History*.

52. Most work focuses on Anglo-American contexts. Katherine D. Watson, *Forensic Medicine in Western Society* (London: Routledge, 2010); Alison Adam, *A History of Forensic Science* (London: Routledge, 2016).

53. Simon A. Cole, *Suspect Identities: A History of Fingerprinting* (Cambridge, MA: Harvard University Press, 2001); Chandak Sengoopta, *Imprint of the Raj* (London: Macmillan, 2003).

54. Katherine Watson, *Poisoned Lives: English Poisoners and their Victims* (London: Hambledon, 2004); Burney, *Poison, Detection, and the Victorian Imagination*; James C. Whorton, *The Arsenic Century* (Oxford: Oxford University Press, 2010).

55. Ken Alder, *The Lie Detectors: The History of an American Obsession* (New York: Free Press, 2007); Geoffrey Bunn, *The Truth Machine: A Social History of the Lie Detector* (Baltimore: Johns Hopkins University Press, 2012); Melissa Littlefield, *The Lying Brain: Lie Detection in Science and Science Fiction* (Ann Arbor: University of Michigan Press, 2011).

56. P. Becker and R. Wetzell, *Criminals and Their Scientists: The History of Criminology in International Perspective* (Cambridge: Cambridge University Press, 2006); Peter Becker, *Dem Täter auf der Spur: Eine Geschichte der Kriminalistik* (Darmstadt: Primus Verlage, 2005); Burney and Pemberton, *Murder and the Making of English CSI*; Neil Davie, *Tracing the Criminal: The Rise of Scientific Criminology in Britain, 1860–1918* (Oxford: Bardwell, 2006).

57. J. Aronson, *Genetic Witness: Science, Law, and Controversy in the Making of DNA Profiling* (New Brunswick, NJ: Rutgers, 2007); Michael Lynch et al., *Truth Machine: The Contentious History of DNA Fingerprinting* (Chicago: University of Chicago Press, 2008).

58. Rosenberg, *Trial of Assassin Guiteau*; Joel Eigen, *Mad-Doctors in the Dock: Defending the Diagnosis, 1760–1913* (Baltimore: Johns Hopkins University Press, 2016); idem, *Unconscious Crime: Mental Absence and Criminal Responsibility* (Baltimore: Johns Hopkins University Press, 2003); Mohr, *Doctors and the Law.*

59. Jeffrey Jentzen, *Death Investigation in America: Coroners, Medical Examiners, and the Pursuit of Medical Certainty* (Cambridge, MA: Harvard University Press, 2009); Ian A. Burney, *Bodies of Evidence: Medicine and the Politics of the English Inquest, 1830–1926* (Baltimore: Johns Hopkins University Press, 2000).

60. Golan, *Laws of Men and Laws of Nature.*

61. Michel Foucault, *Discipline and Punish: The Birth of the Prison*, trans. Alan Sheridan (New York: Vintage, 1979). Cf. Patrick Joyce, *The Rule of Freedom: Liberalism and the Modern City* (London: Verso, 2003).

62. For contemporary global practices, see Douglas H. Ubelaker, *The Global Practice of Forensic Science* (Chichester: Wiley-Blackwell, 2015).

63. A classic study of the consolidation of such expertise is John Law and Michel Callon, "The Life and Death of an Aircraft: A Network Analysis of Technical Change," in *Shaping Technology/Building Society: Studies in Sociotechnical Change*, ed. Wiebe Bijker and John Law (Cambridge: Massachusetts Institute of Technology Press, 1992), 21–52.

# PART ONE

# EVIDENCE AND EPISTEMOLOGY

# CHAPTER ONE

# The Value(s) of Methods

## Method Selection in German Forensic Toxicology in the Second Half of the Nineteenth Century

MARCUS B. CARRIER

The selection of an analytical method that is appropriate to answer a specific question is one of the main tasks of an individual scientist and of the entire scientific community. While this may seem like an uncontested and simple enough statement, the matter might become more complicated once science leaves the "purely" academic realm and enters the context of application. That is because, in these cases, the interests of different social systems collide or at least interact so that a new, specific rationale for method selection regarding scientific expertise emerges. This is also (and perhaps especially) true for the legal context and forensic science on which this chapter focuses. In legal proceedings, the relevant actors who evaluate the suitability of methods are not only the scientific experts themselves, but also judges, attorneys, and juries who have to accept or dismiss these methods to reach a verdict. Therefore scientists must choose what they consider the most suitable method, as well as the one that will convince laypeople, which leads to a particular rationale for method selection being applied in the context of legal matters.

This chapter is divided into four parts. First, I give a short overview of the concept of "epistemic values," drawn largely from the works of Thomas Kuhn and Ernan McMullin. I borrow this concept originating from the philosophy of science in order to explain my understanding of "rationality" in this context and to provide a theoretical basis for my historical analysis. As I explain, however, this concept is understood here in a descriptive way rather than in its original normative meaning. Second, I give a brief (and necessarily superficial) overview of the legal history of the German states in the nineteenth century. In particular, I focus on how criminal procedures changed around the revolution of 1848 and on how these changes affected the status of experts in court trials. Third, using textbooks and archival sources, I present the six values for method selection that I have identified in my historical analysis: sensitivity, selectivity, simplicity, redundancy, minimal sample size, and what I call intuitive comprehensibility. I argue that the former three of these values originated from analytical chemistry, and the latter three came from the legal context. This will bring me, fourth, to the conclusion that the values involved in method selection constitute a particular rationale negotiated between science and law, and that the specific values employed by experts rely on both scientific and local legal culture.

## Values

The underlying theoretical concept of this chapter is based on the philosophical debates over whether—and, if yes, to what degree—scientific judgments (e.g., theory choice) are more similar to some kind of value judgment than to an algorithmic sort of data processing. That is, do guiding values in science exist that are based not on data or experiments but on a (at least tentative) consensus within the scientific community or even within society?[1] Such a consensus on the relevant values does not necessarily lead to only one preferred theory or method; that is, it does not preclude the possibility of dissent and debate within the scientific community. Different scientists can come to completely different evaluations of theories while invoking the same values because these values can be made precise in different ways, emphasized differently, and be traded off against each other. That is exactly the reason Kuhn insisted on calling them "values" and not "criteria."[2]

In my view, such values exist and are not restricted to a consensus within the scientific community. Thus they cannot necessarily be described as "internal" to science. Rather, the belonging of members to the community that is relevant for negotiating the values in question is determined by the context of application of the epistemic practice. This holds true for the specific set of values guiding the choice of methods in a given context, their interpretation, as well as the evaluation of necessary trade-offs between such values.

This concept of ambiguous and debated values and trade-offs between them is useful to get a better understanding of the choices scientists made (and make) in practice. Chemical analysis is not a determined route where the chemist was unable to make any choices. In modern textbooks of analytic chemistry, an abundancy of methods can be found to check a sample for a specific possible constituent. The same is true when a chemist is not looking for a specific substance but rather is employing multiple tests to determine the constituents of a given specimen. In all cases, chemists have to make a choice that is dependent on but not completely determined by the specific circumstances, the questions asked, and the available means. My goal here is to better understand that choice.

In the history of science, this concept of values in science has so far not been the focus of attention. But Lorraine Daston, Peter Galison,[3] and Ernst Homburg have done important work in this regard. In the context of this chapter, the work of the latter is especially interesting. According to Homburg's "The Rise of Analytical Chemistry,"[4] chemical-analytical methods were adopted relatively late in the chemical industry compared to the rapid development of the field of analytical chemistry during the nineteenth century. Moreover, the methods used by the industry for tasks such as product control or purity tests differed from the methods preferred in academic laboratories at the time.[5] Homburg's explanation for this finding can be reconstructed as a value conflict in the terminology employed here: whereas academic analytical chemistry preferred methods with a high level of accuracy and reliability, the chemical industry primarily needed methods that yielded results quickly. Companies could not afford to strive for higher accuracy than necessary at the cost of speed in their production chain. Thus the regular use of analytical chemistry in industrial contexts was dependent on the development of instruments that simultaneously yielded sufficiently high accuracy and speed of analysis, and thus improved efficiency of such tests in the industrial context.[6] Industrial analytical chemists, therefore, had to satisfy the needs and values of two social systems—that is, science and economy—which led them to develop a new value system for method selection or at least led them to reconsider trade-offs between accuracy and speed.

In this chapter, I use this concept of values and their trade-offs in the context of application for understanding forensic analytical toxicology in Germany. I am interested here in what values shaped specific method decisions of forensic experts and how these values and their evaluation are influenced by or dependent on the legal context. Although the history of forensic expertise in recent years has received more interest from the history of science, the focus of studies on that topic lies mainly on the American and English legal systems.[7] Important exceptions to that trend exist but focus their attention in turn mainly on France and its legal context.[8] The history of German forensic expertise, however, has been largely ignored by the history of science, at least after the end of the Holy Roman Empire in 1806, and thus the formal end of validity of the so-called *Carolina*.[9] German social history, by contrast, is interested in scientific expertise in Germany, but—following Lutz Raphael's idea of the "scientification of the social"[10]—it focuses mainly on the social and human sciences (e.g., psychology, criminology, and sociology) and especially on their impact on the development of the modern welfare state.[11] So far, the history of forensic chemistry in Germany has been a research gap.

In the following, I practically demonstrate how I think this field can be approached by making use of the concept of values in science. I am interested here in the practices of chemical experts in German courts of law between the 1840s and circa 1900. Having examined transcripts and records of trials, as well as the available textbook material for forensic chemistry of that time, I have reconstructed six values that I found crucial for guiding the chemists' choice of method. For this reconstruction, I especially used the parts of textbooks where general claims about the duties of forensic chemists are made, as well as sections where different methods are explicitly compared to highlight their advantages and disadvantages in different situations. Such textbooks are important in shaping the community consensus on the values that are to be considered and are at the same time, because they are usually written by practitioners, the result of experiences and negotiations with the legal system. Found in the written expert opinions in the archival sources is whether such values had an actual impact on the chemists' choice. This can be done by examining the expert opinions for explicit and implicit justifications for such choices.

The local legal culture seems of high importance to better understand the choices made. Thus, in the next section, I give a brief overview over the essential changes in the German criminal procedural law that took place in the nineteenth century.

## German Legal History

In the first half of the nineteenth century, most German states adhered to the "inquisitorial process" as defined in the *Carolina* of 1532.[12] The only exceptions were the states on the left bank of the river Rhine, where Napoleon instituted his code of criminal procedure (*Code d'instruction criminelle*) in 1808, which remained valid after Napoleon's defeat and the restoration period.[13] The inquisitorial process was a written-down and secret procedure. It was based on strict rules of proof. Finally, the inquisitorial process did not—or, at least, not for the most part—distinguish between the prosecution and the judge.[14]

In this system, expertise played a relatively subordinated role. The most important piece of evidence in inquisitorial criminal proceedings was the confession of the defendant. Expert testimonies by physicians and midwives were mandatory for some defined crimes to establish the—in modern terms— physical elements of the offense, that is, what happened and whether it is criminal. Importantly, forensic expertise normally did not play a role in conviction, unlike modern forensic identification (fingerprinting, DNA, etc.), for example. Medicine and also toxicology were used by the court for the determination of a crime, not to identify the culprit.

Notably, poisoning was not one of the defined offenses that made experts mandatory. Rather, the corresponding Article 37 of the *Carolina* determined only that the important pieces of circumstantial evidence were someone having bought poison, having had a fight with the deceased, or profiting from the person's death in some other way. Rather complicated rules determined how much circumstantial evidence was needed to justify either a conviction, torture to obtain a confession,[15] or a "punishment for suspicion" (*Verdachtsstrafe*) in case the evidence was not enough to justify the full penalty. If neither was achieved, the charges were dropped but the defendant was not considered not guilty. An acquittal was in need of evidence for the defendant's innocence.[16]

This complicated system of proofs and conditions constituted a kind of special judicial epistemology with—at least in theory—clearly defined rules to establish knowledge about the guilt or innocence of a defendant. Although the use and the status of experts were rather modest, it established a tradition of legal culture that was based on legal comments and revisions of cases. This tradition made it easy to adapt to the use of science in the courtroom, as Catherine Crawford's now-classic article has shown.[17] Nonetheless, the exact status of nonlegal expertise remained underdetermined and was widely debated.[18]

The important changes of the legal procedure in almost all German states took place around the liberal revolution of 1848.[19] The replacement of the inquisitorial process by a reformed criminal procedure, based largely on the French model, was one of the key demands of the revolutionaries. Although the revolution had already failed by 1849, most German states left the revolutionary reforms of criminal procedure in place.[20] The reformed procedure was a public and oral process, and—at least for certain crimes, such as murder—it was a trial by jury. The judge still conducted the main investigation; for example, he posed questions to the witnesses and appointed experts.

In this way, the inquisitorial, comment and revision–based legal tradition stayed intact. But a public prosecutor's office was implemented for the preliminary investigation, and prosecution and judge were thus separated. Finally, the principle of free consideration of evidence on the part of both judge and jury replaced the strict rules of proof.[21] That is, the formal rules of evaluation of evidence—where the testimony of two credible witnesses constituted the point of reference for a so-called full proof—were abolished. The judges and juries were considered able and obligated to find their own standard to evaluate the given evidence including expert testimonies.

Although the legal situation in the German states remained anything but uniform, this reformed criminal procedure was used in most German states. The new unified codes of criminal procedure, which were implemented in Austria in 1873 and in the German Empire in 1877, were both based on this reformed process, thus abolishing for good the Carolinian inquisitorial process in the German states.[22]

Regarding the status of forensic experts in courts of law, the reformed criminal procedure brought two important changes. First, the public and oral procedure opened up new possibilities for the performance of experts. They did not have to convince the judge by their written testimony alone, but could use oral presentations to further strengthen their argument and enhance their credibility.[23] Second, the principle of free consideration of evidence allowed judges and juries to give more weight to expert testimonies than the Carolinian strict rules of proof had allowed before.[24] The eyewitness was no longer the foundation of the modern criminal procedural law, replaced instead by the freedom of judges and juries to evaluate evidence in its context.

## Values in Action

Following this short overview of German legal history, I address the values that I have identified. In my analysis of nineteenth-century written expert testimonies and textbooks for forensic chemistry, I established six values: selectivity, sensitivity, simplicity, redundancy, minimal sample size,[25] and intuitive comprehensibility. Below, I present these values in detail.

None of these values is objectively "better" or "worse" than the others. All of them are important and must be evaluated with reference to the situation and context at hand by the experts. The point is, again, that not all methods can adhere to all values equally at the same time. Method selection is always an evaluation of trade-offs between these values, and it is exactly this kind of decision in which I am interested here.

### Values from the Lab: Selectivity, Sensitivity, and Simplicity

I understand "selectivity" as the quality of tests to only react positively (e.g., by changing colors) if specific substances are present. For example, a test that reacts positively to only one substance—for example, arsenic—would be selective, whereas a test that reacts positively to a whole class of substances—for example, halides[26]—is not selective. The importance of selectivity depended on the details of the case. That is, the relevant degree of selectivity depended on the suspicion as to which poison could have been used in a particular case. To give an example, in 1844,[27] during an investigation in Detmold, the capital of the Principality of Lippe in northwest Germany, the appointed expert deliberately only used tests with very low selectivity. In this case, the local knacker was accused of having poisoned a customer, yet it was not clear which poison he could have used. Therefore the pharmacist Tronlarius employed tests that were not particularly selective in order to narrow down the number of possible poisons. He started with passing hydrogen sulphide into his prepared sample, which failed to produce any sediment. With this, he deduced that neither mercury, copper, lead, antimony, nor arsenic could have been present in the sample. As Tronlarius was ready to admit, no positive result would have proven the presence of any of these metals but could only hint at possible candidates. Yet this negative result was in his mind suited to exclude five possible poisons with one single test. He further supported this first tentative result with other tests and still found nothing. Ultimately, the knacker was acquitted.[28]

By contrast, in another case that also took place in Detmold in 1882–83, in which two brothers were accused of having poisoned their mother, both appointed experts exclusively tested for arsenic using the Marsh test, which is selective. In this case, the preliminary investigation had shown that the poisoning had likely been done with arsenic, so there was no reason to test for anything else if the test for arsenic was successful. Both experts found arsenic, and one of the brothers was subsequently sent to prison.[29]

Recommendations given in relevant textbooks pointed in a similar direction, that is, to use methods whose selectivity was adjusted to the issue at hand. In his textbook for judicial chemistry (*Gerichtliche Chemie*), for example, the Viennese chemist Franz Schneider (1812–97) wrote: "If the task is simply to deliver proof of a specific substance which has been named by the court, it would seem obvious to only examine the solution prepared for analysis with those reagents by which said substance can be detected and synthesized. If, however, the court does merely unspecifically hint at various substances, the inquiry has to be executed in a more general fashion."[30]

Selectivity was largely uncontested as a relevant value. The specific context of an investigation would determine the preferable level of selectivity, but considerations regarding this value always played an important role in method selection. Even when in 1844 the analytic chemists Carl R. Fresenius (1818–97) and Lambert von Babo (1818–99) criticized the nearly exclusive use of the Marsh test on the basis of its high level of selectivity, the importance of this value itself was not questioned. Fresenius and Babo argued that the Marsh test works well only if the arsenic was administered in its oxidized form (i.e., diarsenic trioxide in modern systematic nomenclature). Other arsenic compounds could not be detected using this method, they claimed, and therefore the Marsh test was insufficient to rule out arsenic poisoning. In fact, they recommended the use of the test with hydrogen sulphide used by Tronlarius in the abovementioned case. The Marsh test could then be used as further confirmation if arsenic was found, but a negative result of the Marsh test proved nothing in their opinion, precisely because it was too selective. Selectivity here was not questioned as a determining factor for method selection but rather had to be weighed against the risk of false negatives.[31]

The second value is sensitivity, by which I refer to the capacity of methods to detect even small amounts of certain substances or poisons. High selectivity is intended to exclude false positives, while high sensitivity is supposed to rule out falsely negative findings. As was the case with selectivity before, sensitivity seems to have been nearly universally accepted in the community of forensic toxicologists in the German states. Sometimes it could even be the most important sign of quality of a forensic test. In his textbook, Schneider compared different methods for the detection of arsenic considering only their sensitivities.[32] In general, the more sensitive a test for poison was, the better.

The demand for high sensitivity was especially high when the courts asked for a quantitative analysis. Until now, I have only talked about qualitative chemical analysis, that is, methods that determine whether a certain substance is present in a sample. In many cases, this is only half the work, since the question of *how much* of the substance is present might also be of high importance. But the question about the necessity of a quantitative analysis remained a point of contention between the courts and the community of forensic toxicologists and even within the community of forensic toxicologists. In general, courts demanded a quantitative analysis to establish without doubt that the administered quantity of poison was the cause of death, in other words, that this quantity was sufficient to kill. This principle was not always followed, however, as seen in an 1890 letter from the Prussian minister of justice, Ludwig Hermann von Schelling (1824–1908), in which he underlined the importance of conducting a quantitative analysis in cases of poisoning.[33] This might be because not all chemists were convinced that a quantitative analysis was necessary or indeed useful. Schneider, for instance, argued in his textbook that in most cases a quantitative analysis would be impossible owing to the small sample size, and that it would even be superfluous, since its interpretation would be uncertain owing to the lack of a comprehensive theory of poison absorption by the body.[34] In other words, even if a quantitative analysis could and would be done, Schneider's argument goes, the interpretation of its result would be impossible. If there is no theory of how the body absorbs a specific poison, the found amount of this poison in a certain organ clarifies nothing. Not only would a quantitative analysis be useless in the eyes of some chemists, but it could also actually do harm, as I further explain when talking about the value of minimal sample size, which also directly affects the question of quantitative analyses.

To return to the value at hand—that is, sensitivity—a short look outside of the German states can show how even such a seemingly clear-cut value could be subject to debate. That is, there were situations when an exceeding sensitivity might be problematic. Take the discussion about "normal arsenic" in France as a first example. The Menorca-born French physician and toxicologist Mathieu (or Mateu in its native spelling) Orfila (1787–1853) argued between roughly 1838 and 1841 that the human body naturally contained a certain amount of arsenic. Thus tests with an exceedingly high level of sensitivity might falsely suggest that a person had been poisoned, whereas, in fact, only this naturally occurring normal arsenic was detected.[35] Note that the underlying problem here is similar to the Fresenius-Babo critique of the Marsh test mentioned above. Here, high sensitivity should be weighed against the risk of false positives, but, again, sensitivity as such remains an important property.[36]

As a second example, consider the high-profile case of William Palmer tried in London in 1856. He was famously convicted for poisoning one of his acquaintances with strychnine. But the assigned expert, Alfred Swaine Taylor (1806–80), never found strychnine in the body but deduced strychnine as the killing agent based on the description of the tetanus-like symptoms of the victim shortly before his death. He did find antimony, but in his opinion not enough to kill somebody.[37] Here, the question (in this case, publicly discussed) was what to do if there is no sufficiently sensitive (and, in this case, also selective) test available, a question that must have been pressing, especially for organic poisons in the first half of the nineteenth century, when there was no method known to extract organic compounds as a whole from the body.[38]

"Simplicity," as I understand it here, describes the dependency of the result of the method on the experimental skill of the expert. That is, the simpler the method, the lower the likelihood of experimental mistakes. In the aforementioned Detmold case from 1882–83, for example, one of the two experts, the pharmacist Betzler, used two tests for arsenic on his specimen. First, he used the Marsh test as James Marsh (1794–1846) had described it in 1836.[39] But early on he expressed doubt in the suitability of this method and terminated the test prematurely owing to the strong foaming of his specimen.[40] He then went on to use his second test, which he considered to be much more reliable. Betzler neither revealed the name of this test nor did he specify his sources. Judging from his description, however, it is safe to assume that he was using the Marsh-Berzelius apparatus, which was a modification of the Marsh test by the Swedish chemist Jöns Jakob Berzelius (1779–1848). Chemically speaking, but without going into detail here, the two tests that Betzler executed were identical; Berzelius's modification merely simplified the execution of the Marsh test by eliminating certain possible sources of error.[41] Nonetheless, the higher degree of simplicity of this modification apparently convinced Betzler of its higher trustworthiness. Furthermore, the mere existence of such modifications for analytic tests and the search of chemists for a simpler solution indicate that the value of simplicity was part of considerations about the selection and development of analytic methods.

Still, when compared to the other two values mentioned so far, simplicity played a subordinate role. This is evident, for example, from the general preference for the Marsh test (or its modifications). As early as 1844, the German chemist Hugo Reinsch (1809–99) described a method for identifying arsenic, the Reinsch test, which was much simpler than the Marsh test.[42] But the Reinsch test was also less sensitive[43] and less selective.[44] At least in the German states, the Reinsch test never replaced the Marsh test as the standard method for arsenic detection.

One could be inclined to think that simplicity played a special role in establishing some implicit standard methods for forensic analysis. As Chris Hamlin has shown, British chemists in the nineteenth century were reluctant to give up an easy but questionable method for water analysis (the Wanklyn test) in favor of Frankland's better but more complex combustion analysis.[45] Simplicity, or rather the lack of it, could also be used to consolidate hierarchies within the community of forensic chemists. In 1840, during the Lafarge affair, the Paris-based toxicologist Orfila was called to conduct the analysis and testify only after three committees of (provincial) physicians were deemed inadequate in their skill to be trustworthy experts, which underlines tensions between center and periphery, especially in France.[46] But I am reluctant to consider simplicity as the main cause for some kind of standardizing of forensic methods. Rather, I would stress the point that all the values mentioned in this essay together serve as an implicit equivalent to what is today the Daubert standard in the United States. That is, together, they serve to establish a pool of standard methods without writing them down or legally demanding them. I return to this point of implicit standardizing in the conclusion.

### Values from the Courtroom: Redundancy, Minimal Sample Size, and Intuitive Comprehensibility

So far, I have only addressed values for method selection that can be expected to govern the choice of method in analytic chemistry in general, regardless of the specific context. In fact, what I distinguished here as sensitivity and selectivity can be identified with what Homburg called "accuracy" in his account of the development of analytic chemistry with respect to its relationship with the chemical industry.[47] In the following subsection, however, I describe values for method selection that seem to have been directly linked to the use of these methods in court. An extensive (although admittedly incomplete) comparative analysis of textbooks for academic and forensic analytical chemistry suggests that the values to be considered now have directly evolved from the legal context. The reason is that such values were explicitly justified in textbooks for forensic toxicology by appeal to their relevance for expert testimony in court. In some cases, the suggested significance of these values was even directly contrasted to requirements in the academic laboratory.

The first of these values is redundancy. I use this notion for denoting recommendations in textbooks not to use only one but several, or even all, possible methods to prove (or disprove) the presence of a poison. In this sense, redundancy occupies a peculiar position value for method *selection* since it actually advises to not select at all. For example, Schneider noted in his textbook that "the substance has to be examined with *all* possible reagents."[48] His justification for this recommendation was directly derived from the legal context: "Only by doing that [i.e., using all possible reactions] will his [the chemist's] judgment be exhaustive and, at the same time, will it counter those doubts and objections which might be raised by the court or the attorneys."[49]

This justification could be interpreted cynically by saying that the task of redundancy was to avoid critical questions and thereby to protect the authority of the analyst. Put more positively, one could also argue that the burden of proof in the courtroom was higher than in the academic laboratory and that multiple tests were therefore needed to overcome this threshold. Either way, it seems as though redundancy was born out of the attempt to cope with a specific challenge, namely, the need of the analyst to face laypersons and to convince them.

Redundancy was not only recommended in textbooks but also—at least partly—put into practice. For example, in 1871, a woman and her lover were accused of having poisoned her husband in Springe, near Hannover. They had already confessed to having used potassium chromate, which the lover had brought home from his work in a dye house.[50] The appointed expert, whose task was to confirm this supposed progression of events, precipitated chromium oxide in a series of tests that also served to disprove the presence of several other metals.[51] Having done so, he again dissolved the supposed chromium oxide and used not one but two detection reactions in order to make sure that he had indeed found chromium. First, he used silver nitrate to precipitate red sediment, and, second, he used lead to precipitate yellow sediment.[52] Carried out in this particular combination, these two reactions seem to have been sufficient to prove the presence of chrome, and the couple was convicted.[53]

There is a rather obvious question about what to do if multiple tests do not yield the same result and contradict each other. Are there some tests to be more trusted than others? In more adversarial legal systems—such as in the United States, for instance—the use of multiple tests did not increase trust in forensic expertise but rather was the basis for expert disagreement in court.[54] In more inquisitorial legal systems where the expert witness serves the judge or the court and not the defense or the prosecution, it is not clear how contradicting forensic evidence was handled. In the case of simply disagreeing experts, the court could settle on the hierarchy within the community, as the abovementioned Lafarge affair and Orfila's testimony show. But this does not settle the question of how to handle multiple tests with contradicting results conducted by the same expert who obeyed the value of redundancy. This must remain an open question, since this situation simply did not occur in any of the cases analyzed here.

A different point connected to redundancy is the question of how many tests should be used in practice. Note that in the Springe case, the appointed expert did not use all possible tests, but only two. As I show, this practice is connected to the value of minimal sample size, which stood in direct conflict with redundancy. According to minimal sample size, forensic analysts were advised to prefer methods that consumed the smallest sample possible. Chemical analysis is usually destructive, so it is in most cases impossible to analyze the same sample twice. All the chemical expert can do is use smaller amounts of the sample for different tests. Thus minimal sample size is directly linked to the material circumstances of the investigation in practice. Usually, the amount of a sample—taken from the victim at the postmortem or from the potentially poisoned food or drink—was limited. This is an important difference from other analytic settings, such as water analysis or product control. At the same time, it was possible that either the judge deemed a second opinion necessary, or that the prosecution or defense could appeal the verdict, which again could require a second opinion. Take once more the Lafarge affair, where the body of the presumed victim had to be exhumed because all previously available sample material had been used.[55]

In practice, minimal sample size could be enforced directly by the judge, who could instruct the expert to save evidential material. This occurred in the aforementioned case of poisoning by potassium chromate, in which the expert was instructed to return one-half of the sample to the court after having finished his analysis.[56] Yet even without such direct instructions, experts were allowed to decide not to use their entire sample owing to considerations of minimal sample size.[57] Minimal sample size was recommended in textbooks as well. Friedrich Müller wrote in his *Compendium für Staatsarzneikunde* that the analyst "has to be committed to only use a *minimum* [of the sample] in every single experiment."[58]

Using a minimal sample size stood in direct and obvious conflict with both redundancy and the requests of courts for quantitative analyses. The conflict with redundancy is evident because it is simply not possible to carry out a variety of different detection reactions while using only a small fraction of the sample at the same time. Quantitative analysis posed a similar problem, as analysts were faced with a quantitative analysis that was more reliable with a larger quantity of the sample. Quantitative analysis simply needs more material than qualitative analysis does. In the Detmold case, for example, both appointed experts refused to conduct a quantitative analysis owing to the small size of their respective samples.[59] Also, forensic toxicology textbooks described the need for quantitative analyses as controversial owing to the lack of a theory for poison absorption.[60] By contrast, the Viennese chemist Ernst Ludwig explained that, in most cases, a purely qualitative analysis would be insufficient.[61] But he strongly opposed minimal sample size in general and explained that it would be "foolish" not to give preference to redundancy in favor of minimal sample size.[62] But even Ludwig took minimal sample size into consideration and addressed the conflict between it and quantitative analysis, despite dismissing minimal sample size as a decisive factor for such cases in the end. Before he could arrive at such a conclusion, he needed to compare minimal sample size with competing values.

The value of minimal sample size did not collide with all other values or requests, however. In fact, it was highly compatible with the value of sensitivity. Seeing as highly sensitive methods can detect even minute amounts of a certain substance, they also use small sample sizes.

One way of dealing with the value conflict between redundancy and parsimony, as the example of Ludwig shows, was to ignore one of the opposing values altogether. Another option was to invoke a mediating value, such as the final value I describe here, what I call intuitive comprehensibility. By this, I refer to the courtroom presentation of the substance in question in such a way that everyone present would be able to comprehend. This is articulated most strongly by Schneider, who also explicitly contrasted this practice to practices commonly found in academic analytic chemistry:

> Whereas the analytical chemist will choose such detection methods which are either distinguished by their characteristic chemical behavior or allow for the most accurate quantitative analysis, the forensic chemist must aim to present his substances in such a way that they can easily be recognized and distinguished by everyone. If, for instance, his task is to prove a poisoning by copper or determine the proportion of copper in food products and the like, there is no question that the judge and all laypeople will be much more convinced of the presence of this metal, if iron is used to precipitate it in its elementary form—in which it can be easily recognized by everyone due to its copper red color—than if it is dissolved in excess ammonia, which produces the dark blue color by which every chemist detects the presence of copper in qualitative analyses.[63]

Schneider's argument is simple: since everyone knows what copper looks like, the red color of its elementary form is more convincing to laypeople than the distinct blue color that it displays when dissolved in ammonia. The expert chemist, then, would not have to explain why the blue color is characteristic, but could rather just (perhaps even literally) point to the easily recognizable red color. According to Schneider, intuitive comprehensibility was always preferred and could be used to resolve the value conflict between redundancy and parsimony in the sense that redundancy was only needed if for some reason intuitive comprehensibility could not be achieved.[64] Moreover, in the passage quoted above, Schneider explicitly uses the preference for intuitively comprehensible methods as a means of contrasting the academic analytical chemist with the forensic chemist. The most obvious distinction between the two professions, in his opinion, was that they differed in their modes of method selection.

Intuitive comprehensibility was also put into practice. The most prominent example of this might be the aforementioned preference of forensic chemists for the Marsh test and its modifications over other methods such as the Reinsch test. In the Marsh test, the examined sample is submerged together with zinc in sulfuric acid, and the released gas is ignited. If arsenic is present, the gas consists of arsenous hydride (or arsine), and a mirror of elementary arsenic can be produced on a porcelain dish held over the flame. Otherwise, the gas consists of hydrogen, which burns without producing such a mirror.[65] By contrast, in the Reinsch test, the sample is dissolved and boiled in hydrochloric acid, and a copper strip is held in the solution. If arsenic is present, the copper strip takes on a grayish or black color.[66] The production of pure, metallic arsenic in the Marsh test can be considered more intuitively comprehensible in the terms used here and in the sense of Schneider's recommendations. This also provides a further explanation of the nearly exclusive use of the Marsh test in investigations like the Detmold case of 1882–83.[67] To enhance the persuasive power of his argument, one of the experts even sent the arsenic mirror he had produced to court.[68]

The effect of intuitive comprehensibility can even be observed in cases where it was impossible to synthesize the elementary form of a substance. For instance, the expert in the Springe case of poisoning with potassium chromate sent not only his two precipitations to court but also—in a second expert testimony—comparative probes in which he had executed the same reaction with pure potassium chromate. What he had used for these comparative samples was not just any potassium chromate, but the same potassium chromate he had gathered from the dye house where the defendant had worked and presumably obtained the potassium chromate he had used in the murder. Although in this case he was not able to point to the elementary and easily recognizable form of the metal, as Schneider would have recommended him to do, he demonstrated the decisiveness of the used reaction by directly comparing it to the suspected poison.[69]

There is another important point to be made about intuitive comprehensibility: in Schneider's opinion, this value served the important purpose—as did redundancy before—of convincing laypeople. Much as minimal sample size was an answer to the material conditions of court expertise, redundancy and intuitive comprehensibility are explicitly justified in textbooks with the reference to the presence of laypeople at court and the need to convince them. After all, in the end these laypeople had to decide whether to follow an expert opinion in their verdict.

## Conclusion

In this chapter, I have argued that the concept of epistemic values—as introduced by Kuhn and McMullin—can be used to gain a better understanding of science in the context of application in general, and of nineteenth-century forensic toxicology in particular. In the case of forensic toxicology, the values that governed the selection of methods by experts sprang from both science and the legal context. Whereas the values of selectivity, sensitivity, and simplicity were tied to developments within analytical chemistry in general, redundancy, minimal sample size, and intuitive comprehensibility directly resulted from the material circumstances of court cases and the need to convince the judging laypeople in the legal context of the results of the experiments in particular. When such analytical methods entered a different realm of application, their appropriateness had to be assessed, and their setup modified, according to the standards governing this field.

This brings up the question of how the set of values changes over time, and of how it reacted to changes in both science and the legal system during the nineteenth century. Since I have so far concentrated on the German states in the second half of the nineteenth century—that is, after the midcentury reforms of the legal process—I cannot give a comprehensive answer to this question at this point. But because the number of laypeople in the new process considerably increased owing to the introduction of a jury for some cases, it is reasonable to assume that this reform directly affected the values, since at least two of them were explicitly linked to the problem of convincing laypeople. As a future follow-up question, I would ask how experts dealt with the different roles of laypeople at court. Did they just consider judges, attorneys,

prosecutors, and juries to be equal—namely, equally chemically ignorant—
or did they make distinctions in how to address and convince them?

Moreover, as Schneider's explicit distinction between academic analytical
and forensic chemists indicates, the specific mode of method selection might
be linked to the professionalization of forensic chemistry. Further research
in this direction has yet to be done, but it would seem reasonable to assume
that the values applied in this and other contexts had a direct impact on science and on the establishment of analytical chemistry in public service. For
forensic chemistry in Germany, for instance, the privilege of testifying as expert witnesses in poisoning trials seems to have shifted from pharmacists
to chemists at the turn of the twentieth century. This is at least one possible
explanation for the already mentioned administrative order—or rather
reminder—issued in 1890 by the Prussian minister of justice, to appoint
chemists rather than pharmacists as experts in poisoning cases.[70]

Furthermore, the set of values that emerges from this process might contain conflicting elements. In the case under consideration, minimal sample
size and redundancy do not match well. The community may react to such
value conflicts in various ways. As the example of Ludwig demonstrates,
experts can just ignore such a conflict by strongly prioritizing either one
of the opposing values. The other way is to introduce a mediating value,
such as intuitive comprehensibility. In both cases, practitioners had to
confront the value conflict and find some way of resolving it. In other
words, as McMullin's and Kuhn's understanding of values in science already suggests on the theoretical level, value judgments were subject to constant negotiation and did not clearly predetermine a certain outcome or
choice of methods.

As I hinted above, however, this does not mean that this set of values could
not serve the purpose of implicitly standardizing methods for forensic toxicology. On the contrary, the fact that the Marsh test matched so well the values considered here was partly responsible for its nearly exclusive use when
arsenic poisoning was suspected. Although some chemists, like Fresenius,
asked for it, there existed no set of mandatory forensic methods.[71] Nonetheless,
there was standardization, as the values discussed here can be considered as
an implicit equivalent to the modern Daubert standard. Like the Daubert
standard, the values essentially leave the decision about which method is
appropriate to the scientific community. Yet these values might also have a
conservative effect in that they possibly highly favor widely known and easy
methods over newer ones without focusing on the quality of the respective

tests.[72] Thus further research in this regard should ask when new analytic methods were introduced in academic analytic chemistry and when did these methods then arrive in the courtroom. The interesting point here would then be if there is a delay, and whether this delay can be explained by the conservative effect of these values. Furthermore, it would be interesting to see if, during the switch from pharmacists to chemists as experts, new methods were introduced faster to amplify the chemists' claim to be in possession of superior knowledge.

On a broader scale, the framework I have employed here might have some implications for the history of forensic science and for the history of expertise in general. That is, it further strengthens the idea that local context is of utmost importance to better understanding expertise as a practice. Because the specific context contributes to the specific aims of expert opinions as well as to the means by which these aims are achieved, the values derived from these aims and means will always be inherently local. This corresponds well with the broader concept of forensic cultures, as it emphasizes the strong entanglement of forensic practices with the surrounding culture.[73] By focusing on the cultural embeddedness of choice of methods, it becomes clear that neither the answers to seemingly "hard" questions for forensic science—that is, what happened and what stuff was used—nor the use of the "hard" physical sciences are completely detached from culture.

### ACKNOWLEDGMENTS

This chapter is based on the author's master's thesis, which was submitted at Bielefeld University in September 2015. He would like to thank his supervisors, Carsten Reinhardt and Thomas Steinhauser, as well as Martin Carrier, Julia Engelschalt, and the volume editors for their helpful comments.

### NOTES

1. For classical literature on that topic, cf. Thomas S. Kuhn, "Objectivity, Value Judgment, and Theory Choice," in *The Essential Tension: Selected Studies in Scientific Tradition and Change* (Chicago: University of Chicago Press, 1977), 320–39; Ernan McMullin, "Values in Science," *PSA: Proceedings of the Biennial Meeting of the Philosophy of Science Association* 2 (1982): 3–28; Richard Rudner, "The Scientist Qua Scientist Makes Value Judgments," *Philosophy of Science* 20, no. 1 (1953): 1–6. For an overview of the current state of the debate, cf. also Martin Carrier, "Values and Objectivity in Science: Value-Ladenness, Pluralism and the Epistemic Attitude," *Science & Education* 22, no. 10 (2013): 2547–68.

2. Cf. Kuhn, "Objectivity," 330–31.

3. Lorraine Daston and Peter Galison, *Objectivity* (New York: Zone Books, 2007). In this book, Daston and Galison speak of "epistemic virtues," by which they mean standards for method selection and modes of observation on the one hand and behavioral standards of scientists on the other (39–42). Thus at least the first part of their concept of epistemic virtues is similar or even identical with the concept of values used here. For reasons I cannot discuss comprehensively here, I still prefer the term "values" and would argue for a clearer distinction between attributes of individual scientists (i.e., "virtues") and of methods or theories (i.e., "values").

4. Ernst Homburg, "The Rise of Analytical Chemistry and Its Consequences for the Development of the German Chemical Profession (1780–1860)," *Ambix* 46, no. 1 (1999): 1–32.

5. Homburg, "Rise of Analytical Chemistry," 20–25.

6. Homburg, "Rise of Analytical Chemistry," 21.

7. Cf., e.g., Ian Burney, *Poison, Detection and the Victorian Imagination* (Manchester, NY: Manchester Universitry Press, 2006); Ian Burney and Neil Pemberton, *Murder and the Making of English CSI* (Baltimore: Johns Hopkins University Press, 2016); Tal Golan, *Laws of Men and Laws of Nature: The History of Scientific Expert Testimony in England and America* (Cambridge, MA: Harvard University Press, 2004); James C. Mohr, *Doctors and the Law: Medical Jurisprudence in Nineteenth-Century America* (Baltimore: Johns Hopkins University Press, 1993); James C. Whorton, *The Arsenic Century: How Victorian Britain Was Poisoned at Home, Work, and Play* (Oxford: Oxford University Press, 2010).

8. Cf., e.g., José Ramón Bertomeu-Sánchez and Agustí Nieto-Galan, eds., *Chemistry, Medicine, and Crime: Mateu J. B. Orfila (1787–1853) and His Times* (Sagamore Beach, MA: Science History Publications, 2006); José Ramón Bertomeu-Sánchez, *La Verdad Sobre El Caso Larfage: Ciencia, Justicia Y Ley Durante El Siglo XIX* (Barcelona: Eciciones del Serbal, 2015); Frédéric Chauvaud, *Les Experts du Crime: La Médecine Légale en France au XIXe Siècle* (Paris: Aubier, 2000).

9. For a now-classic study for the influences of the legal culture of the Holy Roman Empire on forensic science, cf. Catherine Crawford, "Legalizing Medicine: Early Modern Legal Systems and the Growth of Medico-Legal Knowledge," in *Legal Medicine in History*, ed. Michael Clark and Catherine Crawford (New York: Cambridge University Press, 1994), 89–116; cf. also Christopher Hamlin, "Forensic Cultures in Historical Perspective: Technologies of Witness, Testimony, Judgment (and Justice?)," *Studies in History and Philosophy of Biological and Biomedical Sciences* 44, no. 1 (2013): 4–15. For a short overview of the *Carolina* and the changes after its abolishment, see "German Legal History," below.

10. Lutz Raphael, "Die Verwissenschaftlichung des Sozialen als Methodische und Konzeptionelle Herausforderung für Eine Sozialgeschichte des 20. Jahrhunderts," *Geschichte und Gesellschaft* 22 (1996): 165–93.

11. Cf., e.g., Peter Becker, *Verderbnis und Entartung: Eine Geschichte der Kriminologie des 19. Jahrhunderts als Diskurs und Praxis*, Veröffentlichungen des Max-Planck-Instituts für Geschichte 176 (Göttingen: Vandenhoeck & Ruprecht, 2002); Kerstin Brückweh et al., eds., *Engineering Society: The Role of the Human and Social Sciences in Modern Societies, 1880–1980* (New York: Palgrave Macmillan, 2012); Christian Müller, *Verbrechensbekämpfung im Anstaltsstaat: Psychiatrie, Kriminologie und Strafrechtsreform in Deutschland 1871–1933*, Kritische Studien zur Geschichtswissenschaft 160 (Göttingen: Vandenhoeck & Ruprecht, 2004).

12. Cf., e.g., Enno Poppen, *Die Geschichte des Sachverständigenbeweises im Strafprozeß des Deutschsprachigen Raumes*, Göttinger Studien zur Rechtsgeschichte 16 (Göttingen: Musterschmidt Verlag, 1984), 60; Crawford, "Legalizing Medicine," 95–100; Katherine D. Watson, *Forensic Medicine in Western Society: A History* (New York: Routledge, 2011), 38–39; Hamlin, "Forensic Cultures," 6–8.

13. Dirk Blasius, "Der Kampf um die Geschworenengerichte im Vormärz," in *Sozialgeschichte Heute: Festschrift für Hans Rosenberg zum 70. Geburtstag*, ed. Hans-Ulrich Wehler (Göttingen: Vandenhoeck & Ruprecht, 1974), 148–62.

14. Poppen, *Geschichte*, 60–74.

15. The *Carolina* provided rules and a context for torture to obtain a confession until the abolishment of torture by the German states between 1740 (Prussia) and 1831 (Bavaria).

16. Poppen, *Geschichte*, 221–22.

17. Crawford, "Legalizing Medicine," 98–100.

18. Poppen, *Geschichte*, 224–31.

19. The only four exceptions to this were the principalities of Lippe and of Schaumburg-Lippe, as well as the Duchies of Mecklenburg-Schwerin and of Mecklenburg-Strelitz. Cf. Arnd Koch, "Die Gescheiterte Reform des Reformierten Strafprozesses: Liberale Prozessrechtslehre Zwischen Paulskirche und Reichsgründung," *Zeitschrift für Internationale Strafrechtsdogmatik* 4, no. 10 (2009): 542–8, here 542n11.

20. Most prominently, however, both the Austrian Empire and the Kingdom of Saxony abolished the trial by jury and thus revoked at least some of the reforms in 1853 and 1855, respectively. Cf. Koch, "Gescheiterte Reform," 547.

21. Poppen, *Geschichte*, 222–23; Rebekka Habermas, *Diebe vor Gericht: Die Entstehung der Modernen Rechtsordnung im 19. Jahrhundert* (Frankfurt: Campus, 2008), 166–73.

22. Poppen, *Geschichte*, 223.

23. In some instances, the expert could even use his testimony to present sometimes shocking visualization, as the case of the Countess Görlitz testifies, where the physiologist Theodor von Bischoff (1807–88) brought the skull of a burned corpse to the witness stand to disprove spontaneous combustion as a possible cause of death. Cf. John Lewis Heilbron, "The Affair of the Countess Görlitz," *Proceedings of the American Philosophical Society* 138, no. 2 (1994): 284–316.

24. Poppen, *Geschichte*, 232–34.

25. Originally, I had called this value "parsimony." I am thankful to Chris Hamlin for pointing out to me that this term might be easily misunderstood and conflated with the principle of using as little assumptions as possible (Occam's razor), which is distinctly not at stake here.

26. An example for this would be the Beilstein test, which detects halides in organic compounds.

27. I am well aware that this is before the 1848 turning point described above. In this context, however, I am only interested in the way (un-)certainty about an allegedly used poison affects the preference of analysts for tests of varying selectivity. The legal reforms most directly affected the values discussed below. Hence I still think that this is a useful example.

28. F. Tronlarius, Expert Testimony in the Harte Case, Detmold, 1844, Archives of the Land North Rhine-Westphalia, Section Ostwestfalen-Lippe (Landesarchiv, Nordrhein-Westfalen, Abteilung Ostwestfalen Lippe [LAV NRW OWL]), L86 Nr. 2020/20, 96–105; the cited test with hydrogen sulphide can be found on pp. 98–99.

29. Carl Betzler, Expert Testimony in the Kruse Case, Horn/Lippe, December 20, 1882, Archives of the Land NRW, Section Ostwestfalen-Lippe (LAV NRW OWL), D21 B Nr. 466, 110–16; Otto Wessel, Expert Testimonies in the Kruse Case, Detmold, November and December 1882, Archives of the Land NRW, Section Ostwestfalen-Lippe (LAV NRW OWL), D21 B Nr. 466, 62–66, 103–5; Regional Court Detmold, Court Record of the Proceedings against the Brothers Kruse, Detmold, March 9, 1883, Archives of the Land NRW, Section Ostwestfalen-Lippe (LAV NRW OWL), D21 B Nr. 466, 159–299.

30. "Soll bloss der Nachweis einer gewissen, vom Gerichte namhaft gemachten Substanz geliefert werden, so versteht es sich wohl von selbst, dass man die zur Analyse vorbereitete

Lösung nur mit jenen Reagentien prüft, durch welche diese Substanz entdeckt und dargestellt werden kann. Lautet dagegen die Frage des Gerichtes unbestimmt und deutet sie bloss vermutungsweise auf einen oder mehrere Körper hin, so muss auch die Untersuchung mehr allgemein gehalten werden." Franz Schneider, *Die Gerichtliche Chemie für Gerichtsaerzte und Juristen* (Wien: Wilhelm Braumüller, 1852), 22. Cf. also pp. 13–14 and Friedrich Wilhelm Böcker, *Memoranda der Gerichtlichen Medicin mit Besonderer Berücksichtigung der Neuern Deutschen, Preussischen und Rheinischen Gesetzgebung* (Iserlohn/Elberfeld: Julius Bädeker, 1854), 290; Ignaz Heinrich Schürmayer, *Lehrbuch der Gerichtlichen Medicin: Mit Berücksichtigung der Neueren Gesetzgebung des In- und Auslandes*, Insbesondere des Verfahrens bei Schwurgerichten 2 (Erlangen: Ferdinand Enke Verlag, 1854), 239–40.

31. Carl Remigius Fresenius and Lambert von Babo, "Ueber Ein Neues, unter Allen Umständen Sicheres Verfahren zur Ausmittelung und Quantitativen Bestimmung des Arsens bei Vergiftungsfällen," *Justus Liebigs Annalen der Chemie* 49, no. 3 (1844): 291.

32. Cf. Schneider, *Gerichtliche Chemie*, 173–74.

33. Cf. Ludwig H. von Schelling, "Die in Strafsachen wegen Giftmordes zur Feststellung des objektiven Tatbestandes erforderliche chemische Untersuchung einzelner Leichenteile," 1890, Archives of the Land Lower Saxony, Section Hannover (Niedersächsisches Landesarchiv [NLA] Hannover), Hann. 173 Acc. 30/87, Nr. 600.

34. Cf. von Schelling, "Die in Strafsachen Wegen," 10–12; cf. also Böcker, *Memoranda der Gerichtlichen Medicin*; Friedrich Müller, *Compendium der Staatsarzneikunde für Ärzte, Juristen, Studirende, Pharmaceuten und Geschworene* (Munich: Joh. Palm, 1855), 262. A different opinion can be found in Ernst Ludwig, *Medicinische Chemie in Anwendung auf Gerichtliche, Sanitätspolizeiliche und Hygienische Untersuchungen Sowie auf die Prüfung der Arzneipräparate: Ein Handbuch für Ärzte*, Apotheker, Sanitätsbeamte und Studirende 2 (Leipzig: Urban & Schwarzenberg, 1895), 144.

35. José Ramón Bertomeu-Sánchez, "Managing Uncertainty in the Academy and the Courtroom: Normal Arsenic and Nineteenth-Century Toxicology," *Isis* 104, no. 2 (2013): 197–225, doi:10.1086/670945; Ian A. Burney, "Bones of Contention: Mateu Orfila, Normal Arsenic and British Toxicology," in *Chemistry, Medicine, and Crime*, 243–59. As a side note, this uncertainty did not prevent Orfila from claiming absolute certainty for his arsenic detections in at least two court cases during which he already held but had not yet published his idea of normal arsenic. Bertomeu-Sánchez, "Managing Uncertainty."

36. A similar problem might be identified with forensic methods of DNA analysis. Today, the sensitivity of DNA testing is so high that contamination that had been under the detection limit so far might now prove to be a serious problem for forensic experts. Cf., e.g., Cynthia M. Cale et al., "Could Secondary DNA Transfer Falsely Place Someone at the Scene of Crime?," *Journal of Forensic Sciences* 61, no. 1 (2016): 196–203; Janine Helmus, Thomas Bajanowski, and Micaela Poetsch, "DNA Transfer—A Never Ending Story: A Study on Scenarios Involving a Second Person as Carrier," *International Journal of Legal Medicine* 130, no. 1 (2016): 121–25; Thomas Kamphausen et al., "Everything Clean? Transfer of DNA Traces between Textiles in the Washtub," *International Journal of Legal Medicine* 129, no. 4 (2015): 709–14.

37. Burney, *Poison*, 118–21.

38. Sacha Tomic, "Alkaloids and Crime in Early Nineteenth-Century France," in *Chemistry, Medicine, and Crime*, 261–96. Unfortunately, I have not yet found a case involving organic poisons in the archives for further analysis of this question, so I have to leave it at this for now.

39. Betzler, Expert Testimony in the Kruse Case, LAV NRW OWL, D21 B Nr. 466, 112; James Marsh, "Account of a Method of Separating Small Quantities of Arsenic from Substances with Which It May Be Mixed," *Edinburgh New Philosophical Journal* 21 (1836): 229–36.

40. Betzler, Expert Testimony in the Kruse Case, LAV NRW OWL, D21 B Nr. 466, 112.

41. Betzler, Expert Testimony in the Kruse Case, LAV NRW OWL, D21 B Nr. 466, 114–15; Jöns Jakob Berzelius, "Paton's, Marsh's und Simon's Methoden, Arsenik zu Entdecken, Nebst Bemerkungen," *Poggendorffs Annalen der Physik und Chemie* 42 (1837): 159–63; Katherine D. Watson, "Medical and Chemical Expertise in English Trials for Criminal Poisoning, 1750–1914," *Medical History* 50 (2006): 193–94. However, the strong foaming up of Betzler's specimen, which forced him to stop the first test prematurely, should not have been solved by Berzelius's modification. Rather, he probably did not properly prepare his specimen for his first test.

42. Hugo Reinsch, "Ueber das Verhalten des Metallischen Kupfers zu Einigen Metalllösungen," *Journal für Praktische Chemie* 24, no. 1 (1841): 244–50.

43. Whorton, *Arsenic Century*, 95.

44. Whereas the Marsh test only reacts to arsenic, antimony, and germanium, the Reinsch test can signify—following Reinsch's own description—arsenic, antimony, lead, silver, bismuth, tin, and mercury. Cf. Reinsch, "Ueber das Verhalten."

45. Christopher Hamlin, *A Science of Impurity: Water Analysis in Nineteenth Century Britain* (Berkeley: University of California Press, 1990), 184–90, esp. 189–90.

46. For a short overview of the case, cf. José Ramón Bertomeu-Sánchez, "Sense and Sensitivity: Mateu Orfila, the Marsh Test and the Lafarge Affaire," in *Chemistry, Medicine, and Crime*, 207–8; on the tensions between center and periphery in nineteenth-century France, cf., e.g., Robert Fox, *The Savant and the State: Science and Cultural Politics in Nineteenth-Century France* (Baltimore: Johns Hopkins University Press, 2012).

47. Homburg, "Rise of Analytical Chemistry."

48. "Die Substanz [muss] mit allen charakteristischen Reagentien geprüft . . . werden." Schneider, *Gerichtliche Chemie*, 6, emphasis in the original; cf. also Ludwig, *Medicinische Chemie*, 143.

49. "Nur dadurch wird sein Gutachten erschöpfend und zugleich jenen Zweifeln und Einwürfen begegnen, welche von Seiten des Gerichtes oder der Rechtsanwälte erhoben werden könnten." Schneider, *Gerichtliche Chemie*, 6.

50. Cf. Public Prosecutor's Office Celle, Indictment against the Widow Fricke and Bernhard Pilz, Celle, March 30, 1871, Archives of the Land Lower Saxony, Section Hannover (NLA Hannover), Hann. 71 C Nr. 127, 2–15.

51. V. Serturnes, Expert Testimony in the Fricke Case, Hameln, February 8, 1871, Archives of the Land Lower Saxony, Section Hannover (NLA Hannover), Hann. 71 C Nr. 127, 447–53.

52. Serturnes, Expert Testimony in the Fricke Case, NLA Hannover, Hann. 71 C Nr. 127, 460.

53. Jury Court Hannover, Judicial Judgments against the Widow Fricke and Bernhard Pilz, Hannover, June 7, 1871, Archives of the Land Lower Saxony, Section Hannover (NLA Hannover), Hann. 71 C Nr. 127, 38–41.

54. Christopher Hamlin, "Scientific Method and Expert Witnessing: Victorian Perspectives on a Modern Problem," *Social Studies of Science* 16, no. 3 (1986): 485–513.

55. Bertomeu-Sánchez, "Sense and Sensitivity," 208.

56. High Court Hameln, Instructions to V. Serturnes for His Chemical Analysis in the Fricke Case, Hameln, December 4, 1870, Archives of the Land Lower Saxony, Section Hannover (NLA Hannover), Hann. 71 C Nr. 127, 50–51.

57. Cf., e.g., Wessel, Expert Testimonies in the Kruse Case, LAV NRW OWL, D21 B Nr. 466, 65–66.

58. "Man muss sich die Pflicht auferlegen, bei jedem einzelnen Versuch nur ein Minimum zu verbrauchen." Müller, *Compendium der Staatsarzneikunde*, 262, emphasis original; cf. also Ludwig, *Medicinische Chemie*, 141.

59. Cf. Wessel, Expert Testimonies in the Kruse Case, LAV NRW OWL, D21 B Nr. 466, 103; Betzler, Expert Testimony in the Kruse Case, LAV NRW OWL, D21 B Nr. 466, 110. Their refusal to execute a quantitative analysis, however, did not prevent them from claiming high certainty for their estimation that the detected amount of arsenic should have been enough to be the cause of death. Cf. Regional Court Detmold, Court Record, LAV NRW OWL, D21 B Nr. 466, 264–65.

60. Cf. Schneider, *Gerichtliche Chemie*, 10–12; cf. also Böcker, *Memoranda der Gerichtlichen Medicin*; Müller, *Compendium der Staatsarzneikunde*, 262.

61. Cf. Ludwig, *Medicinische Chemie*, 144.

62. Ludwig, *Medicinische Chemie*, 143–44.

63. "Während der analytische Chemiker insbesondere solche Formbestimmungen der Körper auswählt, welche entweder durch ihr charakteristisches chemisches Verhalten besonders ausgezeichnet sind, oder welche die Menge der Substanz aufs Genaueste zu bestimmen gestatten, muss der Gerichts-Chemiker seinen Substanzen jene Formen vorzüglich zu geben bemüht sein, nach welchen die Körper von Jedermann leicht erkannt und von ähnlichen unterschieden werden können. Handelt es sich z. B. um die Ausmittelung einer Kupfervergiftung oder um den Nachweis eines Kupfergehaltes in Nahrungsmitteln u. dgl., so wird der Richter und mit ihm alle Laien unstreitig von der wirklichen Anwesenheit dieses Metalles viel bestimmter überzeugt werden, wenn dasselbe durch Fällung mittelst Eisen in seinem elementaren Zustande dargestellt wird, wo es durch seine kupferrothe Farbe von dem gemeinsten Mann gekannt ist, als wenn es, in überschüssigem Ammoniak aufgelöst, die tief blaue Färbung erzeugt, an welcher der Chemiker bei qualitativen Analysen die Gegenwart des Kupfers entdeckt." Schneider, *Gerichtliche Chemie*, 5. Cf. also chap. 3, on representing toxicological knowledge, in Burney, *Poison*, 78–115.

64. Schneider, *Gerichtliche* Chemie, 6.

65. Marsh, "Account of a Method."

66. Reinsch, "Ueber das Verhalten," 245.

67. Cf. Betzler, Expert Testimony in the Kruse Case, LAV NRW OWL, D21 B Nr. 466; Wessel, Expert Testimonies in the Kruse Case, LAV NRW OWL, D21 B Nr. 466. Ludwig criticizes such an exclusive use of the Marsh test in his argument for redundancy (in my terms) as the most important value. Because he subordinated parsimony in every case, he was not in need of a mediatory value. Cf. Ludwig, *Medicinische Chemie*, 143.

68. Cf. Betzler, Expert Testimony in the Kruse Case, LAV NRW OWL, D21 B Nr. 466, 115.

69. V. Serturnes, Expert Testimony in the Fricke Case, Hameln, February 14, 1871, Archives of the Land Lower Saxony, Section Hannover (NLA Hannover), Hann. 71 C Nr. 127, 432.

70. Schelling, "Strafsachen," 1890, NLA Hannover, Hann. 173 Acc. 30/87, Nr. 600. For the use of methods as means to gain authority in certain territories, cf. also Carsten Reinhardt, "Expertise in Methods, Methods of Expertise," in *Science in the Context of Application*, ed. Martin Carrier and Alfred Nordmann (Dordrecht: Springer, 2011), 143–59.

71. Carl Remigius Fresenius, "Ueber die Stellung des Chemikers bei Gerichtlich-Chemischen Untersuchungen und über die Anforderungen, Welche von Seiten des Richters and Ihn Gemacht Werden Können," *Justus Liebigs Annalen der Chemie* 49, no. 3 (1844): 276; cf. also Reinhardt, "Expertise in Methods," 152–56.

72. Christopher Hamlin discusses a situation like this. Cf. *Science of Impurity*, 184–90.

73. Cf. Christopher Hamlin, "Introduction: Forensic Facts, the Guts of Rights," this volume.

# CHAPTER TWO

# The Imperial Serologist and Punitive Self-Harm

## Bloodstains and Legal Pluralism in British India

MITRA SHARAFI

In 1916, news of a conflict between south Indian caste communities reached the pages of the *British Medical Journal*.[1] The Naikers had planned to build a wedding canopy in front of a temple. The Nadars objected because the structure would interfere with the carrying of palanquins in their own processions near the temple. A magistrate held an inquiry to resolve the dispute, but the Nadars remained unsatisfied. They decided to further the dispute by framing their adversaries for murder. Determined to kill one of their own and blame it on the Naikers, the Nadars selected the mistress of one of their members as the victim. Her lover suggested her because she was childless and had no relatives to avenge her death. This unfortunate woman was beaten to death. Her body was then left at the temple with sheep's blood poured on the surrounding ground. The Nadars sent a telegram to the district authorities, reporting that the woman had been murdered at the temple by the Naikers.

This chain of events might have resulted in the conviction of a hapless Naiker for murder. But colonial authorities applied a new form of forensic analysis to the bloodstains in the case. Precipitin testing had been institutionalized in India in 1914–16 through the creation of an official known as the imperial serologist, who ran a laboratory in Calcutta. This figure had no equivalent in Britain.[2] A key part of his job was to test stains for the courts.[3] For bloodstains, his task was to use precipitin testing to assess whether blood was human or animal.[4] Precipitin testing determined the species (or species group) of origin of a bloodstain.[5] In the south Indian case, the imperial serologist determined that the stains at the temple were sheep's blood. Equally, both human and ovine blood were found on a bloodstained loincloth belonging to a Naiker accused of the killing.[6] Because of the odd presence of sheep's blood both at the temple and on the Naiker's clothing, the judge concluded that all the blood had been planted, and that this man was being framed for murder. He acquitted the Naiker defendant for lack of evidence. The *BMJ* published the account so that the empire could learn of precipitin testing's success in India. When animal blood was planted "to obscure or divert the path of avenging justice," the imperial serologist could spot the trick.[7]

Although precipitin testing appeared periodically in criminal trials in Britain in the early twentieth century, it lived a much fuller life in British India.[8] Why? This essay suggests that the technique took hold in India because it spoke to a priority of the colonial state: the detection of fabricated evidence in cases involving south Asians.[9] Detecting dissimulation was a special focus of the criminal justice system in British India. Reflecting anxieties about "native mendacity," the Indian Penal Code (IPC) and Code of Criminal Procedure included extensive provisions on perjury, forgery, and false charges—all manners of manipulating the colonial legal system.[10] One leading textbook on medical jurisprudence called false evidence and fabricated charges "the great difficulty" of judges in India, devoting a chapter to the topic.[11] Fear of the false existed in Britain with strong class- and gender-based associations.[12] In British India, though, suspicions of falsity operated on a different scale: they were racialized, too.[13] Furthering Projit Mukharji's characterization of the late nineteenth and early twentieth century as the age of "serotropicality" in the British Empire, this essay explores how precipitin testing reflected anxieties about truth and trust in the context of empire.[14]

Scholars have attributed the "native mendacity" stereotype to colonial administrators' racism and ended there. This chapter entertains the possibility of dissimulation *without* accepting the racist explanations provided by the colonial sources. By examining precipitin testing, it reaches behind the racialized stereotype of the dissimulating native to propose an alternative explanation. In one important subset of cases, fabricated evidence was the product of what may be called punitive self-harm. The use of precipitin testing by the imperial serologist allows us to read colonial sources against the grain and to catch a glimpse of the ways that certain south Asian notions of collective identity, punishment, and causation remade themselves in the colonial legal context. In this way, the story of precipitin testing is a story about legal pluralism—how state and nonstate normative orders interacted.[15] Specifically, the precipitin cases reveal that a noncolonial mode of disputing (hurting oneself or one's relative to punish an adversary) morphed into something else (framing an opponent for murder by planting animal blood) to harness the punitive powers of colonial criminal law.

Colonial sources referred to the Nadar-Naiker dispute as a feud. Across the British Empire, colonial authorities found themselves pulled into back-and-forth group disputes whose longer histories remained largely illegible.[16] In such cases, the colonial state in India failed to recognize the process of adaptation involving planted animal blood. By one noncolonial model of causation and responsibility, there was little difference between killing a relative and dumping the body at the doorstep of one's adversary in response to a wrong, and doing the same thing but also planting animal blood at the adversary's doorstep and on his clothing. In the former scenario, the adversary would be deemed morally responsible (and cursed) for the death by virtue of the initial wrong. In the latter scenario, he might also be convicted of murder.

This chapter explores an alternative mode of punishment made visible through the bloodstain work of the imperial serologist. Its key theme is the interplay between two forms of disputing: colonial criminal law and punitive self-harm. The latter twisted itself to meet the requirements of the former while subverting the larger aims of the criminal justice system. This chapter proceeds in three sections. The first focuses on the history of science and institutions. It examines the rise of precipitin testing and of the creation of the imperial serologist in British India during the 1910s. The second section turns to legal history. It examines the problem of fabricated evidence, which was considered a leading challenge for the criminal justice system in late colonial India. In the third section, the history of science and disputing meet in the theme of legal pluralism. This final section examines a set of noncolonial disputing practices that may be called punitive self-harm. Precipitin testing revealed planted animal blood in some cases. For us, more significant than the falseness of this evidence is the alternative logic that we can see at play, particularly the notions of causation and responsibility that these cases embodied. Even if a person had been framed, he could still be deemed morally responsible and punishable for a death because he "caused" the death with his original wrong. Through this reasoning, punitive self-harm made the fabrication of evidence legitimate for its proponents. The colonial state saw its serological expertise as exposing mendacity rather than alterity, but its sources revealed the interactions between dispute resolution systems—and two different ways of judging human action.

## Precipitin Testing and the Imperial Serologist

For most of the twentieth century, blood group testing and blood spatter analysis were the best-known methods of forensic blood testing. Karl Landsteiner proposed the ABO blood group system in 1901.[17] This system of blood grouping enabled the expansion of blood transfusion worldwide during the first four decades of the twentieth century.[18] By the late 1930s, it rose to prominence as a method of excluding paternity and maternity in English and Indian courts.[19] It was also used occasionally to determine whether bloodstains could have belonged to the victim or perpetrator of violent crime.[20] Blood spatter analysis developed in the 1950s, emerging as a forensic technique after Indian independence.[21]

Around the same time as Landsteiner's work on ABO blood grouping, immunological researchers developed a form of testing whereby the production of antibodies in blood could be used to determine the species (or species group) of origin of a blood sample. Paul Uhlenhuth proposed in 1901 that precipitin testing could be put to forensic use. Tal Golan has traced the way precipitin testing overtook microscopy in the early twentieth century in the United States.[22] Since the mid-nineteenth century, forensic experts had attempted to determine the species of origin of a blood sample by comparing under the microscope "corpuscles" or erythrocytes from an unidentified sample with those of other species' identified samples. Most of all, one had to compare the size of red blood cells, but the overlap between species' size ranges made the method of limited use in court: one could say that a bloodstain *may* have been human, but not more. The blood sample also had to be relatively fresh.[23]

Precipitin testing could produce clearer results even on an old or small sample, or on a sample that had been exposed to heat up to 50°C to 60°C (122°F to 140°F)—relevant in the Indian context especially.[24] An animal (typically a rabbit or bird) would be injected with blood or serum of another species (e.g., human blood).[25] Over the course of several days, this animal's blood developed antibodies in response to its exposure to human blood. This blood would be extracted from the animal. From it, antihuman serum was preserved. Next, an extract would be created from the suspected stain (known to be blood through other prior testing).[26] This mystery extract would be mixed with the antihuman serum. If the stain was also of human origin, a precipitate (or "precipitin") would form within twenty minutes.[27] If no precipitate formed, the stain extract would then be mixed with a series of antisera for other species, one by one, until a precipitate formed and a match could be declared.[28]

In the first two decades of precipitin testing (from 1901 until the 1920s), there were debates about its ability to distinguish between related species.[29] Could goat's blood be distinguished from sheep's blood?[30] Could the test tell the difference between human and ape blood?[31] Alfred Swaine Taylor's posthumously edited treatise on medical jurisprudence (the leading work in the imperial metropole) oscillated between declarations like "the anti-serum is very definitely specific" (in 1920) and an acknowledgment (in 1928) that in some cases the test produced inconclusive results between similar species.[32] This qualification meant that, on occasion, precipitin testing would be of limited use. But the allied species problem was not enough to prevent the test from establishing itself in the early twentieth-century criminal courtroom.[33]

The empire played a special role in the development of precipitin testing in these early years. One of the leading works in the field was by Cambridge biologist and physician George H. F. Nuttall, who carried out sixteen thousand experiments on 586 different animal species. In the acknowledgments to his *Blood Immunity and Blood Relationship: The Precipitin Test for Blood* (1904), he thanked "seventy gentlemen" who had helped him procure a Noah's Ark of animal blood from around the world. Nuttall could not have created his antiserum reference library without the British Empire. He received animal blood from British central Africa, Canada, the Cape and Lagos colonies, Egypt, Ireland, and Uganda. The largest number of samples came from south Asia. Nuttall thanked colleagues (including many military men) in Bombay, Calcutta, Ceylon, Chitral, Kashmir, Khandesh, South Sylhet, and elsewhere.[34]

British India also played an outsized role in the field of precipitin testing through the figure of Lt.-Col. William Dunbar Sutherland. Like so many Britons in medicine and the empire, he was Scottish.[35] He obtained his medical education in Edinburgh and then at the army medical school in Netley on England's south coast. By the 1890s, Sutherland had proceeded to India as a member of the Indian Medical Service (IMS), which rotated him through south Asia on posts in Burma, Madras, and the Central Provinces. He was chief medical official in charge of a jail and a "lunatic asylum." In the run-up to the creation of the imperial serologist position, Sutherland was stationed in the chemical examiner's office in Calcutta. The chemical examiners were military physicians who ran a network of regional labs across India and tested samples for the state, including the courts.[36] Around 1908, the chemical examiner of Calcutta (at the flagship department) began exploring the potential value of precipitin testing for bloodstains, then sent to his lab for analysis. Sutherland published *Blood-stains* by 1907, and spent 1908–9 working exclusively on the technique at Calcutta's Medical College.[37] In 1910, he published a work on forensic uses of precipitin testing in the "Scientific Memoirs by Officers of the Medical and Sanitary Departments of the Government of India" series.[38] By 1912, he was offering classes in serology to medical officers from across India at the college.[39] The Calcutta chemical examiner begrudgingly shared his lab space with Sutherland and his growing traffic of incoming bloodstains until 1916, when Sutherland set up his permanent serological lab (under orders from the central Indian government) in the new Calcutta School of Tropical Medicine (est. 1914).[40]

In writing *Blood-stains* (his most extensive work on precipitin testing), Sutherland drew upon his own earlier Edinburgh dissertation (which was on blood testing, although not precipitin testing) and methods used in Frankfurt.[41] The short book was aimed at a metropolitan and imperial audience. It included mainly European case studies and issues (it predated the establishment of the imperial serologist in India), but with the occasional reference to tropical settings. Sutherland was in fact presenting German research to an English-speaking audience, noting that little appeared on the precipitin test in Anglo-American textbooks of legal medicine.[42] He was also trying to create a useful resource for "workers in the tropics and the colonies, where libraries are few and but poorly filled with works of reference."[43]

Sutherland made his mark on Indian institutions. His pursuit of precipitin testing led to the creation of the office of the imperial serologist, a figure that seems to have had no equivalent elsewhere in the empire until the 1930s.[44] Sutherland was taking his cue from the German-language research. He noted in *Blood-Stains* Paul Uhlenhuth's recommendation that the state take control of precipitin testing in order to guarantee the quality of various animals' antisera.[45] The department of the imperial serologist did just that.

There had been attempts to institutionalize precipitin testing in India before Sutherland's success in 1914–16. In 1903, E. H. Hankin had unsuccessfully proposed the idea to the central government of British India. Hankin was chemical examiner to government in India's United and Central Provinces, and had been conducting his own experiments with precipitin testing. Following Hankin, another chemical examiner, named F. N. Windsor, had tried to perfect his technique in Burma. Windsor had failed to obtain antiserum from Britain and lacked the time and facilities to make it himself. He could keep neither rabbits nor the necessary supply of blood to create a proper collection of multispecies antisera.[46] (Antiserum could take a month or two to prepare, required special refrigeration, and by Sutherland's account had to be used on-site.[47]) In 1910, the government of India revisited the question. Some officials recommended the creation of an imperial serologist. Others protested that India was "a poor country" that could not afford "fancy" appointments.[48] Most compelling of all was correspondence from Thomas Stevenson and W. H. Willcox, two physicians in London who shared their experiences of offering precipitin-based expert witness testimony in the English courts. They acknowledged that their results were not always reliable enough to be used in criminal trials. In some cases, they could say that a bloodstain was "with a high degree of probability" human blood, but not that the test was absolutely conclusive.[49] They suggested that the precipitin testing fell short of the criminal "beyond a reasonable doubt" standard of proof.

Yet just a few years later, in 1914–16, Sutherland's efforts succeeded. His own research claimed to fine-tune the test such that it produced clearer results in a larger proportion of cases. And yet if India could not afford the "luxury" of a state serologist in 1910, it could surely not afford one in the midst of World War I. The war gave license for state action, however, in a number of domains. Many new industrial and scientific enterprises were created around World War I in India, particularly in areas where India had been hitherto dependent upon German-supplied chemicals.[50] Although antiserum could be purchased in Britain, analysts considered the use of antiserum one had not made oneself to be risky.[51]

No matter what factors led to its ultimate creation in 1914–16, the imperial serologist's department was a raging success in the economic terms that the state understood best. In its first decade of existence, authorities made occasional threats to close the department if it was not self-supporting.[52] Yet not only did the imperial serologist's office support itself, but it also made a sizeable profit for the central government of India through the fees charged to provincial governments and princely states.[53] By the 1920s and 1930s, the imperial serologist's lab was processing over ten thousand bloodstained articles annually on cloth, soil, metal, wood, leather, and other materials. It did blood group testing for blood transfusions and ran a blood bank.[54] The imperial serologist and his assistants also had a lucrative side business in Wassermann testing for syphilis, retaining most of these fees privately.[55] The Calcutta Veterinary College and the municipal government's slaughterhouses provided a steady supply of animal blood.[56] Human blood (which could not be purchased) came from a nearby maternity hospital and the lab's own Wassermann test samples.[57] The imperial serologist taught courses on serology and immunology at the Calcutta School of Tropical Medicine and the All-India Institute of Hygiene and Public Health.[58] By the early 1940s, the department consisted of ten people: the imperial serologist, who was European until S. D. S. Greval's appointment in the late 1930s, and eight others, who over time were increasingly south Asian and consisted of assistant imperial serologists, clerks, and lab assistants.[59]

Early characterizations of the imperial serologist as a "fancy" luxury quickly faded away. In 1918, officials complained that authorities in the Andaman Islands' penal colony had failed to send the bloody clothing of a man being tried for murder to Calcutta for analysis by the imperial serologist. Blood examination was a "very highly specialized and extremely technical proceeding" and because "the life of the accused person frequently depends on its result," it was imperative that bloodstains be sent from Port Blair to Calcutta in future. The man in question was convicted and executed without bloodstain analysis.[60] Other officials echoed these views in the decades that followed.

Writing in 1932, one official commented that the position was more important than that of a High Court judge, routinely involving "questions of life and death as well as matters which affect the efficiency of justice."[61] The director of the Indian Medical Service agreed, calling the imperial serologist's work "an essential wheel in the machinery of the administration of justice."[62] Once the imperial serologist's laboratory was created, precipitin testing was rapidly accepted in British India. This quick endorsement made sense in light of colonial anxieties over dissimulation.

## Fabricated Evidence

Precipitin testing took hold in India earlier and in more institutionalized form than in Britain, where the courts did not consistently accept test results until the early 1930s.[63] Why this difference? Precipitin testing offered a way to allay officials' anxieties over the manipulation of the criminal justice system by south Asians.[64] Forensic science promised relief generally: by privileging scientific experts and tests, courts could depend less on Indian witnesses. The colonial notion of "native mendacity" portrayed Indians as unreliable, and perjury and forgery as rife.[65] More specifically, though, precipitin testing helped officials detect attempts to frame innocent parties through the planting of fabricated evidence. "In some cases," declared Sutherland in 1915, "the fantastic web of lies woven by the accused person or his advisers was torn to pieces by the determination of the true origin of the blood. And above all, in some cases the innocence of a suspected person was established."[66]

Writings for different audiences reflected the fact that the fear of false charges was particularly acute in India. Sutherland's *Blood-Stains* (written for a metropolitan and imperial audience) did not include a single case where an item stained with animal blood was planted in an attempt to frame an innocent person for murder.[67] Taylor's canonical work (also for Britain) similarly offered no such case studies in his section on the precipitin test.[68] By contrast, Sutherland's chapter on bloodstains in Waddell and Lyon's treatise for India included numerous cases—almost a third, in fact—involving fabricated evidence in murder, rape, and sodomy cases.[69] In a 1915 printed report by Sutherland that circulated widely among government officials, precipitin testing suggested that fabricated evidence appeared in half of forty-two case studies.[70] In an article published by Sutherland's successor R. B. Lloyd in 1926, twenty-four of twenty-seven bloodstains submitted in murder or culpable homicide cases tested positive for animal blood, suggesting fabricated evidence.[71]

Across the metropole-facing treatise literature on precipitin testing, the typical case scenario involved a person accused of murder who was found with clothing or a weapon that was bloodstained. The accused would account for the blood by telling authorities that he had recently slaughtered an animal.[72] If precipitin testing revealed that the blood was in fact human, the accused would usually be convicted. But the cases that occasionally appeared in the Indian literature were different. In these cases, a murder suspect would be found with alleged human blood on his clothing, on a weapon, or at his home. Here the suspect made no claim to have recently killed an animal. Precipitin testing revealed animal blood, and forensic authorities assumed that it had been planted to frame the innocent suspect. Colonial authorities used this form of testing to identify cases in which false charges had been made, as in the Nadar-Naiker feud with which this essay began.

In a less common scenario, an individual might try to frame others for her own death. An old woman disappeared from her house in one 1945 case from South India. Bloodstains were found under her bed and elsewhere in her home, but precipitin testing revealed that the blood was avian. After the woman was caught by the police, the local submagistrate reported that she had planted the bird blood to try to "foist a case of murder" on her relatives after "making herself scarce in a mysterious manner."[73]

There were also cases of fabricated evidence involving communal violence. In a 1944 case from a village in Bihar, a group of one hundred Hindus raided Muslim homes, claiming that a Muslim had offended them religiously by slaughtering a cow.[74] Soil and pieces of cow dung were sent to Calcutta for analysis. The soil was stained with sheep and goat blood, not the blood of a cow. The cow dung, in turn, was stained with human blood, presumably from the violence. The claim of cow killing was regarded as a false pretext, and thirty-two Hindu rioters were prosecuted.

A final category of cases involved mixed blood. In these scenarios, both human and animal blood were identified through precipitin testing, suggesting that the alleged crime of murder had occurred, but that there was also some "bolstering" or enhancement of the evidence through the planting of animal blood. Hehir and Gribble's treatise chapter on false evidence addressed this pattern. "Even in cases which are substantially true, there is generally a certain amount of concocted evidence." The job of the judge in India was not to decide which side was true and which was false. Rather, it was to identify the false evidence on each side, and then to decide the case based on what remained.[75] In other words, false evidence did not necessarily mean a false case. In a murder case from Bengal, for instance, a rope and two specimens of soil were submitted to the imperial serologist. Testing revealed bird blood on the rope and one soil sample. The other piece of earth, however, tested positive for human blood.[76] There was mixed blood (human and ovine) in the Nadar-Naiker case, too. For some reason, though, that case was not interpreted as a true case supplemented by false evidence, but as a fundamentally false case.

How did colonial officials make sense of these cases? From its fifth edition in 1905 onward, editions of Taylor's medicolegal treatise included a chapter on India. The author of this contribution was a member of the Indian Medical Service, W. J. Buchanan. Like Sutherland, he was a Scot.[77] He was also editor of the *Indian Medical Gazette* and inspector general of prisons in Bengal. Buchanan spent two pages discussing the "falseness of much of the evidence given by natives of India." He considered whether fabricated cases were the consequence of "inherent Oriental deceit," or "fear, stupidity, apathy, or malice, or to the fact that the witnesses have been 'tutored' by police."[78] The most common impulse among colonial officials was to fall into cultural and racialized stereotypes when explaining the perceived prevalence of false charges and fabricated evidence. By these accounts, south Asians were either incapable of truth telling or did not value it.[79]

Historians of colonial south Asia usually attribute such characterizations of native dissimulation to the "the rule of colonial difference" or racial prejudice.[80] Without downplaying the depth of racism in colonial officialdom, though, there remains more to be said. Racial bias prevented officials from seeing particular patterns in their own sources. It is critical that scholars not replicate the move by also stopping short when interpreting these sources. The last section of this chapter offers an alternative explanation for the phenomenon of fabricated evidence by exploring a set of practices embedded in the precipitin archive.

## Punitive Self-Harm

Colonial reports of Indian mendacity could have been referring to several things. When dissimulation by an Indian witness or accuser was voluntary and intentional, it may have been a form of resistance.[81] Dissimulation by an Indian police officer could have been framed as corruption. Complaints in the medicolegal literature about Indian mendacity extended to Indian police.[82] Certain forms of corruption could have been driven by the desire to resist colonial rule through sabotage from within a state institution—whether through foot-dragging and the unraveling of everyday processes in petty ways or through more significant acts like taking bribes.[83]

At times, though, what colonial sources labeled "native mendacity" was something quite different. In an important subset of false-charge cases like the Nadar-Naiker feud, noncolonial modes of disputing were being bent to meet the new requirements of the colonial legal system. The chemical examiners and imperial serologist regarded these cases as duplicitous perversions of the course of justice. Many of their annual reports ended with *Fiat justitia, ruat coelum*, or "let justice be done, though the heavens may fall," which meant that they had a duty not only to help convict the guilty but also to enable acquittal of the innocent.[84] Unexpectedly, then, the belief in native mendacity produced heightened awareness of the possibility of wrongful convictions, at least in cases among south Asians. But south Asian accusers in these cases probably saw their own actions as consistent with pre- and noncolonial understandings of justice: they were harnessing the colonial system to arrive at the same destination as they would have reached otherwise. "In attempting to introduce British procedural law into their Indian courts," observed Bernard Cohn in 1959, "the British confronted the Indians with a situation in which there was a direct clash of the values of the two societies;

and the Indians in response thought only of manipulating the new situation and did not use the courts to settle disputes but only to further them."[85]

Buchanan's chapter in Taylor's textbook included these examples of south Asian "moral insensibility": "A master murdered his servant and dragged the body to the door of his enemy solely in order that a charge of murder might be brought against the latter; a father murdered his daughter, because his neighbour had slandered her, in order that her blood might be upon the neighbour's head."[86]

These cases reflected a model of collective identity, responsibility, causation, and punishment that was so foreign to Buchanan's own cultural logic that he failed to understand what he was observing.[87] Presumably, notions of family identity and honor (deeply gendered) made the father's killing of his daughter an act of "self"-sacrifice rather than a violent crime by one individual against another. Similarly, the master who murdered his servant may have regarded the household as the crucial social unit, and the killing of the servant as the master's own prerogative. By the logic of punitive self-harm, the original wrongdoer (the enemy in the first case, the neighbor in the second) was directly responsible and punishable for the death of the servant or the daughter. It was he, and not the master or father, who had caused the death. Otherwise put, his original wrong set in motion a chain of "foreseeable" events that culminated in the death of the servant or daughter. The East India Company had made efforts in the late eighteenth century to prohibit punitive self-harm.[88] Curiously, though, early twentieth-century figures like Buchanan seemed to have lost any awareness that a particular mode of disputing underpinned the two cases described. Buchanan attributed these examples to simple cruelty and barbarism, missing—or dismissing—the phenomenon of legal pluralism.

Punitive self-harm had a long and varied history in south Asia, including Gandhian modes of nonviolent resistance (like the hunger strike), political protest suicides, and even some cases of *sati* or ritual widow immolation.[89] At its core was the idea that a person would hurt him- or herself to protest a wrong done by another. Conducting *dharnā* was the most common practice associated with this mode of disputing.[90] It shaded into the fast-unto-death, known in Sanskrit as *prāyopaveśa*.[91] To protest a wrong, a person would sit at the doorstep of the wrongdoer, refusing to eat or leave until the wrong was redressed. If the protester died in the process, the wrongdoer was responsible. This latter person would be cursed, and even haunted by the protester's ghost.[92] In the classic scenario, a creditor would sit *dharnā* at the doorstep of a debtor who had not repaid a loan.[93] But these practices also had broader applications, including protest that state legal processes were misguided, impotent, or absent.[94] *Dharnā* was particularly powerful when carried out by Brahmins, who might increase the pressure by threatening to stab themselves or drink poison. There was no expiation for the sin of causing a Brahmin's death.[95]

Although Hindu law considered such acts legitimate only if inflicted upon oneself, the practice of punitive self-harm was at times extended.[96] In this version, the "self" in self-harm referred not only to the individual wronged, but also to members of the person's family or even caste.[97] These were typically elderly women or little girls—vulnerable females deemed expendable by the patriarchal worldview in question.[98] *Trāgā* was defined by the *Hobson Jobson* as "the extreme form" of *dharnā*, in which a person tortured or killed himself or a relative "for bringing vengeance on the oppressor."[99] In late eighteenth-century Bengal, one Beechuk Brahmin beheaded his own mother in protest during a feud over revenue collection rights, while a lower-caste man named Dhunoo Chamar killed his four-year-old daughter at the house of a village headman after being criticized for drinking and abusing his superiors. Government officials commented that the latter had acted "in conformity with the usage established in Benares with regard to such cases."[100] The practice of establishing a *kurh* also illustrated this expanded version of punitive self-harm. A person resisting arrest or some other action he considered wrongful (like the payment of a tax) would construct a circular enclosure, usually of wood. Inside, he would place a cow or an elderly female relative. He would then threaten to set fire to the wood if touched or forced to undertake the action he was resisting.[101]

Similarly, some communities guaranteed contracts by using this extended version of punitive self-harm.[102] The Bhats and Charans of western India were known for making themselves—and their relatives—available for hire as human security for contracts.[103] Written contracts that were Bhat-guaranteed bore the sign of a dagger.[104] When the relative of a Bhat or Charan was killed to punish another person's breach of contract, the violence was sometimes processed by the state as a prosecution for murder. Bombay High Court judge F. C. O. Beaman offered one remarkably detailed description of such a case, an 1892 murder trial in his *mofussil* or countryside court in western India.[105] Two parties had made a contract and guaranteed it by engaging a third person, a member of the Charan caste. "If the debtor whose debt the Charan had guaranteed, refused to pay, in the last resort his obligation would be enforced by the Charan spilling his sacred blood on the lintel of the offender's house."[106] The debt was not repaid. The debtor laughed at the idea of having the contract enforced "the ancient, grim, terrible way" through the spilling of Charan blood: "You and your curses, he said, no longer have any terrors for me. Are there not the Courts of the Sirkar [colonial state]? If my friend thinks that I owe him money, they are open to him. Let him sue in due form. I am rich. I can engage the best Vakil [lawyer]. Let us see then."[107] The Charan went home to his family, where he, his mother, son, and daughter "each in turn prayed that he or she might be used as the divine scourge." "Charan honour must be maintained," wrote Beaman.

The Charan's elderly mother prevailed. By Beaman's account, she claimed that she had already lived a long life and thus should be the sacrificial martyr. The case reproduced a common gendered pattern in punitive self-harm cases: it was typically young, elderly, or otherwise marginal female relatives who were selected. On the debtor's front door step, the Charan stabbed his mother in the heart and "smeared the lintels of the house with her blood, calling down the curses of the angry Gods on this hardened sinner." Beaman was the judge who had to try the Charan (and his son) for murder, but he was also a defender of tradition and caste.[108] "Of course they were not hanged." In the judge's words, "It is ill to set up new Gods in place of the Old, too hastily, to flout immemorial usage, and incur the stain of Charan blood."[109] Beaman's sympathies were unusual for a British judge at the turn of the twentieth century. As a judge who came up through the Indian Civil Service (ICS), however, Beaman spent more time in the *mofussil* (and from an earlier age) than did his judicial colleagues who had been trained as barristers in London. Although ICS judges were looked down on by their legally trained colleagues on the bench, they exhibited greater familiarity with Indian languages and practices.[110] Beaman's sympathetic view of Charan contracts was a good example. The larger point, however, was that the extended version of punitive self-harm revealed itself through this murder trial.

While colonial criminal law thus operated upon the basic assumption of individual responsibility and punishment, some punitive self-harm practices worked through a collective notion of identity, responsibility, and punishment. Colonial officials (Beaman aside) considered it murder when a man killed his own daughter to punish a person who had slandered her or when the Charan stabbed his own mother over a breach of contract. Within the logic of extended punitive self-harm, however, both men would have seen their actions as legitimate forms of self-harm.[111]

Furthermore, it was not just the human unit—individual or collective—that distinguished these south Asian and European modes of dispute resolution. Conceptions of causation also differed. Within the extended self-harm tradition, the death of the man's daughter (in the Buchanan case) had been caused not by the father, but by the man who had wronged the father. This initial offender would suffer cosmically for the death. For the criminal justice system, though, it was the father, not the wrongdoer, who was solely responsible for the death. Any attempt to attribute the death to the initial wrongdoer would be voided by the father's actions, which broke the chain of causation. Unusually (among Britons), Beaman sympathized with the logic of Charan contracts, noting that "notwithstanding his western veneer, the real offender" almost instantly fell ill and died. He was "looked on askance by his former friends, pitied by none."[112] Causation here was clear: the fault lay with the debtor, not the Charan who had stabbed his own mother.

What did practices of extended punitive self-harm have to do with the work of the imperial serologist? It was only one step from killing another person in protest to then making that death look like it had been committed by the original wrongdoer through the planting of animal blood. In other words, the shift from dumping the Charan mother's body at the recalcitrant debtor's doorstep to also planting animal blood to frame the adversary for the death was not unimaginable. The Nadar-Naiker feud was an example. The master who killed his servant and left the body at his enemy's doorstep could have been another—had animal blood been planted at the doorstep or on the enemy's clothing. Fabricating evidence was hardly illegitimate on the logic of this tradition of private disputing and punishment, in other words. It was simply an extension of preexisting practices.

To be sure, these practices changed while adapting to the colonial justice system. One would no longer receive the publicity or social "credit" for committing extended self-harm when framing the original wrongdoer for murder. The effects of a successful framing were also more earthly than otherworldly. Cases that went undetected would produce no longer just a good haunting, but also perhaps an execution. Actors planting animal blood would have rejected the colonial state's attempt to monopolize dispute resolution and punishment. They must also have derived satisfaction from using criminal law for their own ends. The point was that planting blood-stained evidence, when successful, would achieve the same goal as punitive self-harm: punishment of the original wrongdoer. Precipitin testing enabled officials to identify these cases.

At the same time, there were limits on what precipitin testing and the imperial serologist could achieve. What happened, for instance, when news of the test's abilities spread? Did fabricators of evidence simply start planting human blood at crime scenes, instead of animal blood? In one 1938 murder case, the Lahore High Court ruled that a bloody spearhead had no value as evidence. Although the blood was human, other aspects of the murder made the item suspect. Crucially, all twenty-eight of the victim's wounds were cuts or incisions, not puncture wounds. The court held that no spear could have been used in the murder. The bloody spearhead was probably false evidence.[113] Here the historian must acknowledge the limits of not only the test, but also the archive. The historian of deception requires *imperfect* trickery: it is precisely when ruses failed that they became visible in the primary sources. There may have been successful countermoves or adaptations to the new technology, in other words, but such instances remain almost always hidden from the historian. It is possible that the imperial serologist's usefulness diminished as popular awareness of the precipitin test grew. Equally, precipitin testing may have come to play a secondary role in court over time, providing important information but in conjunction with other forms of forensic analysis (as in the Lahore case), rather than on its own. What is clear is that the imperial serologist's heyday lasted for several decades at least—from the 1910s until the 1930s. The fact that other colonies started appointing state serologists from the 1930s suggests the continuing usefulness of precipitin testing.

## Conclusions

In 1914, a new edition of I. B. Lyon's *Medical Jurisprudence for India* appeared.[114] Its editor, L. A. Waddell, explained in the preface that the most important addition was a chapter on precipitin testing. The chapter's author, W. D. Sutherland, had perfected the technique such that the "great climatic difficulties hitherto experienced" in carrying out this test in India had now been overcome. The test could now be used "with as absolutely certain and trustworthy results as in Europe," and Waddell predicted that the test would now be used in all murder cases in India.[115] For Waddell, precipitin testing in India was now on an equal footing with precipitin testing in Britain. As we have seen, though, the technique played a larger role in the colony than in the metropole.

Why did precipitin testing take hold in India in special ways—like the creation of the imperial serologist? This form of forensic testing gave colonial officials a tool that allowed them to respond to a central obsession of the colonial criminal justice system: fear of dissimulation, particularly the creation of fabricated evidence by colonized subjects. Reading colonial sources against the grain reveals that both stereotypes of native mendacity and scholarly explanations grounded in the "rule of colonial difference" overlook practices like punitive self-harm that operated beyond the blinders of colonial racism. Ending the story with racial difference, in other words, means missing a whole other sphere of activity, namely, noncolonial modes of disputing as they adapted to the new colonial rules. It is possible that some dissimulators responded to the arrival of precipitin testing by planting human instead of animal blood. Yet, still, the question is not so much whether colonizer or colonized had the better moves in this delicate dance of trickery and its detection. In other words, the question is not "who won?" but what did the interaction reveal about each normative system? The story here is about legal pluralism—how layers of disputing norms interacted, with one twisting itself into the other's categories in order to use the other's tools.

When bloodstained evidence was planted near an enemy, colonial officials saw an attempt to frame an innocent party. But if the group, not the individual, was the fundamental unit—and if causation was widely construed—was the framed party undeserving of punishment? Planting animal blood made the killing of one's own relative punitive in a new way. When it succeeded, it produced a criminal conviction, in addition to punishing the adversary socially and psychologically. Planted animal blood drew the criminal courts into complex, ongoing disputes between groups. Through precipitin testing, these cases revealed something other than perversion of the course of justice intended as simple sabotage of the system. The work of the imperial serologist exposed not so much schemes to destroy colonial law as attempts to make it work for what Beaman would have called the "old gods" of punitive self-harm. Through the planting of animal blood to make punitive self-harm look like murder by a rival, this noncolonial mode of disputing adapted itself to produce punishment in the colonial idiom. Colonial analysts may not have recognized it, but precipitin testing exposed the interaction between normative systems. Beyond colonial attributions of fabricated evidence to native mendacity lay a world of noncolonial disputing.

## ACKNOWLEDGMENTS

Earlier versions of this research were presented in 2015–16 at the Locating Forensic Science and Medicine Conference (London); the Centre for the Study of Law and Governance, Jawaharlal Nehru University (Delhi); the University of Minnesota's Legal History Workshop (Minneapolis); and the University of Wisconsin (UW) History of Science Department, Law School, and Center for South Asia (Madison). The author thanks participants at all these sessions for their questions and comments. She also and especially thanks Pratiksha Baxi, Binyamin Blum, Susanna Blumenthal, Donald R. Davis Jr., Aparna Kapadia, Projit B. Mukharji, Nicole C. Nelson, Radhika Singha, Barbara Welke, Heather Wolffram, and Nurfadzilah Yahaya for their insights and references; and the UW Center for South Asia and Law School for research funds. She is grateful to UW law librarians and Alexandra Fleagle for their research assistance, and to the editors of this volume for their rigorous feedback. The author remains responsible for the views expressed here.

## NOTES

Shelf marks beginning with IOR indicate sources from the India Office Records of the Asia, Pacific, and Africa Collections, British Library, London. NAI indicates material from the National Archives of India, Delhi. *BMJ* stands for the *British Medical Journal*, *IMG* for the *Indian Medical Gazette*, IPC for the Indian Penal Code, *TI* for *Times of India*, and *TL* for *Times of London*.

1. "Blood Stains," *BMJ* 1, no. 2877 (February 19, 1916): 283–84. See also "Note by Lieut.-Col. W. D. Sutherland, IMS, on the legal Serological Work Carried Out by him" (undated), p. 8 in "Appointment of an Imperial Serologist for the Whole of India," Bhopal Agency: General, Proc. No. 5, 1914, NAI.

2. There were experts who used forensic serology in English and Scottish criminal trials during the early twentieth century, but there was no official dedicated exclusively to this field. The senior official analyst to the Home Office, county analysts in England, and university professors of medical jurisprudence in Scotland covered both forensic serology and toxicology. See, e.g., "Doctor's Death from Drugs," *TL*, September 21, 1921, 7; "Double Murder Charge," *TL*, November 1, 1928, 11; "Doctor's Suspicions of Arsenic," *TL*, August 20, 1929, 14; "Trunk Murder Charge," *TL*, August 8, 1934, 7; and M. Anne Crowther and Brenda White, *On Soul and Conscience the Medical Expert and Crime: 150 Years of Forensic Medicine in Glasgow* (Aberdeen: Aberdeen University Press, 1988).

3. The imperial serologist also did serological research on tropical and venereal diseases, and worked on blood banking and transfusion. See note 54 below.

4. "Chemistry and Crime: The Detective in the Laboratory," *The Leader*, April 8, 1914, 9.

5. Strictly speaking, precipitin testing was not a blood test but a serum-based "specific protein" test that could be used on blood, bone, skin, muscle, semen, or "albuminous urine." R. B. Lloyd, "The Serological Analysis of Bloodstains in Criminal Cases (Illustrative Cases)," *IMG* 61 (1926): 220; Alfred Swaine Taylor, *Principles and Practice of Medical Jurisprudence*, ed. Sydney Smith and W. G. H. Cook (London: J. & A. Churchill, 1928), 1:499. Nonetheless, I refer to it as a blood test in this chapter because bloodstains were the usual objects of analysis. In this chapter, references to Taylor pertain to editions of the treatise revised and issued after the original author's death in 1880.

6. "Note by Sutherland," 33–34.

7. "Blood Stains," 183.

8. E.g., "Bungalow Crime," *TL*, June 7, 1924, 9; and Donald Carswell, *Trial of Ronald True* (Edinburgh: William Hodge, 1925), 71. See also note 70 below.

9. The term "south Asian" describes affiliation with the region that today includes Afghanistan, Bangladesh, Bhutan, India, the Maldives, Nepal, Pakistan, and Sri Lanka. For the colonial period (1757–1947), "Indian" refers to British India, the territory that would later split into Bangladesh, India, and Pakistan. This chapter uses the broader terms "south Asian" and "European" as well as the more specific labels of "Indian" and "Briton."

10. See, e.g., IPC ss. 191–229 and 463–77 in M. H. Starling, *Indian Criminal Law: Containing the Indian Penal Code and Other Indian Acts Relating to Offences, and Also Acts of Parliament and Orders in Council Relating to Offences Triable in India, Including Act X of 1886* (Bombay: Education Society's Press, 1886); and s. 98 of the Code of Criminal Procedure in H. T. Prinsep, *The Code of Criminal Procedure, Being Act V of 1898* (Calcutta: S. K. Lahiri, 1901), 72–74.

11. Patrick Hehir and J. D. B. Gribble, eds., *Outlines of Medical Jurisprudence in India* (Madras: Higginbotham, 1908), 28–35.

12. On false claims by women relating to abortion, maternity, and rape in Britain, see Taylor, *Principles and Practices of Medical Jurisprudence*, 2:205, 2:238–42, 2:435, 2:450, 2:466. On the view that female testimony was unreliable, see Wendie Ellen Schneider, *Engines of Truth: Producing Veracity in the Victorian Courtroom* (New Haven, CT: Yale University Press, 2015), 206–8. On the association of mendacity with lower-class Britons, see Schneider, *Engines of Truth*, 5, 10, 18, 21, 29–31, 201–3.

13. See Vinay Lal, "Everyday Crime, Native Mendacity and the Cultural Psychology of Justice in Colonial India," *Studies in History* 15 (1999): 154–66.

14. Projit Mukharji, "Sero-Tropicality: Blood and the Reinvention of Tropical Medicine, 1930–50" (presented at the History of Science Brown Bag Series, University of Wisconsin–Madison, March 20, 2015).

15. See Sally Engle Merry, "Legal Pluralism," *Law & Society Review* 22, no. 5 (1988): 869–96, and Mitra Sharafi, "Justice in Many Rooms since Galanter: De-Romanticizing Legal Pluralism through the Cultural Defense," *Law and Contemporary Problems* 71 (2008): 139–46.

16. In support of the creation of the imperial serologist, one administrator in the North-West Frontier Province wrote that "the need for such an appointment is particularly felt in this Province, where the blood feud flourishes and excites so large a degree of public sympathy leading to homicidal crime." Letter from Chief Commissioner, North-West Frontier Province to Secretary to Government of India, Home Department (January 6, 1914), in "Proceedings of the Home Dept. 1914," 822 (IOR/P/9457). For other examples, see "Not Guilty of Murder: Five Villagers Acquitted," *TI*, July 8, 1929, 17; and "Government of Palestine: Annual Report of the Department of Public Health for the Year 1925," *Analyst* 52 (1927): 232 (on the cutting down of adversaries' olive trees as "a favorite form of spite or revenge," like the poisoning of animals).

17. On aspects of ABO blood group testing in India during the first half of the twentieth century, see Projit Bihari Mukharji, "From Serosocial to Sanguinary Identities: Caste, Transnational Race Science and the Shifting Metonymies of Blood Group B, India c. 1918–1960," *Indian Economic and Social History Review* 51 (2014): 143–76.

18. See note 54 below.

19. See the *IMG* articles by S. D. S. Greval: "The Use of Blood Tests in Excluding Paternity and Maternity," 74 (1939): 388–91; and 80 (1945): 204–7.

20. See, e.g., D. P. Lambert, "A Preliminary Report on the Medico-Legal Value of the Finding of Blood on Nail Parings," *IMG* 74 (1939): 745.

21. Eduard Piotrowski published a book proposing blood spatter analysis in 1895, but the forensic study of the physics of blood patterns emerged as a distinct field only in the 1950s with

the Sam Sheppard case in the United States. See Ian Burney, "Spatters and Lies: Contrasting Forensic Cultures in the Trials of Sam Sheppard, 1954–66," chap. 4, this volume.

22. Tal Golan, *Laws of Men and Laws of Nature: The History of Scientific Expert Testimony in England and America* (Cambridge, MA: Harvard University Press, 2004), 174–75. For the earlier nineteenth-century history including the struggle between the "smell test" and microscopy to establish species of origin, see J. R. Bertomeu-Sánchez, "Chemistry, Microscopy and Smell: Bloodstains and Nineteenth-Century Legal Medicine," *Annals of Science* 72, no. 4 (2015): 490–516.

23. *Report of the Chemical Examiner to Government, Punjab, for the Year 1879* (Lahore: Government Civil Secretariat Press, 1880), 17–18 (IOR/V/24/418); Taylor, *Principles and Practice of Medical Jurisprudence* (1894), 1:594–600; W. D. Sutherland, *Blood-Stains: Their Detection, and the Determination of Their Source* (London: Ballière, Tindall and Coz, 1907), 48–68, esp. 55; P. C. Gane, "The Serum or Precipitin Test for Blood: Its Forensic Aspect," *Criminal Law Journal of India* 5, no. 8 (1907): 89; and L. A. Waddell, *Lyon's Medical Jurisprudence for India* (Calcutta: Thacker, Spink, 1921), 176–79.

24. G. S. Graham-Smith and F. Sanger, "The Biological or Precipitin Test for Blood Considered Mainly from Its Medico-Legal Aspect," *Journal of Hygiene* 3, no. 2 (1903): 268–71, and 3, no. 3 (1903): 354–56; George Nuttall, *Blood Immunity and Blood Relationship: The Precipitin Test for Blood* (Cambridge: Cambridge University Press, 1904), 117–18; Gane, "Serum or Precipitin," 93; and Waddell, *Lyon's Medical Jurisprudence* (1921), 185.

25. By the late 1930s, rabbits were used to make bird antiserum, while birds were used to make the antisera of humans and other mammals. *Annual Report on the Working of the Imperial Serologist's Department, Calcutta for the Year 1938–39* (Calcutta: Government of India Press, 1940), 26, in "Report: Imperial Serologist's Department, Calcutta, 1938–39," Department of Education, Health and Lands: Health Branch, 1939, File No. 33-5/39-H (NAI).

26. G. Roche Lynch, "The Technique of the Precipitin Test and Its Forensic Value," *The Analyst* 53 (1928): 8. On the use of chemical and spectroscopic analysis to determine whether a stain was blood at all, see Sutherland, *Blood-Stains*, 11–47.

27. According to Sutherland, a precipitate would form between mammalian serum and antiserum after twenty minutes (what Nuttall called the "mammalian reaction"). Only test results obtained within the first twenty minutes could thus be of use for forensic purposes. *Blood-Stains*, 117.

28. For detailed descriptions of precipitin testing, see Sutherland, *Blood-Stains*, 116–17; Taylor, *Principles and Practice of Medical Jurisprudence* (1928), 1:499–500; and *Annual Report on the Working of the Imperial Serologist's Department*, 27.

29. See "Blood Relationship and the Precipitin Test," *BMJ* 2, no. 2341 (November 11, 1905): 1304–5. A subsequent study asked whether precipitin (or blood group) testing could distinguish between human races. Among most populations, neither test could do so. "The British Association: Races of Mankind," *TL*, September 6, 1932, 7.

30. See Waddell, *Lyon's Medical Jurisprudence* (1921), 187. Distinguishing between ruminants became important in cases involving the killing and maiming of domesticated animals and in Hindu-Muslim violence involving the blood of cows. On the former scenario in a different part of the British Empire, see Binyamin Blum, "From Bedouin Trackers to Doberman Pinschers: The Rise of Dog Tracking as Forensic Evidence in Palestine," chap. 7, this volume, 212–13.

31. Albert S. F. Grünbaum, "Note on the 'Blood Relationship' of Man and the Anthropod Apes," *Lancet*, January 18, 1902, 143; Sutherland, *Blood-Stains*, 109–14; Gane, "Serum or Precipitin," 92, 95–56; "Blood Stains," *BMJ*, February 19, 1916, 283; Waddell, *Lyon's Medical Jurisprudence* (1921), 183; and K. Landsteiner and C. Philip Miller, "Serological Studies on the Blood of the Primates," *Journal of Experimental Medicine* 42, no. 6 (1925): 841–52. See also C. P. Lukis, "Explanatory Note on Major W. D. Sutherland's Investigations into the Applicability to

Medico-Legal Work in India of Bio-Chemical Methods for the Detection of Blood Stains," 10, in "Report by Major W. D. Sutherland on His Investigation in Connection with Blood Stains: Proposed Appointment of a Government Serologist," Home Department: Medical Branch—A. Proceedings, March 1910, Nos. 149–52 (NAI).

32. Compare editions of Taylor's *Principles and Practice of Medical Jurisprudence*, ed. Fred. J. Smith (London: J. & A. Churchill, 1920), 2:155; and (1928), 1:500–501.

33. See Roche Lynch, "Technique of the Precipitin Test," 12–13.

34. Nuttall, *Blood Immunity*, 411–13.

35. On Scots in British India, see T. M. Devine, *Scotland's Empire 1600–1815* (London: Allen Lane, 2003), and John M. MacKenzie and T. M. Devine, eds., *Scotland and the British Empire* (Oxford: Oxford University Press, 2011). Scottish law, which was distinct from English common law and grew out of Roman law, seems to have exerted a disproportionate influence on Indian medical jurisprudence because of the many Scots in medicine and the empire. See, e.g., Waddell, *Lyon's Medical Jurisprudence* (1921), 2n2.

36. Chemical examiners tested the purity of alcohol, ghee, milk, water, kerosene, cocaine, and opium for various state agencies, including Customs and Excise. They dabbled in anesthetics, explosives, chemical warfare, and bacteriology (involving the plague, tuberculosis, typhoid, cholera, syphilis, and leprosy). They also tested for the presence of poisons in food and viscera, both animal and human. See *Report of the Chemical Examiner for Punjab 1879* and David Arnold, *Toxic Histories: Poison and Pollution in Modern India* (Cambridge: Cambridge University Press, 2016), 111–17.

37. *Annual Report of the Chemical Examiner's Department, Bengal, for 1908* (Calcutta: Bengal Secretariat Press, 1909), 2 (IOR/V/24/422).

38. W. D. Sutherland, *The Applicability to Medico-Legal Practice in India of the Biochemical Tests for the Origin of Blood-Stains* (Calcutta: Superintendent Government Printing, India, 1910).

39. Cover letter to *Annual Report of the Chemical Examiner's Department, Bengal, for 1912* (Calcutta: Bengal Secretariat Press, 1913), and *Annual Report of the Chemical Examiner's Department, Bengal, for 1913* (Calcutta: Bengal Secretariat Book Depot, 1914), 1 (both IOR/V/24/422).

40. "Report from Maj. Windsor to Inspector-General of Civil Hospitals, Bengal, Calcutta, 31 Jan. 1913," in *Annual Report of the Chemical Examiner's Department 1912*, 1; and cover letter to *Annual Report of the Chemical Examiner's Department, Bengal, for 1916* (both IOR/V/24/422).

41. Untitled obituary for W. D. Sutherland, *BMJ*, July 31, 1920, 189; and Sutherland, *Bloodstains*, 129.

42. Sutherland, *Blood-Stains*, ix–x. Sutherland's fluency in German was important not only for his serological work but also his lifelong interest in psychoanalysis. He was a founding member of the British Psycho-Analytical Society and visited Freud in Vienna. Sutherland had "a remarkable knowledge of Indian Folk-Lore" and occasionally published articles in the racy German-language journal *Anthropophyteia*, a publication on erotic ethnology from which Freud drew material. "Obituary," *International Journal of Psycho-Analysis* 1 (1920): 341.

43. Sutherland, *Blood-Stains*, x–xi.

44. In Ceylon, Egypt, southern Nigeria, and Palestine, officials like government analysts and official chemists oversaw the testing of both poisons and bloodstains during the first few decades of the twentieth century. Richard B. Pilcher, "A List of Official Chemical Appointments, p. 166," *Proceedings of the Institute of Chemistry of Great Britain and Ireland* 36 (1912): H001–H246; "Government Analyst's Report for 1918," *Ceylon Observer*, May 13, 1919, 699; "Analytical: Local Administration Report," *Ceylon Observer*, May 18, 1921, 15; "Government of Palestine: Annual Report of the Government Analyst for the year 1928," *Analyst* 55 (1930):

48–49; "Egypt: Report on the Work of the Chemical Laboratory, Ministry of Justice, for the Years 1929–34," *Analyst* 64 (1939): 353–54. Outside of India, state serologists began to appear by the 1930s. On South Africa, see "Public Appointments," *TL*, December 28 1933, 3. On the Federated Malay States, see "The Work Done by the Institute of Medical Research," *Straits Times*, September 26, 1937, 13. See also note 3 above.

45. Sutherland, *Blood-Stains*, 133. The German medicolegal model may also have been adopted elsewhere in the Indian criminal justice system. See *Report of the Chemical Examiner for Punjab 1879*, 13–14.

46. Letter from F. N. Windsor to Inspector-General of Civil Hospitals, Burma, Rangoon, April 22, 1910, "Proposed Appointment of a Serologist for India," Proceedings of the Home Department, August 1910, Nos. 966–75 (August 29, 1910) (IOR/P/8443).

47. Letter from F. N. Windsor to Inspector-General of Civil Hospitals, Burma; letter from W. H. Willcox, MD, St. Mary's Hospital to Under Secretary of State, Home Office (London, February 15, 1910); and letter from H. Wheeler, Secretary to the Government of Bengal, Municipal Government to Secretary to the Government of India, Home Department (Darjeeling, June 20, 1910), No. 566-T.-Med., in "Proposed Appointment of a Serologist for India," 39 (IOR/P/8443).

48. L. Jenkins (August 15, 1910) in a letter from the Government of Bombay, no. 3290 (July 14, 1910), 4, in "Proposed Appointment of a Serologist for India," Home Dept., Medical branch, August 1910, Part A, Proceedings No. 135–46 (NAI).

49. Letter from Thomas Stevenson to the Under Secretary of State, Home Office (London, January 12, 1907), in "Proposed Appointment of a Serologist for India," No. 135, 30 (IOR/P/8443). In the same file, see further correspondence from Stevenson and Willcox (September 12, 1904 to February 15, 1910), 29–32.

50. See, e.g., the history of the Pioneer Magnesia Works in Mitra Sharafi, "Parsi Life Writing: Memoirs and Family Histories of Modern Zoroastrians," in *Holy Wealth: Accounting for This World and the Next in Religious Belief and Practice: Festschrift for John R. Hinnells*, ed. Almut Hintze and Alan Williams (Wiesbaden: Harrassowitz, 2016), 265. I also thank Radhika Singha for her observation that government laboratories proliferated in British India circa World War I.

51. On the sale of antiserum in Britain, see Letter from Willcox (February 15, 1910). I have encountered no similar references to antiserum being available for purchase in India. Roche Lynch advised against the use of commercial antiserum even two decades later: "Note: The Precipitin Test for Blood," *Analyst* 53 (1928): 435a.

52. See, e.g., "Recommendation by the Indian Retrenchment Committee," 23, in "Blood Stained Cases: Creation of the Post of Imperial Serologist for the Examination of Blood Stained Cases for the Whole of India (Baghelkhand Agency), 1914," Baghelkhand Political Agency: English Files, 1914, Proceeding No. 304 (NAI).

53. See, e.g., "Transfer of the Imperial Serologist's Department to the Direct Control of the Government of India and Revision of Emoluments of the Imperial Serologist," Finance: Expenditure-I, 1933, Proceedings Nos. 83-Exl, 1933 (NAI); and cover letter from Education, Health and Land Department to Director-General, Indian Medical Service, u. o. No. 29-13/39/9 (February 12, 1940), 2, in *Annual Report Imperial Serologist 1938–39* (NAI).

54. *Annual Report Imperial Serologist 1938–39*, 3, 3a, 27–28; R. B. Lloyd, "Report of the Professor of Serology and Immunology," in *Annual Report of the Calcutta School of Tropical Medicine Institute of Hygiene and the Carmichael Hospital for Tropical Diseases for the Year 1924* (Calcutta: Bengal Government Press, 1925) (IOR/V24/754). See also Projit Mukharji, "Between Empire and Nation: War, Competitive Philanthropy and the Birth of Blood Banking in British India" (presented at the American Association for the History of Medicine Annual Meeting, Minneapolis, MN, April 29, 2016).

55. The Wassermann test for syphilis was much in demand in the early twentieth century. See, e.g., the judgment notebooks of the Parsi Chief Matrimonial Court, 1893–1947, Bombay High Court, Mumbai.

56. *Annual Report Imperial Serologist 1938–39*, 5; untitled statement by J. W. D. Megaw, April 1, 1933, Department of Education, Health and Lands, 13, in "Transfer of the Imperial Serologist's Department to the Direct Control of the Government of India and Revision of the Emoluments of the Imperial Serologist," Finance: Expenditure-I, 1933, Proceedings Nos. 83–Exl, 1933 (NAI).

57. "Question Asked by the General Purposes Sub-Committee of the Retrenchment Advisory Committee Regarding the Imperial Serologist's Dept.," Dept of Education, Health and Lands. Health branch. 1932, 14, File No. 103-4/32-H (NAI).

58. See Lloyd, "Report of the Professor of Serology and Immunology," and *Annual Report Imperial Serologist 1938–39*.

59. S. D. S. Greval was a Punjabi IMS officer with medical, surgical, and public health degrees from the University of Liverpool. See "Question of the Selection of an Officer for Appointment as Chemical Examiner . . . Appointment of Major S. D. S. Greval, IMS, as Officiating Imperial Serologist," Dept. of Education, Health and Lands: Health Branch," 1934, File No. 11-2/34-H (NAI).

60. Letter from Secretary to Government of India, Home Department, to Superintendent, Port Blair (No. 726-C, February 5, 1918), 15, in "Confirmation of the conviction of the sentence of death passed by the Additional Sessions Judge, Port Blair, upon life-convict Johir Ghazi, no. 21374. Decision that blood stains requiring expert examination in connection with criminal cases in the Andamans should be sent to the imperial serologist in Calcutta," Home Department, Branch: Port Blair—A. Proceedings, February 1918, No. 10-14 (NAI).

61. Untitled statement by J. W. D. Megaw, October 13, 1932, in "Transfer of the Imperial Serologist's Department," 6.

62. Director of the Indian Medical Service quoted in statement by Ram Chandra, November 17, 1932, in "Transfer of the Imperial Serologist's Department," 9.

63. The first prominent English murder trial to use precipitin testing was the 1911 Clapham murder case. "The Clapham Murder," *TL*, February 9, 1911, 7; "The Detection of Human Bloodstains," *TL*, February 14, 1911, 10. But the test was not generally accepted as reliable in the English or Scottish courts until the early 1930s. See Crowther and White, *On Soul and Conscience*, 61; Roche Lynch, "Technique of the Precipitin Test," 13; and "Body in Blazing Car," *TL*, January 28, 1931, 6. By the high-profile Buck Ruxton case of 1935–36, in which John Glaister Jr. analyzed the bloodstains, precipitin testing had become generally accepted or "black-boxed." "Ruxton Case," *TL*, December 12, 1935, 9; John Glaister and James Couper Brash, *Medico-Legal Aspects of the Ruxton Case* (Baltimore: William Wood, 1937), 199–225, 264–72.

64. Note, for instance, the focus on fabricated evidence and false charges in reports on precipitin testing in India for the imperial metropole. See the 1916 *BMJ* article "Blood Stains."

65. Elizabeth Kolsky, *Colonial Justice in British India: White Violence and the Rule of Law* (Cambridge: Cambridge University Press, 2010), 108–19, 217–18; Bhavani Raman, *Document Raj: Writing and Scribes in Early Colonial South India* (Chicago: University of Chicago Press, 2012), 137–60; Schneider, *Engines of Truth*, 103–42. See also Projit Bihari Mukharji, "Handwriting Analysis as a Dynamic Artisanal Science: The Hardless Detective Dynasty and the Forensic Cultures of the British Raj," chap. 3., this volume.

66. "Note by Lieut.-Col. W. D. Sutherland, IMS, on the Legal Serological Work Carried Out by him," May 29, 1915, No. 79 H. D.–625—S.N., 8, in "Appointment of an Imperial Serologist for the Whole of India," Bhopal Agency: General, 1914, Proceedings No. 5, 1914 (NAI).

67. In *Blood-Stains,* Sutherland included two cases in which a person fraudulently claimed to be suffering from an illness or injury. Both involved benefit fraud rather than false murder charges, however (136, 138).

68. See Taylor, *Principles and Practice of Medical Jurisprudence* (1920), 1:153–55.

69. Sutherland in Waddell, *Lyon's Medical Jurisprudence* (1921), 200–212.

70. "Note by Lieut.-Col. W. D. Sutherland," 8–13.

71. Lloyd, "Serological Analysis of Bloodstains in Criminal Cases," 220–22.

72. Many Indians were not vegetarian, particularly those who were not upper caste or Hindu.

73. S. D. S. Greval, ed., *Lyon's Medical Jurisprudence for India* (Calcutta: Thacker, Spink, 1953), 323.

74. On cows in Indian legal history, see Matthew Groves, "Law, Religion and Public Order in Colonial India: Contextualising the 1887 Allahabad High Court Case on 'Sacred' Cows," *South Asia: Journal of South Asian Studies* 33, no. 1 (2010): 87–121, and Rohit De, *A People's Constitution: The Everyday Life of Law in the Indian Republic* (Princeton: Princeton University Press, 2018), 123-68.

75. Hehir and Gribble, *Outlines of Medical Jurisprudence,* 28.

76. Lloyd, "Serological Analysis of Bloodstains in Criminal Cases," 222 (case no. 24).

77. See note 35 above.

78. W. J. Buchanan, "A Chapter on Medical Jurisprudence in India," in *Principles and Practice of Medical Jurisprudence,* 2:853–54.

79. For a critique of this colonial stereotype from the Parsi perspective, see Mitra Sharafi, *Law and Identity in Colonial South Asia: Parsi Legal Culture, 1772–1947* (New York: Cambridge University Press, 2014), 71–72.

80. Partha Chatterjee, *The Nation and Its Fragments: Colonial and Postcolonial Histories* (Princeton, NJ: Princeton University Press, 1993); Kolsky, *Colonial Justice in British India.*

81. I am using "accuser" to mean a person who accused another of a crime, rather than an informant cooperating with the prosecution.

82. See, e.g., Sutherland in Waddell, *Lyon's Medical Jurisprudence* (1921), 208.

83. See Jonathan Saha, *Law, Disorder and the Colonial State: Corruption in Burma c. 1900* (Houndmills: Palgrave Macmillan, 2013).

84. *Report of the Chemical Examiner to Government, Punjab for the Year 1935* (Lahore: Superintendent, Government Printing, Punjab, 1936), 13 (IOR/V/24/420); letter from Inspector-General of Civil Hospitals, Bengal, February 22, 1904, 1, in *Report of the Chemical Examiner to Government, Bengal for the Year 1903* (Calcutta: Bengal Secretariat Press, 1904) (IOR/V/24/422).

85. Bernard Cohn, "Some Notes on Law and Change," in *An Anthropologist among the Historians and Other Essays* (Delhi: Oxford University Press, 1990), 569.

86. Buchanan, "Chapter on Medical Jurisprudence in India," 853.

87. For a study of collective identity and responsibility elsewhere in legal history, see Avner Greif, "Impersonal Exchange without Impartial Law: The Community Responsibility System," *Chicago Journal of International Law* 109 (2004–5): 109–38. I thank James S. Krueger for this reference.

88. See Jörg Fisch, *Cheap Lives and Dear Limbs: The British Transformation of the Bengal Criminal Law, 1769–1817* (Wiesbaden: F. Steiner 1983), 50–51, and Radhika Singha, *A Despotism of Law: Crime and Justice in Early Colonial India* (Delhi: Oxford University Press, 1998), 97–100.

89. For a case of *sati* committed in protest during a feud, see Singha, *Despotism of Law,* 93n54. On Gandhian modes of protest, see Ranajit Guha, *Dominance without Hegemony: History and Power in Colonial India* (Cambridge, MA: Harvard University Press, 1997), 57; Ishita Banerjee-Dube, *A History of Modern India* (Delhi: Cambridge University Press, 2015),

262–63, 278–79; and Durba Ghosh, "Gandhi and the Terrorists: Revolutionary Challenges from Bengal and Engagements with Non-Violent Political Protest," *South Asia: Journal of South Asian Studies* 39, no. 3 (2016): 564, 571. On suicide protest in south Asia today, see Simanti Lahiri, *Suicide Protest in South Asia: Consumed by Commitment* (New York: Routledge, 2014).

90. See Waddell, *Lyon's Medical Jurisprudence* (1921), 29; J. Duncan M. Derrett, *Religion, Law, and the State in India* (Delhi: Oxford University Press, 1999), 216–17.

91. See Washburn Hopkins, "On the Hindu Custom of Dying to Redress a Grievance," *Journal of the American Oriental Society* 21 (1900): 146–59. In South Indian languages like Tamil and Malayalam, the term for the fast-unto-death was *paṭṭini*. I thank Donald R. Davis Jr. for sharing his knowledge on this point.

92. In late eighteenth-century Bengal, death rites were not performed for such a protester, making the person's spirit restless. Rather than being cremated or placed in a river, the corpse would be buried at the disputed site (for land or temple disputes) or buried outside the wrong-doer's house. The ghost or *bhut* would later be summoned by the beating of a drum. If the dispute was resolved between the wrongdoer and the deceased's heirs, the body would be removed and given proper death rites to put the spirit at rest. Singha, *Despotism of Law*, 91–92, 95–96.

93. See H. R. Fink, "The Hindu Custom of 'Sitting Dharna,'" *Calcutta Review* 62 (1876): 37–52.

94. See, e.g., Donald R. Davis Jr., *The Boundaries of Hindu Law: Tradition, Custom and Politics in Medieval Kerala* (Turin: CESMEO, 2004), 115–17; Donald R. Davis Jr. and John Nemec, "Legal Consciousness in Medieval Indian Narratives," *Law, Culture and the Humanities* (2012): 7–13; and Neil Rabitoy, "Sovereignty, Profits, and Social Change: The Development of British Administration in Western India, 1800–1820" (PhD diss., University of Pennsylvania, 1972), 315–45.

95. Fisch, *Cheap Lives and Dear Limbs*, 88; and Singha, *Despotism of Law*, 88.

96. On Hindu law's prohibition on killing one's mother in protest, e.g., see Singha, *Despotism of Law*, 100.

97. On relatives as targets, see Waddell, *Lyon's Medical Jurisprudence* (1921), 26–29, 142. For a company-era example of fellow caste members being killed in protest in western India, see Singha, *Despotism of Law*, 96.

98. See Singha, *Despotism of Law*, 98–100.

99. Henry Yule, *Hobson-Jobson: A Glossary of Colloquial Anglo-Indian Words and Phrases*, ed. William Crooke (London: J. Murray, 1903), 937.

100. Singha, *Despotism of Law*, 92, 99.

101. See Fink, "Custom," 48; Fisch, *Cheap Lives and Dear Limbs*, 50; and Singha, *Despotism of Law*, 87.

102. For example, see Hopkins, "Hindu Custom of 'Sitting Dharna,'" 157.

103. On Bhat contracts, see Alexander Kinloch Forbes, *Rās-Mālā: Hindu Annals of Western India with Particular Reference to Gujarat* (Delhi: Heritage, 1973), 558; Rabitoy, "Sovereignty, Profits, and Social Change," 305–55; and (suggesting influence on Gandhi) Howard Spodek, "On the Origins of Gandhi's Political Methodology: The Heritage of Kathiawad and Gujarat," *Journal of Asian Studies* 30, no. 2 (1971): 363. Relatedly, see R. K. Saxena, *Social Reforms: Infanticide and Sati* (Delhi: Trimurti, 1975), 19–56.

104. Rabitoy, "Sovereignty, Profits, and Social Change," 308; and Forbes, *Rās-Mālā*, xii, 559–60.

105. F. C. O. Beaman, "Eheu Fugaces," *Bombay Law Journal* 3, no. 6 (1925): 211–12. For a similar case, see Mary Frances Billington, *Woman in India* (London: Chapman and Hall, 1895), 249–50.

106. Beaman, "Eheu Fugaces," 211.

107. Beaman, "Eheu Fugaces," 211.

108. For the similar case of Thomas Keate, a judge who was sympathetic to customary Bhat contracts, see Rabitoy, "Sovereignty, Profits, and Social Change," 328–32. On Beaman, see Mitra Sharafi, "Judging Conversion to Zoroastrianism: Behind the Scenes of the Parsi Panchayat Case (1908)," in *Parsis in India and the Diaspora*, ed. John R. Hinnells and Alan Williams (London: Routledge, 2007), 162–63.

109. Beaman, "Eheu Fugaces," 212.

110. Sharafi, "Judging Conversion to Zoroastrianism," 162.

111. For example, see Rabitoy, "Sovereignty, Profits, and Social Change," 311.

112. Beaman, "Eheu Fugaces," 212.

113. *"Ujagar Singh and Others vs. Crown,"* Indian Law Reports 20 Lahore (1939): 211–12.

114. L. A. Waddell, ed., *Lyon's Medical Jurisprudence for India* (Calcutta: Thacker, Spink, 1914), v.

115. Waddell, *Lyon's Medical Jurisprudence* (1914), v.

# CHAPTER THREE

# Handwriting Analysis as a Dynamic Artisanal Science

## The Hardless Detective Dynasty and the Forensic Cultures of the British Raj

PROJIT BIHARI MUKHARJI

Two kinds of studies have hitherto dominated the historiography of colonial science in south Asia. On the one hand are the studies of state science that look at what the colonial state did or did not do to encourage or discourage science.[1] On the other hand are studies of traditional sciences that were often "revived" or "refigured" outside the state's direct control through the initiatives of colonized elites.[2] Somewhere in between, and related more to the former than the latter, are a few studies that explore the "nationalist" appropriation of colonial state science.[3] The history of forensics in British India, however, does not fit into any of these paradigms.

In British India, some of the key figures in the development of forensic sciences in general and handwriting analysis in particular were autodidacts. Though they managed to carve out a niche for themselves within the colonial police for a while, their initial training, success, and popularity were clearly independent of putative support from the colonial state. Yet they did not practice any "traditional" or "indigenous" science, nor were they part of the colonized elite. The only parallel to this that I can imagine are the indigo planters studied by Prakash Kumar who operated between the state and the market and independently developed their science through individual efforts.[4] But the scale, nature, object, and chronology of the indigo science were distinct from the forensic graphologists at the center of this study. Their entanglement with nationalism, too, was distinct and extremely complex. These founding figures in the history of forensic graphology were neither nurtured by the colonial state nor nourished by any "indigenous" tradition. Their commitments to both empire and the nationalism that opposed it were contingent, complex, and inchoate. In short, they were between the largest political institutions of their era.

At the center of this history of forensic graphology were multiple genera-tions of a single mixed-race family: the Hardlesses. The racial intermediacy of the family mirrored the liminal character of the science they practiced. The fascinating story of these self-taught colonial detectives is not just the first chapter in the history of modern forensics in India, but also an important corrective to the stark and continuing binarism that divides modern south Asian history into the "imperial" and the "indigenous." Moreover, in their story we have a testament to a distinctive mode of practicing science that I believe has hitherto escaped the historian's attention. This distinctive mode of practicing science is what I call "dynamic artisanal science." I adapt the no-tion of a "dynamic artisanal science" from economic histories that have looked at the dynamic and creative ways in which artisanal industries re-sponded to the challenges of colonial modernity. The notion highlights the simultaneously dynamic and artisanal character of the science practiced by the forensic graphologists I study.

Locating this dynamic artisanal science within the distinctive "forensic culture" of its time and place generates interesting new insights into more familiar questions tackled by other historians of forensic sciences. Christo-pher Hamlin has helpfully identified four key elements to any forensic culture: (1) technologies of witness, (2) technologies of testimony, (3) tech-nologies of judgment, and (4) the arch anxieties.[5] In each of these key aspects, the dynamic artisanal mode introduced new complexities and contradic-tions. The technology of witnessing, or the "exemplary forensic methods" themselves, was fundamentally reimagined by the insertion of the prover-bial "colonial rule of difference," that is, the imperative for the colonial re-gime to constantly distinguish the colonial from the metropolitan.[6] Thus, precisely when "forensic graphology" was being phased out in Britain around the turn of the twentieth century, it was enthusiastically adopted by the co-lonial state, thereby raising the artisanal practice to new heights. Similarly, the sensitive combination of moral authority and expertise that empowers technologies of testimony or "exemplary professions" evolved through the co-lonial rule of difference along its own distinctive quasi-racialized pathway. The tension between the need for public accreditation of professions and the kinship-oriented, secretive artisanal mode of knowledge transmission gen-erated novel resolutions in the form of racialized modern professional iden-tities. The colonial variation of the technologies of judgment was perhaps the most striking.

Whereas expert disagreement was at the heart of the adversarial legal system of the British world,[7] in the colonial context, there was a constant effort to present science as unequivocal. This meant that the legitimacy of expert disagreement was never really recognized. In turn, this led to distinctive calibrations of the "objective" and the "subjective" and encouraged a steady push toward mathematization and mechanization of forensic work. Finally, the arch anxieties that motivated colonial forensics, aside from poisoning and blood frauds, were documentary frauds.[8] In a state that has been described as a "document raj," it is hardly surprising to find a plethora of anxieties around documentary fraud.[9] What is surprising, however, are the claims that it is unimportant for the expert to know the language of the document to detect fraud. But this was yet another one of the distinctively negotiated results of the specifically colonial location of the dynamic artisanate of forensic scientists. All these multiple tensions and complexities were held together by the Hardless dynasty of detectives.

## A Dynasty of Detectives

Thanks to Sherlock Holmes, the Victorian detective of our collective Anglophone imagination is a solitary man devoid of family life. Despite having many other differences, most famous literary detectives have tended to follow Holmes in bachelorhood. Neither the dandy Belgian, Hercule Poirot, nor the mildly alcoholic Oxford dropout Endeavour Morse ever tasted marital bliss. So the idea of a close-knit dynasty of detectives is surprising.

The founder of the dynasty of graphologists was Charles Richard Hardless (1866–1944). Though there is a possibility that he learned some of his earliest skills in detection from an uncle, John H. Hardless, it was C. R. Hardless who transformed detection into a full-time profession. In 1908, despite much hesitation from metropolitan authorities in London, he was appointed the first "Government Handwriting Expert" attached to the Criminal Investigation Department (CID) of the Calcutta Police. Within less than a decade however, Hardless resigned his job and went into private practice, though the police continued to call upon him well into the 1930s.

Though attached to the Calcutta Police, from the beginning, Hardless was deputed throughout the British Raj to solve difficult cases. He traveled widely from Rangoon to Peshawar and from Colombo to Bombay on official work. Later, in private practice, he took on a number of commissions from Indian monarchs such as the Nizam of Hyderabad. Despite this busy life, C. R. Hardless managed to marry four times and raise a large family.[10]

C. R. Hardless, C. E. Hardless, and photographer Raja Deen Dayal at King Kothi Palace, Hyderabad, 1912. From the author's personal collection; courtesy of the late Susan Creedon

From within this large family, at least three of C. R. Hardless's sons went onto become eminent detectives in their own right. These included Charles Edward Hardless (1888–1966), Philip William Ravenshaw Hardless (1900–?), and Harold Richard Gordon Hardless (1903–?), all of whom wrote textbooks on forensic matters. Between them, the Hardlesses pioneered many of the key forensic techniques that might today be called forensic graphology, forensic chemistry, and ballistics.

Their achievements look even more spectacular when seen in the context of C. R. Hardless's own origins. While details of his early life are unclear, it is almost certain that he was born into a poor Eurasian family living in and around Calcutta.[11] In 1885, the local press mentioned one "Mr Hardless" (most likely John H. Hardless) as attempting to organize a central bureau to try to help poor and unemployed Eurasians and Europeans.[12] Yet years later, the plight of poor Eurasians remained dire, and one relative, George Hardless, a pipe-layer by profession, being unable to find regular employment, was driven by poverty to commit suicide in 1896.[13]

C. R. Hardless was a superintendent in the Accountant-General's Office in 1904 when Viceroy Curzon requested his appointment to the CID. Lord

Curzon's letter to the secretary of state for India explained that Hardless's work brought him into daily contact with handwritten documents and he got interested in handwriting and its individuality.[14] According to Frank Anthony, educationist and leader of the Eurasian community, leaving school, Hardless had joined the Telegraph Department at a monthly salary of 50 rupees. There he had risen to the position of superintendent at a remarkably young age. Having done so and looking for fresh challenges, he began dabbling in handwriting analysis, and it was within the Posts and Telegraphs Department that his skills were first noticed. The department then began officially requisitioning his services to catch frauds. Later, the central government took notice of his abilities, and the police began consulting him. According to Anthony, by the time of his death, Hardless was called the "Father of the Handwriting Profession" in India.[15]

During an interrogation at a trial in Colombo in 1908, C. R. Hardless mentioned that he had been studying handwriting since roughly 1888 and had been practicing his science since around 1893. He said he was initially trained under his uncle (possibly John H. Hardless); later, he depended on the magnanimity of local governments to acquire for him different scripts and texts for him to study. He also independently studied any book on the matter that he could find. Subsequently, he also corresponded with international experts on a regular basis and exchanged case notes with them. By 1908, Hardless claimed to have testified in more than a thousand cases, excluding reference cases where he did not appear in court.[16]

While the patriarch's main skills were in the realm of handwriting analysis, his sons went much further. C. E. Hardless, the eldest son, did much to standardize his father's techniques and thus transform them from "knack" or "talent" to "science." It was P. W. R. Hardless, though, the second son, who ventured into the domain of chemistry. He discussed a range of inks, pens, and papers whose identities could be determined by chemical analysis. Finally, there was the young H. R. Hardless, who developed early ballistics. By the end of the dynasty, what had commenced as an expertise grounded in a single skill—handwriting analysis—had grown into a much more complex and heterogeneous set of detective techniques.

The family's evolving repertoire of techniques underwrote a striking social ascent. In 1909, the Calcutta press, for instance, reported on a burglary at the Hardless residence on 1 Ripon Street. Not only did the very fact of burglary suggest wealth, but the same report also mentioned that the burglars had missed jewelry worth 4,000 rupees in a particular almirah.[17] In 1915, the Hardless family, then residing at the fashionable 17 Park Lane, was burgled once again.[18] Soon after, the Hardlesses left Calcutta and moved to a substantial estate at Chunar, in the United Provinces. The Nizam of Hyderabad, Mir Osman Ali, most likely gifted the estate at Chunar to the Hardlesses in recognition of their having foiled a plot to dethrone the young monarch in 1912.[19] From then on, the family operated out of the grand house they built in Chunar called the Sanctuary and, more occasionally, neighboring Allahabad.

Though the exact wealth or income of the family is difficult to gauge, a divorce court in 1932 estimated the youngest son, H. R. Hardless, to be earning at least 3,000 rupees per year from private practice (*Olivia Hardless vs. HR Hardless*). It is certainly safe to assume that C. R. Hardless and C. E. Hardless would have been making much more than what young Harold could command. By the early 1930s, the Hardlesses were enormously successful. So successful and so well known were the Hardlesses as a family, in fact, that at one time impostors began to claim the name. In 1933, C. E. Hardless wrote to the editor of the *Times of India* to disclaim any connection or relationship with an individual using the Hardless name and posing to be a handwriting expert in Madras.[20] For a family that a mere fifty years earlier was part of the impoverished Eurasian underclass, this was a remarkable success story.

## Technology of Witness

As is clear from the foregoing introduction, at the core of the Hardless dynasty's success was handwriting analysis. In fact, "forensic science" as such did not exist at the time as a discrete discipline. As the introduction to this volume observes, prior to the emergence of specialized professionals, the most august authorities in forensic science had been the professors of medical jurisprudence. By the second half of the nineteenth century, however, the story increasingly became one of the emergence and struggle for acceptance of new modes of specialist professional expertise tied to single techniques, sets of tools, or occasions for interventions. Often, the new professions were closely tied to a single prima donna specialist who claimed to have invented or developed some forensic art to a high and even inimitable degree. Handwriting analysis emerged as a specialized set of techniques within this dispersed late nineteenth-century field of forensic sciences. But its histories in Britain and India were quite distinct from each other.

The origins of the post of government handwriting expert in India illustrate these distinctive trajectories. Initial discussions about the post commenced in 1904. A letter dated June 22, 1905, from Viceroy Lord Curzon to Secretary of State St. John Brodrick summarized the arguments for appointing Hardless. The letter described how C. R. Hardless, who "had acquired a considerable reputation in the study of handwriting" and had appeared as an expert witness in a number of criminal cases, was initially appointed by the Bengal government for a period of four months. "On consideration of the good work" done by Mr. Hardless, Curzon wrote that he had extended the period of employment "experimentally" for another eight months. Now at the culmination of this period, he thought that Mr. Hardless ought to be permanently employed by the government at a salary of 500 rupees plus perks.[21]

Brodrick, playing it safe, initially appointed C. R. Hardless for a period of one year and stipulated that further extensions would only be made if the CID director reported favorably on Hardless's work and thought such a man was necessary.[22] The following year, in 1906, referring to Brodrick's directions, the Bengal government forwarded a report from CID Director H. A. Stuart and once again sought to make Hardless's position a permanent one. Stuart's report stated that in the year 1905–6 alone, Hardless had worked on sixty-eight cases for the government. One of these was a famous fraud case involving forged government promissory notes worth 100,000 rupees. It also mentioned that his services had frequently been of "great assistance to the police in conducting their investigations." The report ended by stating that "there are no private handwriting experts in India of the same skill and standing as Mr Hardless, and if his services are not made available it will be impossible to obtain the assistance he has given during the past two years."[23] Faced with such strong support, the metropolitan authorities had little option but to confirm Hardless's position as a permanent one.

Notwithstanding the eventual appointment, we must ask why London dragged its feet for so long despite Calcutta's repeated efforts to employ Hardless. The reasons for the reluctance become clear from another internal memo. Upon receiving Viceroy Curzon's request in 1905, the Judicial and Public Department forwarded the request to the Finance Department, which in turn sent a "D.O. (Demi-Official) Letter" to the Home Office in London asking for its view on the matter.[24] A. L. Dixon from the Home Office responded in detail about the use and status of "handwriting experts" in England at the time.[25]

To begin, Dixon clarified that "there are no such experts holding definite posts under the Govt." When the police required the services of such experts, he continued, they simply "consult some person with a reputation." Most often this person was "Mr Gurrin" (most likely Thomas H. Gurrin). The chief utility of this expert, according to Dixon, was not so much to solve the crime but to "impress the court with expert opinion—'call in the expert.'" Dilating on the matter further, he stated that the handwriting expert was utterly dispensable for all purposes "apart from the question of making an impression on the minds of the jury or others." So far as Dixon was concerned, the actual business of solving crimes by comparing handwritings could be done by anyone "who has a lot to do with various handwritings and has a sort of 'analytic eye.'"[26] Thus Dixon was not only saying that handwriting experts did not hold official positions in England, but he was also impugning their claims to specialized knowledge. The little utility they had was limited to being a sort of a theatrical accouterment in jury trials.

Even more damningly, Dixon stated, "it will probably be within your knowledge that 'handwriting experts' are somewhat at a discount in the public estimation at the present time—no doubt the circumstances of the Beck case have contributed to it."[27] The so-called Beck case was the wrongful conviction of Adolph Beck on a charge of fraud in 1896. Gurrin had been involved in the trial and had testified against Beck. The trial, however, did not turn simply on handwriting evidence. It was much more crucially reliant on eyewitness testimony, and the eventual wrongful conviction also had much to do with the willful suppression of facts by the judiciary and the prison system. Yet, interestingly, it would appear from Dixon's comments that of all the diverse parties to blame for the miscarriage of justice in the Beck case, it was handwriting experts who had taken the worst hit to their credibility.[28]

In contrast to this British situation, in India, the CID sought Hardless's expertise to help them solve crimes, not merely play the part of the expert at trial. There was also no significant public suspicion of expert testimony in British India at this point. When Brodrick permitted the initial temporary appointment of C. R. Hardless, he did so having taken cognizance of Dixon's note. Capitalizing on this difference, "government handwriting experts" or "examiners of questioned documents," as they were variously called, became an integral part of police forces in British India in the last four decades of colonial rule. It was thus that forensic graphology (though it was still not called that) emerged as a legitimate field of forensic science in British India by working through the rule colonial difference.

## Technology of Testimony

In yet another departmental memo written in the late 1940s, F. W. Kidd, a retired officer of the Indian Intelligence Bureau, questioned the utility of handwriting experts.[29] Impugning their claim to specialized expertise and echoing some of Dixon's opinions nearly half a century later, Kidd said these handwriting experts were only "so-called experts." Kidd gave four reasons why these men were not true experts comparable to those who worked with fingerprints. First, while these men were frequently able to point out differences between handwritings of different people, their knowledge was based on mere "experience" and not upon "more solid and convincing grounds," like fingerprint experts.[30] Second, he knew of handwriting experts who regularly disagreed with each other. Third, he knew of no reference books on the subject. Finally, he knew of no interesting cases that had been solved by handwriting experts.

Much of what Kidd said was based on misconceptions and ignorance. There were, for instance, many major cases—including high-profile political ones—that had turned on handwriting evidence. There were also several books on the subject of forensic graphology, including those by the Hardlesses. Most importantly, Kidd's idea that expert disagreement somehow signaled the inadequacy of the science was patently misplaced. Hamlin has shown that not only were expert disagreements rife in Victorian courts but also that scientists recognized such disagreement as a legitimate part of scientific work, rather than seeing it as pernicious sign of scientific inadequacy, as Kidd did.[31] What Kidd shared with Dixon, however, was an idea of "expertise" as something more than merely ability grounded in experience. Dixon had said anyone handling documents and with an "analytic eye" could do the job that handwriting experts did. Similarly, Kidd acknowledged that handwriting experts were able to distinguish handwritings, but refused to call them experts because their ability was grounded in mere experience and not on more "solid or convincing grounds."

How exactly does ability founded on experience differ from expertise properly so-called? This is clearly a historical question whose answer turns upon the way the three terms "ability," "experience," and "expertise" were defined in specific historical contexts. In his book *Identification of Handwriting and the Detection of Forgery* (1914), C. E. Hardless distinguished between the "lay method" of comparing handwriting and the "scientific method" of the expert. The lay method, according to him, was based on "comparison by formation" whereby the shapes of particular words or letters were compared.[32] In complete contrast to this stood the "comparison by characteristics." Unlike the former, this method focused on the "personal or individual writing habit."[33] Whereas the formation of letters could change, be deliberately disguised, and anyway prove useless when dealing with samples written in two different scripts—like, say, English and Bengali—the comparison by characteristics remained largely consistent. Describing this new scientific method, Hardless Jr. wrote, "Two clever artists could produce two similar drawings or paintings of the same object . . . and the drawings or paintings of each could be similar in all details as to form and colour and made to resemble closely. Careful scrutiny would, however, discover that, in his own particular drawing or painting, each artist has his own particular habit of pulling or drawing his pencil or brush and his own peculiar touch and mode of shading and colouring. By this means each could be identified although the pictorial effect of both drawings or paintings may be the same."[34]

This description is remarkably similar to what art historians know as the "Morellian method," developed by the Italian Giovanni Morelli in the late nineteenth century for detecting artistic authorship of antique paintings. In a fascinating article, Carlo Ginzburg draws out a family resemblance between the methods of Morelli, Sherlock Holmes, and Sigmund Freud. In each, Ginzburg argues, "individuality" is seen to reside in small, negligible, but telltale signs that come to be called "clues." Clues are, for Ginzburg, a new epistemic paradigm emerging at the end of the nineteenth century and connected explicitly to the new anxieties about social control and criminality.[35] While I remain agnostic about Ginzburg's larger claim, I do find it remarkable that the new scientific method of Hardless not only resembled the Morellian method but also described itself explicitly through the hypothetical case of art forgery.

One of the reasons I am agnostic on Ginzburg's larger project is because in his telling, this new epistemic paradigm is clearly opposed to quantification. For the Hardlesses, there was no such contradiction. Referring to the work of Alphonse Bertillon, C. E. Hardless wrote, "In the Bertillon system of measurements, for the identification of criminals, only eleven specific measurements are taken and by the means of these few measurements, with a general description of the individual, thousands of persons have been positively identified. It may appear strange, but no two individuals have yet been found possessing equal measurements in only eleven respects . . . The several individual habits in writing number a great deal more than eleven and it will be readily understood how utterly impossible it therefore is to find the writings of two different persons exactly agreeing in such a considerable number of details."[36]

Sociologists of science Harry Collins and Robert Evans distinguish between two different approaches toward historicizing expertise—the relational and the realist. In the former approach, expertise is always a matter of relationships between experts and nonexperts, and expertise is something attributed by others. By contrast, the realist approach understands expertise as something positively acquired by the expert through membership and socialization in a particular group.[37] In their view, both specialist and more ubiquitous forms of knowledge include a component of tacit knowledge that cannot be fully explicated or verbalized. Individuals are possessed of both various types of ubiquitous knowledge available in the social groups of which they are part as well as more specialist forms of knowledge, including its tacit aspects, possessed by particular communities of expertise.

The social structure of artisanal communities often leads to a close overlap between family, kin, and a community of experts. In south Asia, caste provides an additional basis for enclosing expertise within a discrete social group. This overlap means that distinguishing forms of widely available ubiquitous knowledge in these communities from specialized knowledge becomes ever more challenging. Artisanal groups such as weavers, potters, carpenters, and the like usually constitute distinctive caste groups, and the relevant expertise is widely disseminated within the families and extended social networks that make up these caste communities. Much of the forensic knowledge of the Hardlesses, too, was similarly ensconced in familial and caste-like communitarian networks. In a 1910 case, the second in which C. E. Hardless appeared in court, the defense lawyer sought to mine precisely this problematic familial context to undermine his claims to expertise. The lawyer asked Hardless directly, "You have never met a live expert except your father"? When Hardless replied in the negative, the lawyer further mocked Hardless by asking how can one become an "expert" at the age of fifteen? With remarkable chutzpah, the young Hardless deployed humor to deflect the line of questioning by saying that like a "born poet," he was a "born expert." As the court burst out laughing, Hardless won a small but ambiguous victory. The lawyer, however, persisted in attacking the senior Hardless's reputation, possibly in a bid to provoke the young man indirectly throughout the larger part of the cross-examination.[38] Defense attorneys clearly considered the familial nature of C. E. Hardless's training to be a soft spot.

This tension between the familial and professional is tangentially visible in the prefaces to the books C. E. Hardless authored. In the preface to *The Identification of Handwriting*, he wrote, "The youngest and the least recognized experts in handwriting ventures to issue a book on the subject which has been taught him from childhood and which, owing to the circumstances prevailing in India, he was called upon to help and even practice in when scarcely out of his teens."[39] Continuing further, he wrote that "The book is the outcome of direct training and experience in the Office of an Expert in Handwriting of official rank and recognized standing, admittedly possessing the most extensive, varied, as well as the most constant practice anywhere in the profession and who has dealt with the largest case of handwriting ever brought to trial in the World."[40] In all this, not once does he mention that this great expert of official rank under whom he has learnt his craft is his own father!

The remarkable circumlocution was clearly motivated by implicit protocols of expertise. Collins and Evans are clear that any form of knowledge acquisition involves immersive socialization, and yet modern societies do not admit that a son of, say, a doctor, merely by closely observing his father at work for a long time, might acquire specialist medical knowledge. Herein lies the breach between the artisanal and the modern professional modes of expertise. While both require group socialization and immersive experience, the latter insists on a clear differentiation between the groups that provide access to ubiquitous and specialist knowledge, which the artisanal mode often scrambles.

The Hardless family continuously struggled with this fundamental contradiction. On the one hand, they constantly strove to develop recognizable modern professional institutions, while on the other, these institutions repeatedly became family enterprises. Two instances illustrate my point. First was the effort of the Hardlesses to start a private pedagogical institution exclusively devoted to a forensic science.[41] In 1938, the Hardlesses were inviting applications for admission into the School of Document Investigation.[42] The full course of education was to cover "practical lessons in handwriting and fingerprint identification; analysis of inks and paper; lessons on photography, microscopy etc., etc.," all of course at "moderate charges."[43] The following year, fingerprint identification was available as a separate, independent course. This latter course included practical work as well as its theory, history, and case law. Individual lessons could be had for 5 rupees each, and the entire set of twelve lessons was worth 50 rupees. Upon completing the entire course, one obtained a diploma.[44] Later still, another diploma course titled "A Complete Handwriting Identification" was also commenced.[45]

Yet for all the seemingly formal trappings of the school, it was housed at Pennville, 19 Hastings Road, Allahabad—the residence of CE Hardless. There was nothing to suggest that anyone but members of the extended family were involved in running the school. What appeared, then, to be a formal school with a fixed syllabus and standard fees was in effect more like a shortened apprenticeship being run out of the family home of the eldest Hardless son, with possible help from other members of his family.

My second illustrative example is the professional journal *Document Investigation*, which was published by the Hardlesses from 1933 until at least 1938. Unfortunately, the journal has not been preserved in any library that I know of. The only copies I have been able to consult are ones I have in my personal collection. Because so few issues of *Document Investigation* are available today, it is impossible to undertake a full examination of the contents. Yet social historians of scientific disciplines always point to the emergence of journals as one of the key institutions that consolidate and standardize any discipline. It is therefore perfectly legitimate to see *Document Investigation* as an important step toward the formalization of science of "handwriting analysis" as a distinctive form of forensic science.

The journal commenced in 1933 and was in fact the second such venture undertaken by the Hardlesses. Earlier, newspaper advertisements from the early 1920s inform us of a previous journal titled *The Handwriting Expert's Review*, which was also published by "Hardless and Hardless, Handwriting Experts, Chunar, UP." Unfortunately, no copies of this earlier journal seem to have survived, and all I have been able to gather about it is from advertisements. The journal was described in the advertisements as "An Illustrated Quarterly dealing in Fraud and Forgery" and was priced at 8 annas an issue. An annual subscription was worth a rupee and 8 annas.[46] That the failure of this venture did not dissuade the Hardlesses from publishing another similar journal a decade later once again demonstrates the importance they must have attached to such a journal. When *Document Investigation* first appeared in 1933, it carried the same tagline that had described *Handwriting Expert's Review*: "An Illustrated Quarterly Dealing with Fraud and Forgery." Yet the success of the later venture soon transformed the journal from a quarterly to a monthly.[47]

Despite the journal's clear mission to establish forensic graphology as a professional and scientific field of study, once more we find that in practice the publication was an in-house, family affair. Its contributors were either family members or people closely associated with the Hardless family, the advertisements it carried were entirely devoted to other Hardless ventures or those of their close associates, and even the press that published it was a Hardless family venture. Edited by C. E. Hardless and published by the Penn Press, the journal had all the formal accouterments of a scientific periodical. Yet upon looking closer, we find that the Penn Press was in fact located at Hardless's home at 19 Hastings Road, Allahabad, and the manager of the press was none other than Hardless himself. Contributing authors, when not members of the family, such as William Tuck, another Eurasian gentleman from Patna, or the longtime Hardless correspondent H. P. Wunderling of Seattle, Washington, were almost certainly close acquaintances of the family since they also frequently advertised in the journal.

My intention here is not to impugn the objective scientific status of the Hardless dynasty. Rather, what I am trying to illustrate is the Janus-faced character of their science. On the one hand, they were repeatedly at pains to enact a dubious distance between the intimate space of the family and the allegedly formalized space of the expert community. On the other hand, each of their ventures necessarily relied on the mobilization of intimate familial and familiar affective networks. Lengthy circumlocutions such as not naming C. E. Hardless's eminent teacher as his own father or holding separate designations as both editor of *Document Investigation* and the manager of the press that published it were all performative gestures that enacted the distance between "family" and "profession," and yet every professional move relied more and more heavily on the resources and labor of the same family and friends. It is this Janus-faced dynamic that characterizes the figure of dynamic artisanal science.

## Technology of Judgment

Hamlin describes the *technology of judgment*—the institutional assemblage of justice in colonial India—simply as the "colonial rule of incomprehensible others."[48] The otherness and the incomprehensibility were of course predicated upon a deeply embedded racialized view of colonial society that had crystallized in the second half of the nineteenth century. Legal disputes— such as the Ilbert Bill controversy in 1883–84 that gave the right to Indian judges to try Europeans, or the Age of Consent protests in 1891 over the determination of minimum legal ages of marriage—served to racialize the legal apparatus in stark and public ways.[49]

As a Eurasian family, the Hardless dynasty occupied a strange and liminal position in this Manichean colonial world. They were neither white nor brown. They had to straddle a curious path between the alleged "incomprehensibility of the Other" upon which the judicial apparatus—just as much as the nationalist opposition to it—was based and the detective's ability to know the truth.

That some of C. R. Hardless's most significant successes had come in high-profile political cases did not endear him to the increasingly vocal nationalist press of the day. Hardless's work had been crucial to the Alipore Bomb Case (1908–9), which put major extremist leaders such as Aurobindo Ghose in prison, and as a result, the nationalist press remained ever ready to dilate upon every one of Hardless's failures or reversals at trial. Partha Chatterjee has described how one of the minor dramas in the hugely influential case of the Princely Impostor of Bhawal (1935), with its thinly veiled nationalist underpinnings, had involved a challenge to the neutrality of C. E. Hardless.[50] The dispute arose after Hardless published an advertisement in *Document Investigation* stating that he had been involved in the Bhawal case on behalf of the government.

The incident in the Bhawal case was actually the culmination of a long history of nationalist suspicion of the Hardlesses. As early as 1907, defense lawyer Mr. Norton had described C. R. Hardless as "a hound tied to the chair of the Government of Bengal, let loose every now and then to track a man to death or loss of liberty."[51] Subsequently, in 1911, Justice Sundara Aiyar, while presiding over a case in the Madras High Court, wrote, "I by no means doubt that Mr. Hardless carried out his comparison with perfect bonafides, but it is unfortunate that the expert knew what the prosecution wished to be proved and that circumstance must, in my opinion, detract to some extent from the weight to be attached to the ex-part's [sic] testimony."[52] Despite the different idioms in which Mr. Norton and Justice Aiyar expressed their suspicions, at the heart of both was the same distrust of an expert who was in government employ. The nationalist sentiment served to further deepen this suspicion.

Some sections of the nationalist press were so predisposed to disliking the Hardlesses that when in 1933 the Madrasi impostor, operating under the false name "Bertrand Hardless," was convicted, many nationalist dailies rejoiced that the handwriting expert had been jailed. The *Zamindar* of Lahore, for instance, ran a headline announcing that the "Famous Lahore Expert of Forged Documents Convicted" (at the time, C. E. Hardless briefly lived in Lahore). The *Bombay Chronicle* equally gleefully proclaimed, "Hardless Gets

Six Months." The unseemly haste and glee in denouncing the family without waiting to check the name of the man convicted betrayed the nationalist press's latent dislike of the Hardlesses.[53]

Yet C. R. Hardless was politically a committed moderate nationalist. C. R. Hardless became in 1921 a staunch and close supporter of the Anglo-Indian leader Col. Sir Henry Albert John Gidney and his Anglo-Indian and Domiciled European Association. Col. Gidney asserted that "We must demand recognition as equals of Britisher and Indian though, in so doing, we must remember that in the India of today and the India to be, we have to live on terms of reciprocity and friendship with all races and communities."[54] At the same meeting, Hardless declared, "let us each think kindly of our Anglo-Indian brothers and let us do our duty to God, the King, the Government, our country and our community."[55] Though couched in a loyalist idiom, it is significant that Hardless chose to mention a distinct duty to the "his country." At a time when many Anglo-Indians still preferred to call Britain "home," this mild and inconspicuous insertion was telling.[56] By the following year, however, Hardless was much more forthright in his assertion of nationalist sentiment. In 1922, Hardless, in his capacity as general secretary of the association, led the office bearers in putting out a public call to the Anglo-Indian community to think of themselves as "citizens of India first." The call also encouraged the community to wholeheartedly support moderate nationalists and help them in creating a "sane and proper form of democratic government." Hardless and his allies emphasized that "India was their motherland and they had to live and die in India."[57]

That Hardless's politics was no hollow boast is also proved by the fact that when religious riots broke out in 1917 in Mirzapur, the district in which the Hardlesses lived, Hardless and others actively mediated between the rival groups of Hindus and Muslims to restore peace. The conflict arose over the route for the annual Dussehra and Muharram processions. This was a familiar trope through which northern Indian towns in the last decades of the Raj came to be "communalized."[58] A handful of local gentlemen came together to actively mediate and diffuse the violent conflict. Among these men, predictably, most were either Hindus or Muslims themselves. Charles Hardless was the only non-Hindu and non-Muslim to attempt to make peace in Mirzapur. His efforts were recognized in the local press, which thanked him for his efforts to build bridges between the communities.[59]

Unfortunately, whereas the nationalist press often labeled C. R. Hardless as a "hound of the empire" for his role in the prosecution of nationalist "terrorists," they did not as often discuss his attempts to carve out a legitimate space for Anglo-Indian nationalism. While I have not been able to find clues about the political orientation of the Hardless sons, the fact that they all chose to stay on in India after 1947 suggests that they shared their father's commitment to trying to find a place for themselves in an independent India. Being betwixt and between the empire and the nation did not mean the Hardlesses rejected the language of race. Instead, they embraced it and spoke consistently of the "Anglo-Indian race." Here it is worth noting that many Anglo-Indian families sought to deny the mixed Anglo-Indian identity and claimed to be pure Europeans settled in India. Hence the proud assertion of Anglo-Indianness is in itself significant.[60] The mere fact of their espousal of such an identity also immediately complicated the black-and-white politics of empire and opened up the *technology of judgment* based on the alleged "incomprehension of the Other" to an implicit critique. Their very presence, not to mention their vocal embrace of their biracial identity, mocked the alleged absolute otherness organized neatly along the colonial divide.

Race, as David Arnold has pointed out, was never a "homogenous set of ideas and practices, driven by material greed and social anxieties in the West, and capable of delivering social power and political authority to the Whites across the globe." Instead, Arnold reminds us that race was a "far more nebulous and self-contradictory concept, and rather than being the voice of White authority alone, could form part of an interactive process."[61] The figure of the Eurasian is of course the consummate embodiment of the interactive process.

Moreover, the figure of the Eurasian opened up what Ann Stoler calls the "affective grid of power," where one might locate the "microphysics of colonial rule."[62] The transracial intimacy that produced the community once more forestaged the intimate space of the family as a subtle but eloquent counterpoint to the abstract identities of the white colonizer and the brown colonized who were enacted in legal controversies and nationalist politics. The very "affective grid" within which a son learned his artisanal science from a father was where the "microphysics of colonial rule"—reliant so often on sexual and personal relations across the racial boundaries—became visible.

## Arch Anxieties

While Hamlin is right in pointing out that the main anxiety that inspired the forensic formations of British India was the inscrutability of the Other, an even bigger problem arose from the fact that to run the empire, the colonial state had to substantially depend on that very Other. Unlike settler colonies or the African colonies where Indians played the intermediate roles, in south Asia, the British state was crucially dependent on Indian participation within the state. This meant that one of the key anxieties animating the state was the potential corruption within it.

C. E. Hardless mentioned several instances of documentary fraud where subordinate officials of the state were the culprits. One type of fraud involved railway employees deliberately sending goods to the wrong station and then, having declared them officially lost, selling them privately. Another type of railway fraud involved ticket clerks selling tickets for shorter journeys to passengers who could not read English and pocketing the surplus fare. The misuse of free railway passes was yet another type of fraud involving railway employees. A further major set of frauds involved tampering with government records. Police officers adding, subtracting, or amending case records well after the incident were a frequent phenomenon, as were Patwaris altering village records of proprietary rights. Clerks were known to alter the dates on medical certificates to stretch their leave. Employees also tried to alter the birthdate on their service records in a bid to obtain extensions. Most remarkably, corrupt court clerks were known to change documents in the court's own possession. In one case, a court clerk helped a forger attempting to pass off a forged will as legitimate by inserting a spurious reference to the will in an older judgment passed by the court.[63]

Bias or corruption of subordinates, especially those tasked with the day-to-day working of the Document Raj, was therefore a pervasive source of anxiety for the colonial administration. As is frequently the case, such suspicion of subordinates encourages a process of mechanization. The constant effort by the Hardlesses to develop mechanical tools for their work enacted this suspicion of "native subordinates" even as they arguably occupied a position fairly close to that of the native subordinate. Their unceasing quest for mechanization through the 1920s and 1930s must therefore be understood as a bivalent gesture. On the one hand, machines had the aura of cutting-edge science about them and hence further helped the Hardlesses claim scientific status for their techniques. On the other hand, machines, by enacting a "mechanical objectivity," served to implicitly shore up the moral authority of the Hardlesses as experts within a regime that deeply and increasingly mistrusted its subordinates.[64]

One such early instrument was known as the Terror to the Forger, a combination of a camera and a microscope that produced scaled enlargements of documents called "photomicrographs." Describing these photomicrographs, William H. Tuck, writing in *Document Investigation*, stated, "Although the microscope takes us into a strange world of its own, the glimpses it gives to a single observer are but fleeting, but fortunately photomicrography enables those glimpses to be permanently recorded so that they are always at hand for everybody to see."[65] This promise of transparency was actually enacted in trial courts. Blown-up images demonstrated step by step how the handwriting expert had come to his conclusions. Tuck's felicitous phrase "at hand for everybody to see" was mere code for "skeptical juries and judges to see."[66]

In the same issue of *Document Investigation*, another regular, possibly Anglo-Indian, ally of the Hardlesses, Russell Gregory Afzal, described a slew of new mechanical instruments to aid handwriting analysis.[67] Among these

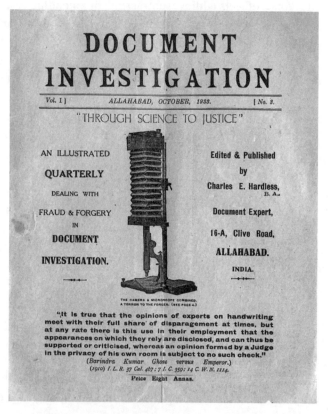

The Terror to the Forger was a machine that combined a camera and a microscope. It produced scaled enlargements of documents called "photomicrographs." From the author's personal collection

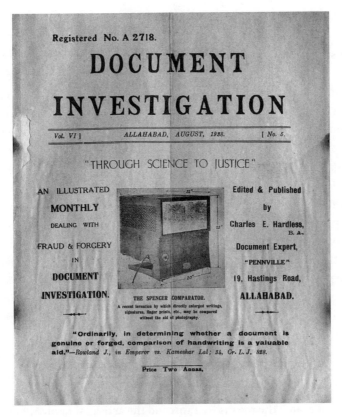

Registered No. A 2718.

# DOCUMENT

# INVESTIGATION

Vol. VI ]        ALLAHABAD,  AUGUST, 1938.        [ No. 5.

"THROUGH SCIENCE TO JUSTICE"

AN ILLUSTRATED

**MONTHLY**

DEALING WITH

FRAUD & FORGERY

IN

**DOCUMENT**

**INVESTIGATION.**

THE SPENCER COMPARATOR.
A recent invention by which directly enlarged writings, signatures, finger prints, etc., may be compared without the aid of photography.

Edited & Published

by

**Charles E. Hardless,**
B. A.,

Document Expert,

"PENNVILLE"

19, Hastings Road,

**ALLAHABAD.**

"Ordinarily, in determining whether a document is genuine or forged, comparison of handwriting is a valuable aid."—*Rowland J., in Emperor vs. Kameshar Lal; 34, Cr. L. J. 828.*

Price Two Annas,

The Spencer Comparator was a type of projector that allowed for enlarging text or fingerprints without having to take a photograph first. From the author's personal collection

was the Whistler, an "electrical apparatus based on the inventions of Marconi." The machine featured a loudspeaker with wireless valves. Upon applying it to normal paper and ink, characteristic "music" ensued. When moved over the disputed part, despite being indistinguishable to the eye, the sonic vibrations were apparently altered and a new tune was emitted, thus signaling a forgery.[68] Besides the Whistler, Afzal described other new machines such as the Chromoscope—an instrument with four lenses equipped with different-colored light filters and connected to an electronic apparatus that allowed the color filters to be combined. This instrument was useful in detecting differences between different types of inks.[69] The use of ultraviolet rays was yet another popular technological invention deployed in document investigation. Both Afzal and the American correspondent,

H. P. Wunderling, who often wrote in the journal, elaborated on the use of these rays.[70]

In a subsequent issue of the journal, Tuck contributed a lengthy piece on another machine known as the Spencer Comparator. This was essentially a projector of sorts that allowed the writing or the fingerprint to be enlarged without going through the process of photography. Similarly, Wunderling contributed an article in a later issue on the various modifications one could make to the standard microscope to use it more effectively for document investigation.

With the solitary exception of the Whistler, all the machines were visual apparatuses promising greater and greater transparency. This quest for transparency was as much about peering into the unknown as it was about "seeing for one's self" unaided by subordinates. These machines, however, became yet another trope by which to enact the distance between the familial milieu and the professional identity. In an editorial in *Document Investigation*, C. E. Hardless wrote, "The time is fast approaching when the arts and devices of the handwriting expert will be set to naught if he does not rise to the occasion and depend more on his laboratory than the former methods employed by his grandfather."[71] Modern crime required modern methods, and in so doing, one had to go beyond the methods of one's ancestors. The laboratory rather than the family became the space that engendered the new science just as science became the means to transcend the artisanal nature of the expertise of the handwriting experts. This was the message sounded again and again on the pages of *Document Investigation*. The prominently displayed photographs of curious machines on the cover further emphasized this core message of technomodern expertise, as did the boldly emblazoned motto: "Through Science to Justice."

## Dynamic Artisanal Science

I draw the idea of the dynamic artisanate from the economic historian Tirthankar Roy.[72] Roy has argued that traditional Indian craftsmen, far from being wiped out by the onslaught of modernity, were frequently able to control the impact of social, political, and economic changes by their dynamism. Master craftsmen in these artisanal communities led the way in calibrating the community's response to the changes taking place. Though our detectives are a different type of artisans, there are also some remarkable similarities.

Like artisanal communities, family and kinship ties played an important part in the development and transmission of knowledge amongst the

Hardlesses. Like artisans, too, a single master dominated the craft. Finally, despite the efforts to mechanize the process, for most of its career, handwriting analysis remained a fairly hands-on job that required the expert's experienced judgment.

Yet, like the artisanate described by Roy, the Hardlesses were not stuck in some ahistorical swamp called "tradition." They were dynamic and engaged with both the state and the market around them. While drawing upon both the opportunities of the market and the state, they continued to try to control their destinies. Despite coming from extreme poverty, the Hardlesses showed remarkable courage in foreswearing the safety offered by state employment and in carving out a more lucrative career in between state and market.

Thinking of handwriting analysis as a dynamic artisanal science not only allows us to appreciate the inventiveness and agency of the Hardless dynasty, but also opens up a new avenue from which to explore certain types of colonial science that were neither developed by the colonial state nor domiciled by the nation nor belonging entirely to the unregulated space of the market. Dynamic artisanal science while engaging with each of these major institutions of modernity in British India—namely, the colonial state, the anticolonial nation, and the market—was driven and shaped above all by a tightly knit family and its extended network of associates.

ACKNOWLEDGMENTS

The author is deeply indebted to the descendants of the Hardless family in the United Kingdom who have generously helped me with their family photographs, time, and memories. The author would like to thank Sue Creedon, Wendy Duke, Shahila Mitchell, Amir Shaikh, and, above all, Karin Tearle, without whose help he would never have met the rest of the family. He is also thankful to Ian Burney for inspiring and encouraging him to think of the "real" detectives, rather than the fictional ones loved by the author. Discussions with Gyan Prakash were helpful in shaping the arguments in this chapter. Earlier versions of this chapter were presented at the ICHSTM Conference in Manchester, the Locating Forensics Conference in London, and the South Asia Seminar at Princeton, and the author remains thankful for the many comments, suggestions, and encouragement he received at all three venues. The author is also deeply indebted to Sanjay Sircar for reading and commenting at length on an earlier draft.

## NOTES

1. David Arnold, *Science, Technology and Medicine in Colonial India* (New York: Columbia University Press, 2004); Deepak Kumar, *Science and the Raj: 1857–1905* (Oxford: Oxford University Press: 1995); Gyan Prakash, *Another Reason: Science and Imagination of Modern India* (Princeton, NJ: Princeton University Press, 1999), chaps. 1 and 2; Pratik Chakrabarti, *Western Science in Modern India: Metropolitan Methods, Colonial Practices* (Delhi: Permanent Black, 2004), chaps. 2–4.

2. Kavita Sivaramakrishnan, *Old Potions, New Bottles: Recasting Indigenous Medicine in Colonial Punjab* (Hyderabad: Orient Longman, 2006); Seema Alavi, *Islam and Healing: Loss and Recovery of an Indo-Muslim Medical Traditions, 1600–1900* (Basingstoke: Palgrave Macmillan, 2008); Guy Attewell, *Refiguring Unani Tibb: Plural Healing in Late Colonial India* (New Delhi: Orient BlackSwan, 2007).

3. Prakash, *Another Reason,* chaps. 3 and 4; Chakrabarti, *Western Science in Modern India,* chaps. 5–8; Projit Bihari Mukharji, *Nationalizing the Body: The Medical Market, Print and Daktari Medicine* (London: Anthem, 2009).

4. Prakash Kumar, *Indigo Plantations and Science in Colonial India* (Cambridge: Cambridge University Press, 2012).

5. Christopher Hamlin, "Forensic Cultures in Historical Perspective: Technologies of Witness, Testimony, Judgement (and Justice?)," *Studies in History and Philosophy of Science Part C: Studies in History and Philosophy of Biology and Biomedical Sciences* 44 (2013): 4–15.

6. Partha Chatterjee, *The Nation and Its Fragments: Colonial and Postcolonial Histories* (Princeton, NJ: Princeton University Press, 1993).

7. Christopher Hamlin, "Scientific Method and Expert Witnessing: Victorian Perspectives on a Modern Problem," *Social Studies of Science* 16, no. 3 (1986): 485–513.

8. On poisoning anxieties, see David Arnold, *Toxic Histories: Poison and Pollution in Modern India* (Cambridge: Cambridge University Press, 2016). On blood frauds, see Mitra Sharafi, "The Imperial Serologist and Punitive Self-Harm: Bloodstains and Legal Pluralism in British India," chap. 2, this volume.

9. Bhavani Raman, *Document Raj: Writing and Scribes in Early Colonial South India* (Chicago: University of Chicago Press, 2012).

10. I am indebted to Wendy Duke and Karin Tearle for helping me work out the details of C. R. Hardless's marriages.

11. The term "Eurasian" was used until 1911 to refer to people of mixed ancestry with at least one European, usually British, ancestor on the paternal side. Many of those referred to by this label, however, found the term offensive and succeeded in 1911 in getting the official designation changed to "Anglo-Indian." Significantly, prior to 1911, the term Anglo-Indian referred to European families settled in India but without any Indian ancestry. See Satoshi Mizutani, *The Meaning of White: Race, Class and the "Domiciled Community" in British India, 1858–1930* (Oxford: Oxford University Press, 2011).

12. "[Poor Eurasians]," *Amrita Bazar Patrika* (hereafter *ABP*), February 5, 1885, 7.

13. "Suicide," *ABP,* July 4, 1896, 5.

14. Letter from Lord Curzon to St. John Brodrick, June 22, 1905, Police Proceedings, Government of India, July/December 1905, Vol. 7061, 955, Asian and African Studies Collection, British Library (hereafter BL).

15. Frank Anthony, *Britain's Betrayal in India: The Story of the Anglo-Indian Community* (Bombay: Allied Publishers, 1960), 67.

16. "The Indian Expert's Evidence," *Supplement to the Ceylon Observer,* August 28, 1908, 3.

17. "Lady's Struggle with Burglar," *ABP,* September 10, 1909, 4.

18. "European Lady Molested by Burglars," *ABP,* March 30, 1915, 9.

19. See Denison Berwick, *A Walk along the Ganges* (New York: Javelin, 1986), 136. See also Zahir Ahmed, *Life's Yesterdays: Glimpses of Sir Nizamat Jung and His Towns* (Bombay: Thacker, 1945), 262–63.

20. Charles R. Hardless, "A Disclaimer," *Times of India* (hereafter *TOI*), August 29, 1933, 16.

21. Letter from Lord Curzon to St. John Brodrick.

22. Letter from St. John Brodrick to Lord Curzon, September 1, 1905, Judicial Despatch No. 39, Judicial Department, Vol. 2039/05, BL.

23. Letter from H. A. Stuart to The Secretary to the Government of India, February 5, 1906, Financial Despatch No. 121, Judicial Department, Vol. 1330/06, BL.

24. D. O. Letter, July 14, 1905, Judicial Department, Vol. 2039/05, BL.

25. Arthur Dixon (1881–1969) would go on to play an important role in reforming British police forces and pushing for a more scientific approach to forensic work. He served as secretary to the powerful Desborough Committee, whose report formed the basis of the Police Act of 1919, and then headed the new Police Department at the Home Office. Between 1933 and 1938 as home office assistant secretary, he presided over a departmental committee on detective work and procedure that came to be known as the Dixon Committee. The Dixon Committee was crucial in introducing new scientific methods in British forensic work and setting up a network of forensic laboratories. Dixon's illustrious and influential later career is amply documented by a number of historians. Alison Adam, *A History of Forensic Science: British Beginnings in the Twentieth Century* (London: Routledge, 2015); Ian Burney and Neil Pemberton, *Murder and the Making of English CSI* (Baltimore: Johns Hopkins University Press, 2016); Norman Vincent Ambage, "The Origins of the Home Office Forensic Science Service, 1931–1967" (PhD thesis, University of Lancaster, 1987).

26. Letter from A. L. Dixon to the Secretary of Finance, July 17, 1905, Judicial Department, Vol. 2039/05, BL.

27. Letter from A. L. Dixon to the Secretary of Finance.

28. Interestingly, Frank Brewester, who eventually took over as government handwriting expert after C. R. Hardless resigned, in an article on the Beck case in *Document Investigation*, gave advanced a spirited defense of Mr. Gurrin. Brewester alleged that Gurrin was misled because the standard writing with which Gurrin had compared Beck's samples were not actually Beck's. According to Brewester, the moment Gurrin learned that the police had misled him, he withdrew his previous testimony. Frank Brewester, "The Beck Case," *Document Investigation* 1, no. 3 (1933): 6–8.

29. F. W. Kidd, "Examiner of Questioned Documents," MSS/Eur F 161/186, BL.

30. Simon A. Cole has described the historical trajectory by which fingerprint experts overcame identical tensions between claims of "expertise" and "objectivity." The fact that when it first emerged the vast majority of those claiming such expertise were law enforcement officers meant that their claims were much more readily accepted without being subjected to the kind of "organized scepticism and careful scrutiny that is supposed to be inflicted upon scientific and legal facts." *Suspect Identities: A History of Fingerprinting and Criminal Identification* (Cambridge, MA: Harvard University Press, 2011), 186.

31. Christopher Hamlin, "Scientific Method and Expert Witnessing: Victorian Perspectives on a Modern Problem," *Social Studies of Science* 16, no. 3 (1986): 485–513.

32. Charles Hardless Jr., *The Identification of Handwriting and the Detection of Forgery* (Calcutta: Hardless Press, 1914), 14.

33. Hardless, *Identification of Handwriting*, 28.

34. Hardless, *Identification of Handwriting*, 29.

35. Carlo Ginzburg, "Clues: Roots of an Evidential Paradigm," in Carlo Ginzburg, *Clues, Myths and the Historical Method*, trans. John and Ann Tedeschi (Baltimore: Johns Hopkins University Press, 1989), 96–125.

36. Hardless, *Identification of Handwriting*, 36.

37. Harry Collins and Robert Evans, *Rethinking Expertise* (Chicago: University of Chicago Press, 2007).

38. Special Reporter, "Twelfth Day's Hearing," *ABP*, November 21, 1910, 8.

39. Hardless, "Preface," in *Identification of Handwriting*.

40. Hardless, "Preface," in *Identification of Handwriting*.

41. The only other private educational institution devoted entirely to a forensic science at the time, so far as I have been able to gather, was the Henry Bennett School of Handwriting Experts in neighboring Cawnpore (Kanpur). This school, however, only seems to have run correspondence courses and did not hold any classes as such. "Forgery: Become an Handwriting Expert," *TOI*, January 21, 1936, 2. Incidentally, like the Hardlesses, Henry Bennett was an Eurasian.

42. Charles E. Hardless, "The School of Document Investigation," *Document Investigation* 6, no. 5 (1938): inside cover.

43. "If You Wish to Become a Document Expert," *Document Expert* 6, no. 5 (1938): 13.

44. "A Complete Fingerprint Course," *Document Investigation* 5, no. 4 (1937): full-page advertisement after p. 14.

45. "A Complete Handwriting Identification Course," *Document Investigation* 6, no. 8 (1938): 14.

46. "Handwriting Expert's Review," *The Leader*, August 9, 1922, 2.

47. Exactly when the transition took place is difficult to tell, since few issues have survived, but by the fifth volume, published in 1937, the journal was described as an "Illustrated Monthly."

48. Hamlin, "Forensic Cultures."

49. Mrinalini Sinha, *Colonial Masculinity: The "Manly Englishman" and the "Effeminate Bengali" in the Late Nineteenth Century* (Manchester: Manchester University Press, 1995).

50. Partha Chatterjee, *A Princely Impostor? The Strange and Universal History of the Kumar of Bhawal* (Princeton, NJ: Princeton University Press, 2002), 187–88.

51. "Counsel and Expert," *The Tribune*, January 17, 1907, 1.

52. Sundara Aiyar, "In Re: Basrur Venkata Row vs. Unknown, on 29th December 1911," Sessions Case No. 19, Sessions Court of South Canara, accessed June 7, 2016, https://indiankanoon.org/doc/781684/.

53. C. E. Hardless, "Whats in a Name?," *Document Investigation* 1, no. 3 (1933): 3.

54. "European Association: Col. Gidney's Speech," *TOI*, January 12, 1921, 13.

55. "European Association."

56. According to Alison Blunt, Anglo-Indian attitudes toward India only began to change in the 1930s, when many, including Gidney, started referring to their dual affiliation to India and Britain. But Hardless was clearly a decade ahead of this shift. Alison Blunt, *Domicile and Diaspora: Anglo-Indian Women and the Spatial Politics of Home* (Malden, MA: Wiley-Blackwell, 2005), 43–50.

57. "The Anglo-Indians," *TOI*, February 28, 1922, 4.

58. "Communalism" in the south Asian context refers to the political mobilization of people around their religious identity. Such mobilization usually led to violence and eventually to the partition of British India along religious lines. Muharram and Dussehra processions were particularly notorious sources of such conflict. See, e.g., Gyanendra Pandey, *Construction of Communalism in Colonial North India* (Oxford: Oxford University Press, 1990).

59. Correspondent, "Dasehra-Muharram in Mirzapur," *The Leader*, October 18, 1917, 6.

60. See, e.g., R. B., "Blue-Eyed and Brown-Skinned: Uncovering a Hidden Past," *Kunapipi* 25, no. 2 (2003): 128–36.

61. David Arnold, "'An Ancient Race Outworn': Malaria and Race in Colonial Bengal, 1860–1930," in *Race, Society and Medicine, 1700–1960*, ed. Waltraud Ernst and Bernard Harris (London: Routledge, 1999), 123–43.

62. Ann Laura Stoler, *Carnal Knowledge and Imperial Power: Race and the Intimate in Colonial Rule* (Berkeley: University of California Press, 2002), 7.

63. Hardless, *Identification of Handwriting*, 1–9.

64. For "mechanical objectivity," see Lorraine Daston and Peter Galison, *Objectivity* (Brooklyn, NY: Zone Books, 2007).

65. William H. Tuck, "Photomicrographs," *Document Investigation* 1, no. 3 (1933): 4–5. Quote on p. 4.

66. Visualizations play an important part in translating forensic expertise for judges and juries. As Sheila Jasanoff points out, any trial opens up a "plurality of visions"; it is the judge who possesses the "eye of power" and must authorize the expert's visualization by controlling the skepticism that constantly threatens to deconstruct the expert's way of seeing. Sheila Jasanoff, "The Eye of Everyman: Witnessing DNA in the Simpson Trial," *Social Studies of Science* 28, nos. 5/6 (1998): 713–40.

67. Going by the name alone, Russell Gregory Afzal might either have had an European or Anglo-Indian mother and a Muslim father, or he may have been an Indian Christian. The former, though technically of mixed heritage, were usually not referred to as Anglo-Indians and were usually absorbed into the Muslim community. Indian Christians had a distinct identity, though at times it blurred into the Anglo-Indian identity. For the cultural and affective ties between the Indian Christians and Anglo-Indians, see Sanjay Sircar, "Matters of Language," in *The Way We Are: An Anglo-Indian Mosaic*, ed. Lionel Lumb and Deborah van Veldhuizen (Monroe Township, NJ: CTR, 2008), 257–66.

68. Russell Gregory Afzal, "Obliterations," *Document Investigation* 1, no. 3 (1933): 11–12. Whistler described on p. 12.

69. Afzal, "Obliterations," 11.

70. H. P. Wunderling, "The Use of Ultra-Violet Rays in Scientific Crime Detection," *Document Investigation* 1, no. 3 (1933): 13–16.

71. C. E. Hardless, "The Handwriting Expert versus the Forger," *Document Investigation* 6, no. 8 (1938): 1–3. Quote on p. 3.

72. Tirthankar Roy, *Traditional Industry in the Economy of Colonial India* (Cambridge: Cambridge University Press, 1999).

# CHAPTER FOUR

# Spatters and Lies

## Contrasting Forensic Cultures in the Trials of Sam Sheppard, 1954–66

IAN BURNEY

In this chapter, I analyze the contrasting forensic cultures involved in one of twentieth-century America's most notorious homicide cases, in which Cleveland osteopathic surgeon Sam Sheppard stood trial for the murder of his wife, Marilyn. The Sheppard case unfolded over three cycles, each of which involved distinct forensic regimes with their own particular configurations of techniques, practices, and experts. Here I consider only the first two cycles, the third—a failed 1999 civil action pursued by the Sheppards' son Chip against the state of Ohio for his father's wrongful imprisonment—being too much to handle within the confines of this volume.[1]

The first cycle revolved around the investigation, trial, and conviction of Sheppard in 1954. The driving force behind this cycle was the Cuyahoga County coroner, Dr. Sam Gerber. A well-connected member of the Cleveland political establishment who enjoyed the strong support of the city's main newspapers, Gerber took charge of the scene investigation, conducted a highly publicized inquest that named Sheppard as the murderer, and provided sensational trial testimony featuring a striking interpretation of the blood evidence found at the crime scene. This interpretation was qualitative in nature—Gerber claiming to have recognized the pattern of a "surgical instrument" impressed on the bloody pillow on which Marilyn had lain the night of the fatal attack.

A second cycle began in the weeks following Sheppard's conviction, when the Sheppard defense team recruited the eminent University of California criminologist and research scientist Paul Leland Kirk to review the forensic evidence heard at trial. Kirk's intervention produced an alternative, but equally striking, reading of the blood evidence: where Gerber saw an identifiable shape, Kirk's interpretation was a pioneering (and since celebrated) exercise in spatial reasoning based on the emerging discipline of blood spatter analysis. At the same time, Kirk's work on blood served as a deliberate strategy to try to depoliticize the case and to recast Gerber as the embodiment of a locally interested and outdated model of expertise, one that had overseen a major miscarriage of justice. Ultimately, a second trial, held in 1966 in the wake of a US Supreme Court ruling that overturned his initial conviction, pronounced Sheppard not guilty of Marilyn's murder.

Within this broad framework, I consider a series of linked themes that reveal the complex set of forensic cultures at work both within and across the two cycles of investigation and evidentiary dispute. First, I outline the initial crime scene investigation orchestrated by Gerber, and in particular draw attention to its approach to the blood evidence at the crime scene. This investigation centered on traditional criminalist questions and techniques, including chemical testing to determine identity (blood or not blood) and provenance (human or not human), using blood trails to reconstruct physical movement across the scene, and ultimately Gerber's finding of the distinctive pillow imprint. Equally significant is what the investigation ignored: the potential meaning of the blood shed in and distributed across the murder room itself.

In contrasting the Gerber and Kirk investigations, I indicate how these align with a set of key oppositions: between analytical and experimental criminalistics and one based on commonsense judgment, and between locally generated narratives of truth founded on local interest(s) and those generated by experts at a distance who purported to be at once scientifically objective, cutting edge, and politically disinterested. These oppositions can be tracked historically, first from the newspaper accounts of the Kirk investigation as it was being undertaken, then from the debates about the value of Kirk's results as they were highlighted in a 1955 legal appeal against Sheppard's conviction, then to the grounds for the rejection of Sheppard's appeal by Ohio's Supreme Court, and finally to the circumstances for the US Supreme Court's quashing of Sheppard's conviction in 1966 and his subsequent retrial.

The first three elements of this historical narrative explore the way that Kirk's evidence was treated in the immediate aftermath of the first trial, culminating in an analysis of the circumstances of its 1955 judicial rejection. The fourth is more complicated, and to an extent runs counter to assumptions about the inexorable triumph of new and improved forensic techniques. When Sheppard's cause was taken up again following the rejection of his initial appeal, it was not Kirk's blood spatter evidence that dominated discussion, but the promise of another contemporary truth technology—the polygraph. By examining the dynamics of this second intervention, I both open up the epistemological, political, legal, and cultural underpinnings of polygraph evidence in 1950s America and at the same time consider the relationship between and comparative fortunes of this (from a present-day perspective) "failed" technology of truth and of the (again from a present-day perspective) comparatively "successful" technology of truth—blood spatter analysis.

Ultimately, Kirk's evidence did play a significant role in the Sheppard retrial, where it successfully marginalized what by then had become Gerber's risible exercise in qualitative pattern recognition, and thus by the conclusion of the second cycle under discussion, it does emerge as the dominant forensic pathway to truth. But its marginal status in the public campaign to trigger a retrial by comparison to the allure of the polygraph underscores the contingent and partial nature of forensic evidence as a vehicle for publicly granted authority. As such, by analyzing techniques and practices stemming from a time and place most commonly associated with universalist scientific modernity, this chapter contributes to this volume's central agenda: to recognize through multiple geographic and temporal examples "the reciprocity of forensic cultures and practices as norm rather than aberration."[2]

## Impressions of Blood

The Sheppard case began in the early hours of the morning of July 4, 1954, when local police officers were summoned to a four-bedroom Dutch colonial-style house in Bay Village, a fashionable suburb of Cleveland. The house, sitting on a high cliff above the shores of Lake Erie, belonged to one of the community's most prominent families, the surgeon Dr. Sam Sheppard and his wife, Marilyn, both aged thirty.

Officers discovered a gruesome scene in the master bedroom—the savagely beaten body of Marilyn Sheppard, four months pregnant with the couple's second child. The room was covered in blood, and there were signs of a break-in. According to Sheppard, he had been asleep on a sofa in the ground floor living room when he was awoken in the early hours by sounds of a struggle in the upstairs bedroom. Rushing to the room, he confronted an intruder who knocked him unconscious twice, once during their initial encounter and subsequently on the lake shoreline, where Sheppard had pursued the "figure" after regaining consciousness. When he awoke from this second confrontation, Sheppard telephoned his friend and neighbor Spencer Houk, a local butcher and mayor of Bay Village. Houk and his wife, Esther, arrived at the Sheppard home just before 6:00 a.m. and were joined minutes later by Bay Village patrolman Fred Drenkhan. Around 8:00, Cuyahoga County coroner Samuel Gerber, accompanied by members of the Cleveland Police Department, arrived and proceeded to take charge of the scene. Sheppard was taken under sedation to the Sheppard family–owned and operated Bay View Hospital to recover from his injuries.

In the following weeks, the city's newspapers were dominated by accounts of the murder and its developing investigation. Headline articles accompanied by extensive photographic images of detectives working at the crime scene ran alongside accounts of Sheppard's physical state, and of attempts made by Gerber and his associates to interview him. According to reporters, these efforts were resisted by Sheppard and his family, who seemed to be using their local influence to thwart the inquiry. The papers noted that Sheppard's attorney advised his client not to talk to police, and a few days later widely reported Sheppard's refusal to submit to a police-administered polygraph examination.

Soon the dominant tone of press coverage became one of frustration at the apparently stalled nature of the investigation, and at the Sheppards' tactic of noncooperation. The leading voice in this increasingly hostile atmosphere was Ohio's largest circulation daily, *The Cleveland Press*, and its long-serving and immensely powerful editor Louis Seltzer, dubbed "Mr. Cleveland" in a 1950 *Life Magazine* feature article.[3] On July 21, under the banner headline "Why No Inquest? Do It Now, Dr. Gerber," a Seltzer editorial demanded a public inquest to force Sheppard to break his silence.[4] An inquest immediately followed, staged over three consecutive days in a local high school gymnasium before a large crowd of members of the public, newspaper reporters, and radio and television crews. Gerber questioned Sheppard at length, including repeated demands that he admit to a long-term affair with a former Bay View Hospital medical technologist. Gerber concluded proceedings by delivering a verdict naming Sheppard as Marilyn's murderer. A week later, Seltzer issued another front-page editorial titled "Why Isn't Sam Sheppard in Jail?"[5] That night, Sheppard was arrested and taken into the custody.

The febrile press coverage continued unabated over the ensuing months of police investigation and pretrial legal hearings, and when on October 18 jury selection commenced, the courtroom was packed with local and national media reporters. The presiding judge, Edward Blythin, denied a defense motion lodged the following day—the first of many such pleas—for a delay in court proceedings until the media storm died down, and for a relocation of the trial to a neutral venue.[6]

From late October to late December 1954, the court and the nation heard testimony featuring eighty-seven witnesses and some three hundred exhibits. Here I focus on the testimony of two witnesses only: Coroner Sam Gerber and Henry Dombrowski of the Cleveland Police Department's scientific identification unit. Gerber spent three days on the stand, significantly longer than any other witness testifying to the police and scientific investigation of the murder. He made it clear that he had taken charge of the case from the start, dictating the terms of engagement of the investigators and others— including the press, whom he invited into the Sheppard residence on the first day of the investigation and encouraged to take photographs. He stated that he had personally entered and briefly inspected the murder room. He then ordered the removal of Marilyn's body, and instructed a photographer to document the scene. It was at this point that he made what newspapers characterized as his "bombshell" discovery:[7] "When I turned the pillow over later I saw this stain, and in this portion of the stain I saw this imprint. It is the imprint of what I believe to be a surgical instrument."[8]

A member of Sheppard's defense team, Fred W. Garmone, immediately objected to Gerber's characterization as a conclusion rather than a description, but Blythin was unmoved: "You go ahead, Doctor, and tell us what you saw. Go ahead, Doctor." Gerber readily took up the invitation to deepen his reading: "This impression here represents the blades, and the blade on each side is about three inches long and the two blades together, its widest part, measure about two and three-quarter, and there is a space between the two blades indicating the fact that these blades do."[9] At this point, Gerber was cut off by Garmone's objection to the term "indicating," but with the protest once again ignored, Gerber's testimony entered the court record.

When it came to the defense cross-examination, Sheppard's lead attorney, William J. Corrigan, pursued several strategies to limit the damage caused by Gerber's synechdotal association of the bloody impression with Sheppard's professional identity. He disputed the evidentiary status of the pillowcase and subsequent photographs purporting to represent its status as a pristine scene object, observing that the picture referred to by Gerber had been taken a day after the initial investigation, after it had been handled by Gerber and many others.[10] He noted that Gerber's instructions for the investigation of the pillow was limited to photography, ignoring chemical and microscopic examination. He forced Gerber to admit that none of the medical implements

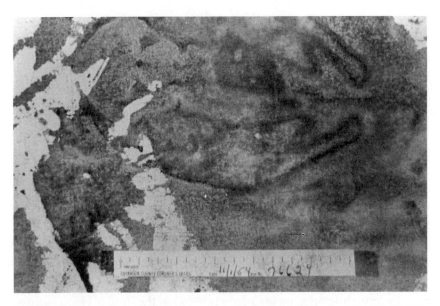

The impression of a bloodstained surgical instrument on Marilyn Sheppard's pillow, according to Sam Gerber. Cuyahoga County Coroner's Office

recovered from the Sheppard home matched the marks that had been shown to the jury. He concluded this line of questioning by suggesting that the pattern was caused by the folding over of the pillowcase while the blood was still wet—much like, he suggestively observed, a Rorschach inkblot.[11]

The Sheppard defense team made further attempts to undermine Gerber's competence as a criminal investigator. They questioned his care in securing the crime scene, in ordering necessary analytical tests of crime scene items, and in examining Marilyn's body at both the scene and the lab. In an exchange foreshadowing future developments, Garmone led Gerber to claim expertise on what should be done at the scene of a murder, encouraging him to confirm that he had written and lectured extensively on the subject. He then asked whether Gerber was familiar the work of Paul Kirk. Garmone's tactic in this question was clear but limited. He was seeking to establish a general laxity in the Gerber regime in contrast to what he was suggesting as a national standard of excellence. He was not, however, seeking to press Gerber on an omission of spatter analysis in particular. Gerber, again significantly in light of subsequent events, bristled at the implication that Kirk's standing as a national expert gave him epistemic priority over his own team of local investigators. Responding to Garmone's characterization of Kirk as "one of the foremost persons in the United States on the matter of criminal investigation," Gerber retorted: "I wouldn't say that. He maybe is in your opinion. He isn't in my opinion. . . . Not out in the East, and locally—locally he is considered by certain people as an authority, locally he is not considered an authority by some people."[12]

The purpose of the challenge to Gerber's "surgical instrument" testimony is evident: Gerber's claim established a commonsense link between Sheppard's professional identity and a key detail of the physical attack. In seeking to deconstruct his claim, the defense was not so much proposing an alternative interpretation based on alternative forensic methods as disputing the legitimacy of the interpretation itself. A similar approach is evident in the examination of the prosecution's main criminalist witness. Henry Dombrowski had been a member of the Cleveland Police Department's scientific identification unit since 1946, and in extended testimony he established himself as a careful and intrepid hunter of blood. His account, however, overwhelmingly focused on his attempts to follow the "trail of blood" across the entirety of the Sheppard home, and his efforts to determine which of the spots were of human origin and what type these were.[13]

In his initial testimony, Dombrowski was invited to share his expertise and the meticulous care he took in all aspects of blood hunting.[14] He revealed his techniques for finding hidden and obscure blood spots by the use of luminol and ultraviolet lighting, marking out with chalk the precise location where the spots lay, and removing blood samples without contamination so they could be tested for source by the use of chemical processes such as the precipitin and benzedine reaction. By using this set of techniques, Dombrowski informed the court that he had managed to reconstruct a blood trail consisting of ninety identifiable spots leading from upstairs to the basement. Through this meticulously conducted and described trace hunting, Dombrowski appeared to have successfully reconstructed the movement of the murderer carrying the bloodied weapon—perhaps Gerber's "surgical instrument"—from the scene of the crime.

The defense's lengthy cross-examination was similarly focused on Dombrowski's tracking work, aiming to seed doubt as to the reliability of his reconstruction and its physical and interpretive integrity. How, for example, did he collect the individual spots? How did he establish the purity of his reagents? How long would a stain remain detectable by the chemical tests he utilized? How specific was the chemical reaction on which he based his conclusion that the spots were of human origin? Only fifteen pages of cross-examination were devoted to Dombrowski's work in the murder room, and for the most part this followed the same lines and purpose. Here, Garmone sought to exploit what appeared to be the comparative lack of attention paid to the blood shed on the bedroom carpet. Despite his having gone down on hands and knees, with bright illumination, "looking for any foreign particles, or any evidence that we thought might be of any value," he did not check for blood spots on carpet "because there was obviously so much blood in the room."[15] Dombrowski revealed that he had not subjected the murder room blood to chemical investigation beyond making a control test to check that his solutions were functional. "It was our opinion that, just from the appearance of the blood in the room, it would add nothing to the investigation," he explained. "It was just there, as far as we were concerned."[16]

## Written in Blood: Paul Kirk and the Story of the Murder Room

On December 21, 1954, on day five of its deliberations, the jury returned a second-degree murder verdict against Sam Sheppard. He was given a life sentence. In early January, his defense team lodged a motion for a retrial, citing forty grounds as justification, including adverse pretrial publicity and denial of a change of venue. Blythin rejected the motion, and in subsequent consultations with Corrigan the Sheppard family, led by his brother Richard, decided on a new tactic: hiring the Berkeley professor of criminology Paul Leland Kirk to reinvestigate the crime scene.

By the mid-1950s, Kirk had established his reputation as one of the nation's foremost experts on the application of science to crime scenes. The recipient of a doctorate in biochemistry, Kirk was a veteran of the Manhattan Project—the exemplar of recent national scientific achievement—and since then had published scores of articles, as well as four authoritative books on the subject of criminalistics. When he arrived in Cleveland in late January 1955, Kirk also brought with him an emergent forensic specialism that promised to use the fundamental laws of fluid physics to reconstruct the precise sequence of violence that had produced the appearance of a specific bloody scene.[17]

Then, as now, the foundational principle underpinning blood spatter analysis is that the shape of a blood spot is determined by the velocity at which it was traveling, distance traveled, the angle of impact, and the type of target onto which it lands. Flying blood starts to coagulate as soon as it is shed and leaves beaded droplets. A drop falling perpendicular to the floor makes a round spot. A drop flying to a flat surface at an angle leaves a bowling pin shape—a bead with a tail. The more acute the striking angle, the longer the tail. The line of the tail reveals the trajectory of flight.

Unlike Gerber and Dombrowski, who treated the murder room blood merely as indicators of the obvious—that it was the scene of Marilyn's death—Kirk recast it as a productive and disciplined site of investigation, in which the blood spattered upon its various surfaces—walls, bed covers, furniture, radiator, doors—could be turned into evidence. To do this, he first measured the dimensions of the room and took note of the placement of the furniture within it. He then turned his attention to the blood spots on the walls, meticulously inspecting, measuring, recording, and photographing their size, shape, and distribution.

In doing so, Kirk made a series of discoveries. Because all the blood spots in the room and the bedsheets radiated in straight lines from the victim's head, he reasoned that this was their source, and that it had been positioned near the center of mattress and had remained there throughout the attack. "Because of the characteristic shape of blood spots striking in different directions and at different velocities," he wrote in his final report,

> it is possible to trace the direction of a drop through the air, and to estimate the velocity with considerable certainty. Utilizing the spots on the defendant's bed, it was noted that all those that gave elongated patterns had originated at a single center of origin which corresponded exactly with the region of Marilyn Sheppard's mattress on which the blood intensity was greatest, and which was occupied by her head at the time she was found. It can therefore be stated with certainty that her head was in essentially the same position during all of the blows from which blood was spattered on the defendant's bed.[18]

With the position of the victim determined, Kirk turned to that of the attacker. All four walls had blood on them, but in the extreme east end of the north wall, for approximately two feet and continuing over a contiguous portion of the east wall, he found no spots of any description. For Kirk, the explanation was clear: "This single region in the entire periphery of the room in which no blood had travelled through the air must by necessity be the region in which the attacker stood, since it is the only place in which the blood drops have been intercepted. . . . Close to the edge of the bed and slightly overlapping it, the width of the cone created by measuring lines would be about 2 feet, which approximates the width of a man's body. It places the attacker very close to the foot of the bed on the east side."[19] The absence of blood spatter on Sam's clothing, in Kirk's view, indicated that he had not created the blood void, which in turn served as evidence that he was not Marilyn's killer.[20]

Having determined the existence and significance of a blood void, Kirk took up the challenge of distinguishing between and reading the bloodstains that hit various surfaces from multiple heights and at different velocities. He was particularly interested in determining the sequence and angle of the blows that had created the precise arrangement of spatter patterns, and what this said about the type of weapon used. To achieve this, however, Kirk had to move—analytically and physically—to a new space of investigation. Returning to California, he constructed a murder simulation room inside his laboratory, within which he would conduct a series of experiments to re-create the various blood spatter patterns he had observed in the murder room. To do this, he soaked a sponge pad with blood as a substitute for Marilyn's head, and placed it on a stool. Around the stool he built a rectangular wall with removable paper strips to collect and record all flying blood. Within this experimental space, Kirk repeatedly struck the blood sponge at different angles, in different ways, and with different weapons. With each successive

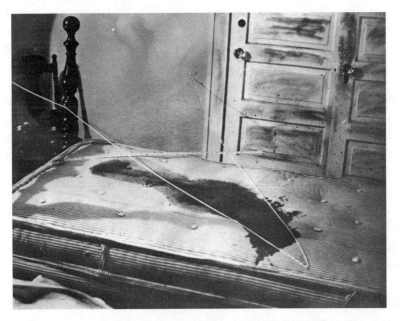

Measuring lines over the victim's bed. The strings extending outward from the front of the picture onto the wall delineate the blood void. Paul Leland Kirk; https://engagedscholarship.csuohio.edu/kirk_photos/6/

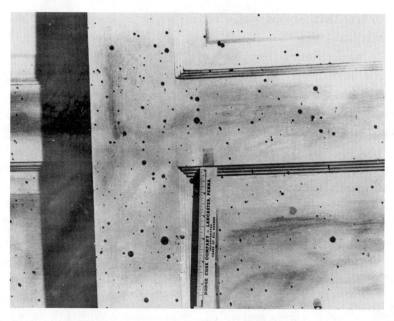

Blood spatter on the east wall (door). "Blood spots on the east wall were exceptional in their indications. Nearly all of them contrasted sharply with other spots in the room in that they were placed by low-velocity drops. Most of the them impacted the wall nearly at right angles to it, as is clearly demonstrated from their essentially round shape." Paul Leland Kirk; https://engagedscholarship.csuohio.edu/kirk_photos/13/

experiment, he replaced the paper strips and re-soaked the sponge, thereby creating a layered data set to establish which of his simulated acts of violence best approximated the blood patterns he had observed in the murder room.

One of the key outcomes of this painstaking work was Kirk's determination of the arc within which the weapon was swung. At the scene, he had observed a series of patterns of large blood spots on a number of doors and walls. In his experiments, he found that these large drops were usually associated with cast-off blood, that is, blood that had accumulated on the weapon and was thrown off as it was swung back and forth. The large drops on the wardrobe, according to Kirk, corresponded with the back-cast-off from a weapon, low-velocity blood spatter, while those on other walls wall were high-velocity spots that corresponded to cast-off patterns from a weapon thrust forward. The position of the attacker and the patterns and distribution of blood spots generated by the weapon cast-off allowed him to determine that the attacker held the weapon in the left hand,

and used a swing like that of a left-handed baseball batter. Sam Sheppard was right-handed.

Kirk's spatter analysis did not end there. He identified an "anomalous" blood spot in the wardrobe spatter that in his view could not have been created by a backswing motion. This was a large round spot measuring about one inch in diameter. In his simulation room, "extensive experiments" would not produce a blood spot of this shape and size, no matter how the various blood-drenched weapons were swung. The only method by which Kirk could reproduce the spot was to take blood into his cupped hand and toss it at low velocity at the paper-lined wall. "This spot could not have come from impact spatter," Kirk concluded. "It is highly improbable that it could have been thrown off a weapon, since so much blood would not have adhered during the back swing for so long a distance, and then separated suddenly at just the right moment to deposit as it did. This spot requires an explanation different from the majority of the spots on the doors. It almost certainly came from a bleeding hand."[21]

To bolster this theory, he turned first to Marilyn's tooth fragments that had been found—but analytically ignored—by the police in their initial investigation of the murder room. If she had been struck by an external blow to the face, Kirk reasoned, the teeth would have been found inside her mouth or in her throat. The fact that they lay in the room suggested to him that Marilyn may have bitten her assailant's hand during the attack with a force sufficient to fracture her teeth and draw blood. Kirk then turned to a complex blood-typing process involving rehydration and the introduction of antiserum to determine the agglutination rate of his samples. The rate observed for large wardrobe spot, he observed, was different from that for known blood samples taken from Marilyn and from Sam. The conclusion, for Kirk, was clear: a third party was the source of the "anomalous" spot.[22]

After two months of experiments in his murder room, Kirk submitted a forty-six-page affidavit to the Ohio judicial authorities that at once detailed the methods and results and drew a sharp contrast to the futile scene work of Gerber and his team: "The presence of blood trails throughout the premises has no bearing whatever on the guilt or innocence of the defendant. Whoever the murderer may have been, these would have occurred to a similar extent and degree."[23] The key to the brutal events of July 4 lay not in trail tracking or shape recognition but in meticulous spatial analysis of the spatter left on the surfaces of the murder room: "It is in this room and only here," Kirk declared, "that the story of the actual murder is written . . . Every blow struck," he later insisted, "placed its signature in the room in blood."[24] A properly scientific reading of this story led to only one conclusion: Sam Sheppard was not responsible for Marilyn's death.

Thus far I have outlined two distinct forensic engagements with blood: the first centered on trails and impressions, the second on spatter patterns. At this stage we might expect Kirk's analytical and experimental intervention to have easily displaced Gerber's commonsense speculation as the authoritative and dominant discourse of proof. But this proved not to be the case, at least initially. Kirk's fresh evidence did not trigger an immediate retrial. It would be more than ten years before one was granted. Moreover, in the intervening decade, it was not Kirk's blood spatter work but another technology of investigative truth—the polygraph—that dominated discussions about the merits of Sheppard's conviction and the grounds for reopening the case.

Several factors help explain the failure of Kirk's conclusions to supersede those reached at trial, not the least of which was the response of the Cleveland media to his challenge. Again led by Seltzer's *Cleveland Press*, Kirk's visits to the city and the crime scene were reported with barely disguised hostility. Kirk was portrayed as an unwelcomed out-of-towner who possessed no authority in local criminal or legal matters, a hired gun willing to sell his expertise to indulge the Sheppard family's ongoing attempt to use its position of influence to subvert the course of justice. In its article announcing Kirk's arrival in Cleveland, for example, the *Press* observed that Kirk "admits" to charging more than $100 per day plus expenses. In its response to the filing of Kirk's affidavit, the paper liberally deployed quotation marks to imply skepticism about Kirk's principal claims. Carrying a cropped photograph of Kirk showing only his eyes, the report repeatedly referred to what they purported to "see": "In the quiet of his laboratory," it began, "Prof Paul L Kirk 'saw' the July 4 murder of Marilyn Reese Sheppard. His 'eyes' were test tubes, slide rules, chemicals, charts, diagrams and photographs. These are the tools of the 'crime professor.'"[25]

Kirk's affidavit met with similar resistance from the prosecution team and the trial judge. In its motion for dismissal, filed on December 2, 1955, the district attorney's office cast Kirk's intervention as a barefaced attempt to circumvent established legal process: rather than presenting newly discovered evidence—the only sanctioned ground for an appeal—Kirk had set himself up as a "reviewing court" for an extralegal weighing of the evidence. He was inserting himself as "a sort of thirteenth juror" whose "self-assumed pose of objectivity" was "utterly absurd" when considered in light "its self-serving declarations, theories, speculations, arguments, conclusion, and misstatements and misrepresentations of the facts."[26] Judge Blythin agreed: in a written statement solicited by the appeals court, he charged Kirk with seeking to "conduct a post-mortem examination of the trial . . . to review the case and to conduct his own trial."[27]

Judge Julius Kovachy, writing the opinion for the 8th Ohio District Court of Appeals, extended these critical judgments. On simple procedural grounds, Kovachy observed that appeals for a new trial founded on new evidence were valid only if that evidence could not have been discovered and produced at trial "with reasonable diligence." This criterion, in his estimation, had not been met, and without it, justice itself was threatened with instability: "If the courts permitted such practices, the inherent certainty of a trial by jury would soon wane, and such function in our system of jurisprudence ultimately disintegrate and disappear."[28]

Kovachy also ruled on epistemic grounds. Citing an 1891 precedent, he declared that, according to state law, for experimental evidence to be admissible at trial, the experiments "must be performed with identical or substantially similar equipment and under conditions closely approximating those existing at the time of the occurrence being investigated."[29] Kirk's spatter analysis failed this standard on multiple levels: his tests using the simulated head were made without any "scientific correlation to the original body whatever"; the array of weapons used to simulate the attack were "selected on the basis of pure speculation";[30] and, because the temperature and humidity in the bedroom at the time of the murder were unknown, his contrived murder room could not be taken as reproducing the original conditions to a legally acceptable standard. "They are interesting and no doubt would be of value in a textbook on the subject," he pointedly observed, "but clearly they would have no probative value in the trial of this cause."[31]

Kovachy's ruling was vigorously challenged by Sheppard's advocates. In his formal legal response, Corrigan protested that the decision amounted to a charge that Kirk had "violated the basic requirements of his profession. Scientists such as Dr. Kirk do not arrive at conclusions by guess work," Corrigan bristled, "and it is a finding by the Court of Appeals that they are more scientifically equipped to determine the results of Dr. Kirk's experiments than the scientist himself."[32] He also drew attention to what he took to be the absurdity of privileging an 1891 legal precedent over scientific progress: "The advance of criminalistics since that day has been extraordinary. . . . If the syllabus in *Perkins v State* is the law of Ohio, it is time it was changed."[33] Kirk's own response was captured in an article in the *Cleveland Plain Dealer*, which reported the "bitter" reaction of "the small, scholarly scientist" to Blythin's initial rejection of his affidavit: "I am not very happy with the people of Cleveland or the people who were directly connected with the case," he stated, adding that in California the case would not have gone to trial on the evidence presented. The public opinion created by the local press was largely responsible for Sheppard's conviction: "Cleveland," he declared, "decided he was the right man."[34]

## The Politics of Conviction: (Cold) Warriors for Truth

Kirk's blunt assessment of the local politics of the Sheppard conviction reflected a criticism of the atmosphere surrounding the case that had been articulated from its outset. As the media demanded that the protective wall of wealth, privilege, and social standing surrounding Sam Sheppard be torn down, a counternarrative emerged that recast Sheppard's treatment in a diametrically opposed—and highly charged—political framework. Reverend Alfred Kreke, the Sheppard family's pastor at the Bay Methodist Church, articulated from the pulpit the strongest version of this challenge. As reported in the August 2 issue of Seltzer's own paper, Kreke told his congregation that, upon returning from a two-week family holiday, he had found his community in the thrall of "an open inquest with a miniature McCarthy at the head. . . . There were all the earmarks not of an inquest but of a medieval inquisition," he continued. "The morbid, sadistic attitude of the audience reminded one of the blood-letting arenas of Nero days."[35]

As striking as his references to the outrages of the distant past may have been, it was Kreke's characterization of Gerber and his inquest as a local counterpart to the inquisitorial excesses of McCarthyism that would have most resonated with his congregation. After years of using Cold War paranoia to run roughshod over civil liberties and the rule of law, at the time of Kreke's denunciation, McCarthy was himself in the process of being exposed—on live national television—at the infamous Army-McCarthy hearings. Instigated by the Wisconsin senator in April of 1954 to reinvigorate his flagging regime of persecution, the hearings had by the summer turned dramatically against him, and as Kreke spoke, Army chief counsel Joseph Welch's June 9 rebuke of McCarthy's "cruelty [and] recklessness" ("Have you no sense of decency, sir, at long last?") would have still been ringing in the national (and congregational) consciousness.

Seltzer's actions were similarly criticized. In a letter to the editor published in his own *Cleveland Press*, Bay Village resident Mrs. James Aspinwall defended her convicted neighbor against the self-serving and hypocritical sensationalism of Seltzer's journalism. Aspinwall recalled having heard the influential editor "speak many times on the fact that people need to 'wake up and be aware of the fact that our rights and freedoms in this democracy are being threatened.' When you left one of Seltzer's speeches you were more aware of these dangers and you, too, were dedicated to doing something." Like the disgraced McCarthy, Seltzer's past calls to the defense of American values seemed hollow in the wake of his persecution of Sheppard: "Now, the paper whose editorial policy you dictate becomes the ringleader of a howling mob. To me it seems like a direct parallel to the lynching crowd, with *The Press* assuming the voice of the hard-mouthed rabble rouser. What has happened to one of our original liberties where supposedly 'a man is innocent until proven guilty?'"[36]

The culmination of this confluence of the excesses of McCarthyism and anti-Sheppard fervor came the following spring, while the appeal based on Kirk's affidavit was failing in a series of preliminary legal hearings prior to Kovachy's ruling. On April 1, 1955, under the headline "Ray Jenkins Hits Dr Sam Verdict," the *Cleveland Plain Dealer* reported that the celebrated Army-McCarthy Senate subcommittee special counsel had publicly criticized the circumstances of the Sheppard trial. Though insisting that he was not passing judgment on Sheppard's ultimate guilt or innocence, Jenkins, as "a firm believer in the American system of legal procedure and trial by jury and evidence measuring up to the standards for a conviction," declared himself— "in the forceful manner familiar to millions of Americans who followed the McCarthy-Army hearings last year"—unsatisfied that these standards had been met. The stakes went beyond the facts of the case, and instead pointed to what Jenkins saw as a fundamental principle of American justice: "It is an enormous thing to think of an innocent man being convicted and having to spend a large part of his life in prison when his guilt was not proven by legal standards. . . . It is far better for many guilty men to go unpunished than for one innocent man to be convicted and punished for a crime he did not commit."[37]

Jenkins's insistence on prioritizing presumptive innocence echoes an established genre of legal scholarship that concerned itself with the causes and potential remedies of "wrongful convictions." Yale law professor Edwin Borchard's 1932 *Convicting the Innocent* was a landmark contribution to this literature, and in turn stimulated many law reformers to pursue the cause of miscarriages of justice.[38] One such (unlikely) reformer was the best-selling crime novelist Erle Stanley Gardner, who, in concert with New York publisher Harry Steeger, founded "The Court of Last Resort" in 1944.

The "Court" was a grouping of self-appointed freelance "experts" in law and criminal investigation charged with investigating possible cases of wrongful conviction. Its work was publicized in feature articles, written by Gardner and other members of its "board of investigators," and published in one of Steeger's magazines, the *Argosy*. This publication has a long and colorful history: starting out as a youth magazine in the 1880s, by the turn of the century it had switched to an adult fiction formula, reaching a circulation of a half million in 1905. By 1940, this figure had dropped to forty thousand, prompting a sale in 1942 to one of America's leading pulp magazine publishers. Steeger was appointed editor, and for the next several years he repositioned the *Argosy* to serve—as the new subtitle dating from 1946 declared—as "The Complete Man's Magazine." Though retaining its long-standing remit as a publisher of original fiction, an increasing proportion of each issue was devoted to features on masculine interests. By 1951, its circulation had recovered to a healthy 1.25 million.[39]

Gardner's Court was part of this strategy designed to capture the attention of the postwar American male. There is a large and impressive recent literature that has detailed the fractured state of Cold War masculinity, forged from an unstable mixture of rugged frontier individuality and outdoor virility, anxiety about the emasculating threats of suburban domesticity and white collar managerialism, a desperation to fulfill the norms of consumerist well-being projected by corporate advertising, and a paranoia about lurking external and internal threats to American liberties.[40] The *Argosy* represented this fractured masculine identity to itself. A lower-brow version of *Esquire*, it addressed itself less to the coastal sophisticates (real or aspirant) than to a reader in Middle America who wanted his cultural fare leavened with a generous helping of red-blooded hunting and fishing stories (with paraphernalia invitingly advertised in its "Men's Mart" section); crime, military, and espionage reportage; titillating cartoons of curvaceous women (in various states of unintended undress); and straight-talking advice (and advertising) about how to be a real, and successful, man.

The imagined landscape of American masculinity constructed by the *Argosy* made the magazine a perfect mouthpiece for Gardner and his enterprise. By the 1940s, Gardner had established himself, in the words of *Saturday*

*Post* reporter Alva Johnston, as "the king of the pocket libraries and the people's choice for America's greatest living writer"—in the first five months of 1946 alone, a single publisher of his mystery novels had sold over thirteen million copies.[41] At the center of the Gardner publishing enterprise sat Perry Mason, the intrepid defense attorney who since his debut in 1932 had built a career exonerating, through determined investigative work and flawless courtroom cross-examination, underdogs who for no fault of their own had been ensnared in false charges of (predominantly) capital crimes. Mason's fictional heroics were for the most part an extension and projection of Gardner's own legal career prior to his turn to writing. Born in Massachusetts in 1889, Gardner moved with his mining engineer father and family to various points west before settling in California. After graduating from Palo Alto High School, he pursued a self-taught program of legal studies, and in 1913, having passed the bar exam, set up a law practice in the Southern California agricultural town of Oxnard. He quickly established himself as a champion of the downtrodden—his work defending the immigrant population earned him the local sobriquet of "the Chinamen's attorney."[42]

Gardner's career as a trial litigator was cut short, not for lack of success, but because "the law kept interfering with his hunting trips."[43] This establishes another key element of the Gardner persona that mapped onto the *Argosy*'s landscape: his self-projection as a man of the outdoors, and the freedoms represented by the American frontier. Success in authorship delivered the frontier to Gardner: "Most writers write for fame and money," Johnston gushed. "Gardner wrote to go hunting."[44] The enjoyment born of autonomy of thought and action—in his case freedom from material constraint—was an essential feature of Gardner's understanding of the Court's mission and at the same time its guarantor of impartiality.

The nerve center of the people's Court was Gardner's aptly named Rancho del Paisano—an isolated, sprawling desert property in the Southern California district of Temecula. Here Gardner could act out his ideal of manly, self-directed, and self-sufficient liberty. But in his account of the origins of the Court of Last Resort, Rancho del Paisano also served as the backdrop for contemplating the plight of the wrongfully convicted: "The more I came to revel in my own liberty to go where I wanted to, whenever I wanted to, the more I found myself thinking of innocent men cooped up in cells."[45] Gardner had recently become involved in a suspected miscarriage of justice case in Southern California, and when Steeger visited his ranch, Gardner introduced him to the details of the case. According to Gardner's account, while the pair were camping out in the wild, Steeger became convinced that the subject could form part of the *Argosy*'s new mandate: to use instances of possible wrongful conviction to test, develop, and harness public interest in American justice and liberty. "So, down there," Gardner recalled, "we began to speculate on the idea of . . . testing the reaction of the American people to find out if they were really interested in the cause of justice and at the same time using space to correct some specific instance of an injustice." Other legal systems had established formal tribunals for reviewing potentially unsafe convictions, but they concluded that the unique conditions of American liberty demanded a different approach: "Out there in the wide open spaces of Baja, California, we came to the conclusion that in a country such as ours no officially organized tribunal ever could be the *real* court of last resort. The real court of last resort, we felt, was the people themselves."[46]

For "the people" to fulfill this guardian role, however, they required access to cases that engaged them as active participants rather than mere followers who might in turn be misled. Gardner accordingly proposed a tried and tested narrative conceit: "We knew that most magazine readers like detective stories. How about letting the readers study the case of John Doe, fact by fact, until they reached an intelligent opinion? That would mean investigators in whom the readers would have confidence, and who could unearth those facts. It would mean that reader interest must be kept alive."[47]

In Gardner's telling, the Court, its aims, and methods, and the *Argosy*'s role in publicizing it, emerged from this campfire discussion. Steeger pledged sufficient column space to pursue, in monthly installments and without knowing the outcome in advance, the twists and turns of ongoing investigations of potential wrongful conviction cases. The quality and integrity of these investigations were to be guaranteed by convening "board of investigators," composed of

> men who were specialists in their line, men who had enough national reputation so readers could have confidence in their judgment, men who would be public-spirited enough to donate their services to the cause of justice (because any question of financial reward would immediately taint the whole proceedings with what might be considered a selfish motivation). We also needed men who had achieved such financial success in their chosen professions that they were in no particular need of personal publicity. Moreover, the aggregate combination must be such that it would be virtually impossible for any prisoner to deceive these men as to the true issues in a case.[48]

"Sheriff" Erle Stanley Gardner, as photographed in the *Argosy*, March 1955. Copyright The University of Manchester

Forged in the crucible of an idealized version of American justice secured by men of skill whose incorruptibility was guaranteed by their transcendence of venal motives—an idealized version that of course depended upon its corrupted other as an aberration that should and could be corrected—the Court of Last Resort offered its services as a modern frontier posse, providing disinterested expertise in the service of a just and humane public. Habitually photographed in western gear, Gardner took on the role of the presiding sheriff, and he selected his deputies with due care to their complementary skill set and their irreproachable integrity. These essential qualities were guaranteed by the sheriff's personal assessment of individual character. In his account of assembling his Court, in selecting cases for investigation, and in identifying uncorrupted allies (prosecutors, reporters, state officials, and police and prison officers), Gardner repeatedly followed his "hunch." Introducing the Washington State Penitentiary warden who assisted the Court in one of its earliest cases, for example, Gardner wrote: "I found Tom Smith to be entirely different from the type of warden I had expected to find. In the intimate association with him which was to come later I learned to know that man's big heart, his almost naïve idealism and his passionate desire for justice."[49]

The human resources of the Court were thus characterized by independence, love of truth, and investigative skill, but for Gardner there was a further technical resource essential to its work: the polygraph. Interrupting his account of how he selected the Court's "associates," Gardner explained that, to be successful, "we'd want to have absolutely accurate information for our own guidance. We had to know whether the men we were talking with were telling the truth."[50] His first recruit, the infamous polygraph entrepreneur Leonarde Keeler, died soon after joining the Court and was replaced by Alex Gregory, whose impartiality was guaranteed by his professional skill and Gardner's faith in his character. Gregory may not always be able to tell if a man is innocent or guilty, Gardner explained, "But I feel that Alex Gregory would never say that an innocent man was guilty. He might say he didn't know. But if he said a man was definitely guilty, I wouldn't want to run against his judgement. And similarly if Alex Gregory assures us that a man who says he is innocent is telling us the truth, I for one am all in favor of going ahead and launching an investigation which may run into hundreds of hours of time spent."[51]

As Ken Alder has demonstrated, the polygraph was another perfect fit both with the vision of American justice that Gardner was pursuing and with the underlying anxieties that this quest reflected.[52] Although since the landmark Frye ruling of 1923 polygraph evidence had been excluded from formal trial proceedings, its extrajudicial use had in subsequent decades proliferated markedly, such that by the 1950s an estimated two million American men and women were subjected to "lie detection" tests to confirm their status as honest, reliable citizens. Employees of large, anonymous commercial organizations were tested for their adherence to company rules, government and military personnel for their allegiance to the state and their moral rectitude, and spouses for their marital fidelity. Crime investigation also availed itself of the promise of instrumental access to guilt and innocence. The polygraph was used in police cells as a humane version of the "third degree" to either support or undermine alibis of suspects and to lay the groundwork for plea bargaining as a low-cost alternative to full trials. This use was ostensibly voluntary, but refusal to submit would itself raise suspicions about a suspect's protestations of innocence. By contrast, individuals could seek a test to "prove" their innocence, and if both the defense and prosecution consented, the results could be admitted as background information at trial.

The polygraph played a significant role in the Sheppard case from the initial 1954 investigation, where it featured in the increasingly shrill newspaper accounts of the hunt for Marilyn's killer. In this initial phase, it was deployed within the inquisitorial framework described by Alder: Gerber, accompanied by a member of the Cleveland Police Department, visited Sheppard in hospital the day following the murder, where during the course of hostile questioning he invited Sheppard to undergo a polygraph examination. Sheppard refused. The press soon latched onto this as another instance of the Sheppard family's attempts to stymie the investigation. In a July 14 article, Seltzer's *Cleveland Press* reported the refusal of Sam's brothers Stephen and Richard to submit to polygraphs as "challenged" by Assistant County Prosecutor Thomas Parrino. "Only the day before," the article continued, "Coroner Samuel R. Gerber had suggested to Dr Richard that he attempt to persuade Dr Sam to take a lie test. Dr Sam had steadfastly refused to do so, claiming that he was 'emotionally upset,' so the polygraph would be inaccurate."[53] Days later, the *Cleveland Plain Dealer* added to the febrile atmosphere of suspicion when it reported—under the banner headline "Mayor Houk Reveals He Voluntarily Took Lie Test in Bay Probe. Friend of Doctor Reported to Have Come Through with 'Flying Colors'"—that the Sheppard family friend had—like any truth-loving innocent—taken the test. "So far as is known," the reporter archly commented, "Mayor Houk is the only figure in the July 4 tragedy to volunteer to submit himself to the scientific evaluation of his veracity."

Despite vigorous objections by his defense team, Sheppard's refusal was highlighted at several points during his seven-week trial—a fact that would later form one of the grounds for his (ultimately successful) appeal. In the wake of his conviction, a letter to the editor in the *Cleveland Press* summed up the mood of inquisitorial suspicion: "If Dr Sam is as innocent as he claims, he surely would have nothing to lose."[54]

The polygraph was also central to the Court of Last Resort's involvement in the case. In early 1957, Sam's brothers were seeking judicial approval for him to be tested in prison, for which they would require the services of a qualified examiner. In February, Stephen Sheppard wrote to Gardner requesting that he consider supervising a polygraph test of the family and publicizing the results in the *Argosy*. A publishing contract was drawn up the following month, and on May 15, the *Cleveland Press* revealed that both Sheppard brothers and their wives had been tested in Chicago earlier that month by Court members Alex Gregory and Lemoyne Snyder, along with two other experts described by Gardner as "the tops in the field of lie detection." The results of this test, Gardner continued, would not be released until the next issue of the *Argosy*—"You don't expect anyone to read the magazine two months later if we announce the results today."[55]

In the June 1957 issue, Gardner authored the first in a series of Sheppard feature articles in which he announced that the family had passed their tests, and that the Court would now seek permission from the governor of Ohio to perform a polygraph examination on Sam.[56] In announcing this initiative, Gardner provided a summary of the information that had been gathered to date, including Kirk's rejected spatter evidence. After summarizing Kirk's key findings—for example, the attacker's left-handedness and the discovery of the "anomalous" blood spot—he immediately marginalized them, using the 1955 Kovachy ruling that the information submitted in Kirk's affidavit did not constitute "new evidence" as his implicit justification: "This is a case where we start out feeling that *unless some new evidence can be discovered* the case is and should be completely closed. Can new evidence be discovered?"[57]

For Gardner, the answer was yes, but only by abandoning any quest to win a rehearing of Kirk's novel but judicially displaced spatial interpretation of the blood-spattered murder room. Instead, the cause of truth could only be successfully pursued by accessing a genuinely new and hitherto untapped resource: Sam Sheppard's mind. There were, Gardner explained, gaps in the original police investigation of the physical evidence—no murder weapon, for example, had ever been found: "Therefore, we have the following interesting situation: If Dr. Sheppard murdered his wife, he must know what the murder weapon was. If Dr. Sam Sheppard is guilty, he probably knows where that weapon now is."[58]

It was here, Gardner insisted, that the objectivity of the Court and its mastery of "scientific interrogation" would make its signal contribution to resolving, one way or another, the doubts surrounding the Sheppard case. Moreover, in publishing developments of its investigation in monthly installments without knowing where the facts might lead them, the Court would also be fulfilling its broader remit of raising public awareness about crime investigation and criminal justice, thereby enabling the American public to take on its responsibility as "the real court of last resort."

> We are going to take you readers right along with us on these investiga-
> tions. We are going to have you sit in with these investigators. We are going to
> let you evaluate the various charts and graphs. . . . In the coming issues of the
> magazine we are going to take you readers behind the scenes on these tests.
> The nation's outstanding authorities will explain to you what the lie detector
> is and how it works, the various types of machines which will be used, the
> various methods by which a person who wishes to do so can attempt with
> varying degrees of success to scramble the lie-detector test. It is impossible to
> beat the machine. . . . By the time we get done with these articles you readers
> who have followed them will know a great deal more about the lie-detector
> examination than the average person ever knows, and there is a chance you
> may have a lot of new information about the case of Dr. Sam Sheppard.[59]

Gardner's campaign unsurprisingly met a hostile reception in the Cleveland press. Reporting an early effort to co-opt the city's police department's homicide captain David Kerr onto the board's Sheppard investigation team, the *Cleveland Plain Dealer* highlighted Kerr's mistrust of the Court's "strictly promotional" nature: "They're selling *Argosy* magazines. If members of the Sheppard family have any new information they can come to us. They don't have to run to Gardner."[60] The next month the *Cleveland Press* carried an interview with Gardner, who claimed to have been "deluged" with mail from Cleveland and across the country in favor of testing Sheppard. He then repeated his agnostic stance on Sheppard's claims of innocence: "It would not surprise me at all that, if he took a test, it would show that he is just filled with guilty knowledge." The Court's pursuit of a state-sanctioned polygraph test gathered momentum over the summer months, and with Governor O'Neill reportedly on the brink of agreeing, the press stepped up its opposition. A Seltzer editorial of July 20 found it "surprising" that the governor was seemingly "falling for a magazine promotion": "Despite its impressive sounding name," it continued, "the Court of Last Resort has no legal standing. . . . Any test by an outsider is decidedly out of place."[61] Two days later, the *Cleveland Plain Dealer* reported Sheppard trial judge Julius Blythin's criticism of the Court's "meddling" in the wake of the legitimate process of review undertaken by the Ohio Court of Appeals and then the Ohio Supreme Court. "Now a group of purely private individuals, who are not even citizens of our state, and who have no evidence to offer, are appearing in the interest of increasing the circulation of a monthly magazine." Denouncing the "entire performance [as] nothing short of fantastic," Blythin urged the attorney general to issue a statewide ban to prevent groups from constituting themselves as a court of last resort.[62]

O'Neill bowed to local pressure and reversed his previously reported inclination to sanction the test. Now it was Gardner's turn to express indignation in print: "We resent remarks that the Court of Last Resort got into this to promote *Argosy* magazine," Gardner told a *Cleveland Press* reporter: "I am a high-priced writer, and these other men (his associates) are pretty well off. We are an impartial group trying to perform a public service."[63] A few days later, he cast his criticism in words more befitting a sheriff who had been run out of town by a corrupt local citizenry: "If you want it in plainer language, we're getting the hell out of here and not coming back unless we're invited back in connection with some future lie detector test—or pressure from our 4 million readers forces us back." But unlike the typical routed lone sheriff, Gardner was able to marshal a nationwide media platform to thwart his opponents: the reporter concluded his account of the interview with Gardner's thinly veiled threat that "he might kick off his television show with an account of his Sheppard case caper."[64]

The Court's final statement on the Sheppard case during this cycle appeared in the November 1957 issue of the *Argosy*, where Gardner repeated his analysis of the broad politics of justice served and justice denied. "*You*, as a member of the American public, are the Court of Last Resort," he insisted. "We represent *you*." It was therefore up to *Argosy* readers to take direct action to redress local powers of resistance galvanized by a media that, to his "shock" had over the last few months seemed to "thoroughly . . . misunderstand what we were trying to do." "Don't write to us," he concluded, "write to Governor O'Neill," promising that his board of experts remained available to return to the cause should the politics of the case shift.[65]

## From Last Resort to Supreme Court: Exonerating Sam Sheppard

The Court would have to wait for another several years, by which time Gardner had distanced himself from its activities. His place as the public face of the Sheppard campaign was taken by another polygraph enthusiast, the young defense attorney F. Lee Bailey.[66] A graduate of Boston University Law School's class of 1960, Bailey during his studies had set up a private "investigative service" offering for hire freelance expertise to local attorneys involved in criminal cases. The hallmark of Bailey's service was its thoroughly modern, scientific methodology, and in this the polygraph was its exemplary technology. In August 1961, while lecturing at the Keeler Polygraph Institute in Chicago, Bailey was approached by the *Chicago Tribune* investigative journalist Paul Holmes, who, on behalf of the Sheppard family, asked whether Bailey would supervise a polygraph examination of Sam Sheppard in the event of the family gaining permission of the new Ohio governor, Mike DiSalle.[67]

Bailey and Gardner in some respects could not have been more different. Gardner, the self-styled frontiersman, held court in his isolated ranch, which for many years, on Gardner's insistence, was without a telephone. Bailey, by contrast, was an East Coast urbanite who luxuriated in the pleasures, and embraced the promises, of 1960s modernity. As a *Time* magazine feature noted, Bailey embedded himself within an "electronic empire" that enabled him to cope with his growing nationwide legal caseload: "Along with electronic gadgets, his jet-age operation includes five office cars and five investigators . . . The whole empire is connected by two-way radios that keep the boss in constant touch as he swoops around the country in his Cessna 310 airplane."[68]

Bailey's embrace of the cause of the wrongfully convicted also differed from Gardner's in the generational idiom within which it was couched. As Bailey's involvement with Sheppard's and other high-profile cases gained publicity in the first half of the 1960s, his embrace of technological modernity was mirrored by his embrace of the liberalizing wind of the Warren Court, which in landmark decisions such as *Gideon* (1963), *Escobedo* (1964), and *Miranda* (1966) ushered in a new set of judicial protections of the rights of the accused. As *Newsweek* observed, this was a change that was sorely needed: "For clients who are poor and obscure, the nation's criminal courts have all too frequently been wretched symbols of justice. Often they are presided over by political hacks as deficient in knowledge of the law as in interest and compassion. Often too, they are overrun with 'dirty-shirt' lawyers who practice law out of a phone booth, recruit bewildered clients in the corridors and automatically enter pleas of guilty."[69]

In the Sheppard case, Bailey's crusade against the forces of oppression and self-serving corruption was personified in his combative approach to Gerber and Seltzer as representatives of 1950s inquisitorialism to be exposed by the tide of 1960s liberation. Writing to a local attorney recruited to serve on his Sheppard defense team, Bailey reported his first impression upon meeting Gerber informally at a 1962 lecture that Gerber had delivered in Boston: "He is an extremely frightened little man who all but begged me to drop the case and stay in Massachusetts. . . . Crushing him into infamy will be a distinct pleasure."[70] Bailey's zeal to expose the corrupt local regime that had presided over the first Sheppard trial applied equally to Seltzer. In his autobiography, he gleefully recounted an exchange with the aging press baron when they met in 1962 to discuss the *Cleveland Press*'s editorial stance on the attempt to seek gubernatorial permission for a Sheppard polygraph test. Met with Seltzer's disdain for the ploy, Bailey claims to have replied, "you have almost every advantage in this case that a man could have . . . But you're an old man, and I'm twenty-eight. You've got the money and the influence but somehow, someday, I'm going to beat you. And when I do, I'm going to hold you up to scorn and ridicule. And you better hope you die before I do it, because then I'll only be destroying your image instead of you as a person."[71]

As Sheppard's new champion, Bailey was in some respects eager to distance himself from Gardner's prior efforts. In correspondence with the Chicago journalist Paul Holmes in which the pair sought to forge a strategy for taking the appeal forward, both voiced skepticism: "I am not in favor of turning the matter over to Argosy Magazine," Bailey wrote, "for I suspect that they are principally interested in sensational reporting and secondarily interested in the drudgery of such hard-core investigation as is necessary to the production of any satisfactory evidence in this case."[72] Holmes replied that he "share[d] your distrust for Argosy's altruism." Yet in one key respect, both agreed with the core feature of the *Argosy* campaign—"The present objective is a lie test," Holmes wrote. "If we get a lie test, and Sam runs a clear test, we can then swiftly take the next step—the federal courts."[73]

Over the next three years, Bailey pursued a dogged campaign to take the Sheppard cause to the federal court level, citing what an earlier Ohio justice, in dissenting from the state's refusal to grant an appeal, descried as the "Roman Holiday" atmosphere surrounding the first Sheppard investigation and trial.[74] In an October 1963 writ of habeas corpus, Bailey drew attention to the elective basis of Ohio judgeships and other key investigative and prosecutorial office holders, and the way that this had been exploited by Seltzer. Among other prejudicial features of the media coverage, Bailey cited Blythin's decision to permit law enforcement officers to testify that Sheppard had refused a lie detector test while Houk had accepted, "despite the fact that such evidence has been ruled incompetent and prejudicial by every jurisdiction in the US."[75]

Bailey's efforts were handsomely rewarded when on June 6, 1966, the US Supreme Court ruled on what is now considered a landmark judgment on the dangers of prejudicial media publicity—*Sheppard v. Maxwell*. The majority opinion, written by Associate Supreme Court Justice Tom C. Clark, was forthright in its condemnation of the Cleveland establishment's actions:

> From the outset, officials focused suspicion on Sheppard. After a search of the house and premises on the morning of the tragedy, Dr. Gerber, the Coroner, is reported—and it is undenied—to have told his men, "Well, it is evident the doctor did this, so let's go get the confession out of him." . . . Later that afternoon Chief Eaton and two Cleveland police officers interrogated Sheppard at some length, confronting him with evidence, and demanding explanations. Asked by Officer Shotke to take a lie detector test, Sheppard said he would if it were reliable. Shotke replied that it was "infallible," and "you might as well tell us all about it now."[76]

Later in his opinion, Clark returned to the prejudicial coverage of Sheppard's persecution at the hands of the polygraph, condemning media reports that "played up Sheppard's refusal to take a lie detector test and 'the protective ring' thrown up by his family."[77] With only the arch defender of press freedom Justice Hugo Black dissenting, the decision concluded by ordering that Sheppard be released from custody "unless the State puts him to its charges again within a reasonable time."[78]

When the second Sheppard trial opened on October 24, 1966, it was immediately evident that it would be tried within a framework that self-consciously repudiated the circumstances that prevailed a decade earlier. Judge Francis Talty delivered a preliminary ruling imposing strict limits on press reporting, and this reflected a wider sense of a generational shift in which the persecutions of the previous decade yielded to a new permissive era. Bailey himself devised his trial strategy to emphasize this shift. The first defense had—apparently at Sheppard's request—refrained from pursuing evidence of Marilyn's own infidelities, thereby reifying her public image as an ideal wife while leaving Sam exposed to puritanical attacks on his sexual misconduct. Bailey would have no such compunctions and, as he advised his client in the lead-up to the trial, he would use the process of jury selection to ensure a more balanced hearing: "Everybody either loved you or hated you in 1954. . . . This time, I'm going to get younger people on that jury, and they won't have those feelings. And they'll acquit you on the evidence—there's no question about it."[79] The selection as foreman of the jury of Ralph Vichell, a quality control engineer who at thirty-three was the same age as Bailey, exemplified this tactic.

The scene was thus set for Bailey to showcase the testimony of Paul Kirk, who, as he later recalled, was "our star." "Working with Dr Kirk was a pleasure," Bailey gushed. "We were like a pair of basketball guards so keyed to each other's moves that it was almost impossible to miss a pass."[80] On the witness stand, Kirk confidently rehearsed the findings of his blood investigations that he had detailed in his judicially excluded 1955 affidavit.

There is no need here to engage with the detail of this testimony, which largely rehearsed his 1955 affidavit findings. Instead, I use the reappearance of Kirk's spatter evidence to make two concluding observations. First, and most straightforwardly, was the way that Bailey used this evidence to fulfill his long-standing ambition of "crushing [Gerber] into infamy." In extensive cross-examination, Bailey repeatedly led the aging Coroner to admit that his sensational 1954 claims to have discovered the impression of a "surgical instrument" on the bloodstained pillow could not be substantiated. In one such exchange, Bailey asked Gerber to "tell the jury what surgical instrument you see impressed in that pillow, the nature of it, please?" "I can't give you the name of it, because I don't know what it is," Gerber replied meekly.[81] Bailey continued to goad, asking for the names of the instruments that Gerber thought might have made the impression, demanding that he draw the one he had in mind, and later asking where one might find such an instrument. "Any surgical store," Gerber ventured, "but I hunted all over the United States and I couldn't find one."[82]

Second, it is worth reflecting briefly on the comparative standing of Kirk's spatter evidence and that of the polygraph in the 1966 retrial. As suggested above, Kirk's testimony was presented as the exemplar of forensic modernity. It was, as Bailey declared in his final summation, "the touchstone to the truth"— "He sees things that should have been seen long before and never were."[83] The jury seemed to concur: on November 16, 1966, after less than a day of deliberation, it found Sam Sheppard not guilty of the murder of his wife Marilyn. The testimony of the polygraph made no direct contribution to this outcome, as there was no mention of it in court.

Does this mean that all the discussion focused on this alternative truth technology—including my own—was to no effect? Yes and no. Yes, if we restrict our attention to what transpired in the courtroom. As historians have conclusively established, however, the power of the polygraph as an instrument of forensic inquiry in twentieth-century America did not stem from its formal role in direct trial evidence. Indeed, it was its formal exclusion from the courtroom that created the conditions for its cultural salience. That is, it was the polygraph's extrajudicial uses—its capacity to activate and promote the promise of technological divination—that made it significant. In Alder's evocative phrasing, its purpose was to serve its vast and expectant public as "an oracle whose judgements descended on the scene of human entanglement like a mechanical god descending from the rafters: condemning the guilty, absolving the innocent, and making retrospective sense of [a] drama."[84] In the Sheppard case, it was the Court of Last Resort's persistent and public pursuit of this "oracular" access to truth that ultimately led to the 1966 retrial, which showcased spatter evidence. Without this intervention, in short, Kirk's triumphant "story" of blood would have remained confined within the walls of his distant laboratory. As such, the trials of Sam Sheppard provide a cogent illustration of an essential characteristic forensic authority, wherever and whenever located: that is, its historically and culturally contingent standing as a conduit for truth and justice.

One final observation indicates my broader intentions for engaging with this singular instance of forensic cultures at work. We live in an age of recovered innocence. Featured in the high-profile work of the Innocence Project, and dramatized by compulsively watched documentaries like Netflix's *Making a Murderer*, the righting of wrongful conviction has emerged as a powerful, prevalent, and from most legal and ethical perspectives salutary feature of our times. It is commonly seen as the product of forces peculiar to our present moment: principled and informed legal critique of the workings (and biases) of the criminal justice system, dogged investigative journalism, and improvements in the methods and standards of forensic investigation—associated most notably with DNA typing. Yet recovered innocence has a history. Take the Court of Last Resort: here we have a symmetrical counterpart to the present era of innocence, forged in a markedly different political, legal, cultural, and scientific context but sharing essential structural features: reformist ideals, media advocacy, and the promise of technical revelation. As such, this material may serve as a means of opening up a critically informed history of innocence, past and present.

NOTES

1. There is an extensive literature on the Sheppard cases, most of which has been written by investigative journalists or authors with a direct involvement in them. The most comprehensive account, on which much of the background information for this chapter is based, is by the *Chicago Tribune* reporter Paul Holmes, *The Sheppard Murder Case* (New York: David McKay, 1961). Other works include Cynthia Cooper and Samuel Reese Sheppard, *Mockery of*

*Justice* (Lebanon, NH: Northeastern University Press, 1995); Jack P. DeSario and William D. Mason, *Dr. Sam Sheppard on Trial: Case Closed* (Kent, OH: Kent State University Press, 2003); and James Neff, *The Wrong Man* (New York: Random House, 2001). The forensic dimension of the case is discussed briefly in Jeffrey M. Jentzen's *Death Investigation in America: Coroners, Medical Examiners and the Pursuit of Medical Certainty* (Cambridge, MA: Harvard University Press, 2009), and by Paul C. Giannelli in his "Scientific Evidence in the Sam Sheppard Case," *Cleveland State Law Review* 49 (2001): 487–98. Another good source of background information can be found at "The Sam Sheppard Case, 1954–2000," EngagedScholarship @ Cleveland State University, http://engagedscholarship.csuohio.edu/sheppard/. This exceptional collection is also the source for much of the archival material used in this chapter. I am deeply indebted to the staff who created and manage this resource for their assistance in my research, especially its curator Beth Farrell.

2. Christopher Hamlin, "Introduction: Forensic Facts, the Guts of Rights," this volume.

3. Richard L. Williams, "Mr. Cleveland," *Life*, March 13, 1950, 130–42.

4. Louis Seltzer, "Why No Inquest? Do It Now, Dr. Gerber," *Cleveland Press*, July 21, 1954.

5. Louis Seltzer, "Why Isn't Sam Sheppard in Jail?," *Cleveland Press*, July 30, 1954.

6. Blythin maintained his apparent indifference to what would later be denounced by the US Supreme Court as the "Roman Holiday" atmosphere surrounding the trial, for example, refusing to sequester the jury and to sanction authors' continuous press reports that repeatedly published the names and photographs of jurors.

7. "Imprint on Sheppard Bed Made by Surgical Device," *New York World Telegram*, November 16, 1954. The report noted that Gerber's testimony "hit the courtroom like a bombshell."

8. "Volume 06, 1954 Trial Transcript: State's Witness Dr. Samuel R. Gerber, Cuyahoga County Coroner," Cuyahoga County Court of Common Pleas, 1954 Trial Transcript, Book 6, 1392. Available at the EngagedScholarship @ Cleveland State University website, http://engagedscholarship.csuohio.edu/sheppard_transcripts_1954/6.

9. "Volume 06, 1954 Trial Transcript," 1396.

10. "Volume 06, 1954 Trial Transcript," 1400, 1726.

11. "Volume 06, 1954 Trial Transcript," 1742. The discussion of the Rorschach blot did not explicitly engage with the test's broader function of eliciting fantasy projections on the part of the viewer.

12. "Volume 06, 1954 Trial Transcript," 1707–8.

13. Of 260 pages of typed transcript evidence given by Dombrowski, only 23 were devoted to questions about his engagement with the blood found in the murder room: 8 by the prosecution, 15 by the defense.

14. It was this blood-hunting work that was the most extensively covered aspect of the initial police investigation. *Cleveland Press* reports in particular looked to "the magic of modern science" to foil the killer's attempts to cover the track of blood that he had presumably made in a panicked state following the brutal attack. "Find Killer's Bloody Trail: Science Cuts Through Cover-up," *Cleveland Press*, August 3, 1954.

15. "Volume 08, 1954 Trial Transcript: State's Witnesses; Recross of Fred Drenkhan; Recall of Coroner Gerber; Defense Motions," Cuyahoga County Court of Common Pleas, 1954 Trial Transcript, Book 8, 2699. Available at the EngagedScholarship @ Cleveland State University website, http://engagedscholarship.csuohio.edu/sheppard_transcripts_1954/8.

16. "Volume 08, 1954 Trial Transcript," 2771–72.

17. In practitioner accounts, the modern history of blood spatter analysis is traced back to the work conducted by Eduard Piotrowski at the Institute for Forensic Medicine in Krakow, published in *Über Entstehung, Form, Richtung und Ausbreitung der Blutspuren nach Hiebwunden des Kopfes* (Vienna: März, 1895). Another acknowledged landmark was Victor Balthazard

et al., "Étude des gouttes de sang projeté," *Annales de Médecine Légale de Criminologie et de Police Scientifique* 19 (1939): 265–323. Kirk's work in the wake of the Sheppard case is generally taken as the next significant development. For an overview of the history and science of spatter analysis, see Herbert L. MacDonald, *Flight Characteristics and Stain Patterns of Human Blood*, National Institute of Law Enforcement and Criminal Justice Project Report 71-4 (Washington, DC: US Department of Justice, 1971), 1–30, and Stuart H. James and William G. Eckert, *Interpretation of Bloodstain Evidence at Crime Scenes* (Boca Raton, FL: CRC Press, 1999).

18. Paul L. Kirk, "Affidavit of Paul Leland Kirk," in *1954–1955 Post-Trial Motions and Ohio Eighth District Court of Appeal*, 15. Available at the EngagedScholarship @ Cleveland State University website, http://engagedscholarship.csuohio.edu/sheppard_eight_district_1950s/10.

19. Kirk, "Affidavit of Paul Leland Kirk," 14.

20. "The comparative absence of blood on the clothing of the defendant is highly significant. It is entirely certain that the actual murderer received blood on his person, and no portion of his clothing that was exposed could have been exempt from blood staining." A significant portion of Kirk's laboratory experiments was devoted to proving that Sheppard could not have washed the blood that must have spurted from Marilyn's body to the degree that no traces would be detectable. Kirk, "Affidavit of Paul Leland Kirk," 11.

21. Kirk, "Affidavit of Paul Leland Kirk," 21. Emphasis original.

22. Kirk, "Affidavit of Paul Leland Kirk," 21–22. Kirk's blood-typing results were among the most hotly contested of his findings, and though the terms of the contest are interesting, space does not permit a discussion of them here.

23. Kirk, "Affidavit of Paul Leland Kirk," 6.

24. Kirk, "Affidavit of Paul Leland Kirk," 6, 8.

25. "Crime Expert Checks Sheppard Death Home," *Cleveland Press*, January 24, 1955; "Eyes of Dr Sam's Expert See Left-Handed Sex Attacker," *Cleveland Press*, April 27, 1955.

26. Frank T. Cullitan et al., "Motion to Dismiss Appeals as of Right and Brief in Support of Motion to Dismiss and Opposing Motions for Leave to Appeal," 1955. Available at the EngagedScholarship @ Cleveland State University website, https://engagedscholarship.csuohio.edu/sheppard_ohio_supreme_court_1950s/3, 153.

27. Cullitan et al., "Motion to Dismiss," Appendix B, Opinion of the Court of Common Pleas on the Motion for a New Trial on the Ground of Newly Discovered Evidence, May 9, 1955. Available at the EngagedScholarship @ Cleveland State University website, https://engagedscholarship.csuohio.edu/sheppard_ohio_supreme_court_1950s/3, 206.

28. *State v. Sheppard*, 100 Ohio App. 399, 128 N.E.2d 504 (Ohio Ct. App. 1955), 408.

29. *State v. Sheppard*, 412.

30. *State v. Sheppard*.

31. *State v. Sheppard*, 413.

32. William J. Corrigan, Fred W. Garmone, and Arthur E. Petersilge, "Assignments of Error and Brief of Defendant-Appellant," in *1955–1959 Ohio Supreme Court and US Supreme Court Direct Appeal*, 483–84. Available at the EngagedScholarship @ Cleveland State University website, http://engagedscholarship.csuohio.edu/sheppard_ohio_supreme_court_1950s/8.

33. Corrigan et al., "Assignments of Error and Brief of Defendant-Appellant," 486.

34. "Kirk Berates Blythin on 'Face-Saving' Trial Ruling," *Cleveland Plain Dealer*, May 10, 1955.

35. "Bay Pastor Hits Methods Used in Sheppard Probe," *Cleveland Press*, August 2, 1954.

36. Mrs. James Aspinwall, letter to the editor, *Cleveland Press*, August 3, 1954.

37. "Ray Jenkins Hits Dr Sam Verdict," *Cleveland Plain Dealer*, April 1, 1955.

38. Edwin M. Borchard, *Convicting the Innocent: Sixty-Five Actual Errors of Criminal Justice* (New Haven, CT: Yale University Press, 1932). Borchard's book was based on his 1913 article in the *Journal of the American Institute of Criminal Law and Criminology*. The successor

to Borchard's critique was Judge Jerome Frank's *Not Guilty* (London: Victor Gollancz, 1957), which provided a newly energized context within which Sheppard's cause was debated. Frank's book featured the work of Erle Stanley Gardner's "Court of Last Resort," which is discussed below. For an overview of the history of wrongful conviction literature, see Jon B. Gould and Richard A. Leo, "'Justice' in Action One Hundred Years Later: Wrongful Convictions after a Century of Research," *Journal of Criminal Law and Criminology* 100, no. 3 (2010): 825–68. For insightful analytical engagement with wrongful conviction past and present, see Richard A. Leo, "Rethinking the Study of Miscarriages of Justice," *Journal of Contemporary Criminal Justice* 21, no. 3 (2005): 201–23, and Jay D. Aronson and Simon A. Cole, "Science and the Death Penalty: DNA, Innocence, and the Debate over Capital Punishment in the United States," *Law and Social Inquiry* 34, no. 3 (2009): 603–33.

39. Information in this paragraph is derived from Tom Pendegrass's *Creating the Modern Man: American Magazines and Consumer Culture, 1900–1950* (Columbia, MO: University of Missouri Press, 2000), 238–42. See also Kenon Breazeale, "In Spite of Women: *Esquire* Magazine and the Construction of the Male Consumer," *Signs*, 20 no. 1 (1994): 1–22, and Elizabeth Fraterrigo, *Playboy and the Making of the Good Life in Modern America* (Oxford: Oxford University Press, 2009). I am deeply grateful to Phil Stephensen-Payne, who not only gave me access to his full-run collection of postwar *Argosy* magazines, but also generously donated the collection for the University of Manchester's library so that this rare material can be consulted by other scholars.

40. The most insightful recent study is K. A. Cuordileone's *Manhood and American Political Culture in the Cold War* (Abingdon: Routledge, 2005). See also Barbara Ehrenreich, *The Hearts of Men: American Dreams and the Flight from Commitment* (New York: Anchor Press / Doubleday, 1983); Garbriele Dietze, "Gender Topography of the Fifties: Mickey Spillane and the Post-World-War-II Masculinity Crisis," *American Studies* 43, no. 4 (1998): 645–56; Thomas Doherty, *Cold War, Cool Medium: Television, McCarthyism, and American Culture* (New York: Columbia University Press, 2003).

41. Alva Johnston, *The Case of Earle Stanley Gardner* (New York: William Morrow, 1946), 9. Johnston's book was based on her series of feature articles first published in the *Saturday Evening Post*.

42. Johnston, *Earle Stanley Gardner*, 31.

43. Johnston, *Earle Stanley Gardner*, 10.

44. Johnston, *Earle Stanley Gardner*, 9.

45. Erle Stanley Gardner, *The Court of Last Resort* (New York: William Sloane, 1952), 14.

46. Gardner, *Court of Last Resort*, 15. Emphasis original. For more on the allure of "frontier justice" in the postwar period, see Richard Slotkin, *Gunfighter Nation: The Myth of the Frontier in Twentieth-Century America* (New York: Atheneum, 1992); Stanley Corkin, *Cowboys as Cold Warriors* (Philadelphia: Temple University Press, 2004); and Andrew C. Isenberg, "The Code of the West: Sexuality, Homosociality, and Wyatt Earp," *Western Historical Quarterly* 40 (2009): 139–57.

47. Gardner, *Court of Last Resort*, 16.

48. Gardner, *Court of Last Resort*, 17. Gardner repeated his reformist project in numerous law journal articles, e.g., "Adventures in Justice: A Call to Arms," *American Bar Association Journal* 37 (1951): 5–8, 84–85; "The Court of Last Resort," *Cornell Law Quarterly* 44 (1958–59): 27–37; "Need for New Concepts in the Administration of Criminal Justice," *Journal of Criminal Law and Criminology* 50 (1959): 20–26.

49. Gardner, *Court of Last Resort*, 22. Gardner's Washington recruits also included, as reviewing judge, "a patient, a kindly individual who is prone to try to get at the facts of a case," "one of the most fair-minded attorneys general I have ever met," and "a very alert, able reporter" (47, 54, 55).

50. Gardner, *Court of Last Resort*, 18.

51. Gardner, *Court of Last Resort*, 18–19.

52. Ken Alder, *The Lie Detectors: The History of an American Obsession* (New York: Free Press, 2007); see also Geoffrey Bunn, *The Truth Machine: A Social History of the Lie Detector* (Baltimore: Johns Hopkins University Press, 2012), esp. chap. 7; Michael Pettit, *The Science of Deception: Psychology and Commerce in America* (Chicago: University of Chicago Press, 2013), esp. chap. 5; and Joseph Masco, "Lie Detectors: On Secrets and Hypersecurity in Los Alamos," *Public Culture* 14, no. 3 (2002): 441–67.

53. "Doctor's Brother Agrees to Lie Test, Then Refuses," *Cleveland Press*, July 14, 1954.

54. "Why No Lie Detector?," *Cleveland Press*, December 28, 1954.

55. "Sam Got Fair Trial, Says Investigator," *Cleveland Press*, May 15, 1955.

56. Erle Stanley Gardner, "The Dr Sam Sheppard Case: A Dramatic New Turn," *Argosy* (June 1957): 15–17, 56–58. As the state had no jurisdiction over the Sheppard family, no such permission was required for their tests. Instead, a private contract was drawn up, stipulating each subject's consent "that you may publish in ARGOSY Magazine the facts and circumstances surrounding such examination, the results thereof, and the opinions with respect thereto. . . . I also agree that the Court of Last Resort may make any other uses of such examinations, results, opinions, and my name, picture and photograph." March 16, 1957. Available at the EngagedScholarship @ Cleveland State University website, https://engagedscholarship.csuohio.edu/sheppard_habeas/33.

57. Gardner, "Dramatic New Turn," 56. Emphasis original.

58. Gardner, "Dramatic New Turn," 57. The "Court" confirmed its strategy of marginalizing Kirk's judicially excluded evidence in favor of an instrumental interrogation of the unexamined terrain of Sheppard's mind in its next instalment on the case: "The Court of Last Resort accepts that verdict as being the final determination of the case *according to the facts adduced at the trial.* It is, however, going to explore the possibility of uncovering additional facts by the use of scientific interrogation." "Nothing but the Truth: The Dr. Sam Sheppard Case," *Argosy* (July 1957): 41, 74–76, 41. Emphasis original.

59. Gardner, "Dramatic New Turn," 58.

60. "Kerr Sees Sales Stunt in Argosy Move for Dr. Sam," *Cleveland Plain Dealer*, May 17, 1957.

61. "No Special Favors," *Cleveland Press*, July 20, 1957.

62. "Private Unit's 'Meddling' Hit by Trial Judge," *Cleveland Plain Dealer*, July 22, 1957.

63. "Gardner Indignant at Cancellation," *Cleveland Press*, July 22, 1957.

64. "'Court' Keeps Door Open," *Cleveland Press*, July 23, 1957. At the same time that Gardner was engaging with this round of the Sheppard saga, he was also preparing for the debut of the NBC television show *The Court of Last Resort*, which was co-produced by his own company, Paisano Productions, and ran in weekly episodes from October 4, 1957, to April 11, 1958.

65. Erle Stanley Gardner, "The Other Side of the Sheppard Case," *Argosy* (November 1957): 22–23, 58–62. Emphasis original.

66. As his standing as the nation's emergent celebrity defense attorney grew over the course of the decade, Bailey paraded his credentials as a master of the polygraph on national television, appearing on Mike Douglas's and Jack Parr's shows to administer tests to guests. In the early 1980s, Bailey hosted a short-lived syndicated daily program called *Lie Detector* in which he again polygraphed guests. Kerry Seagrave, *Lie Detectors: A Social History* (Jefferson, NC: MacFarland, 2004), 151.

67. The author of an excoriating journalistic exposé of the first Sheppard trial, Holmes had become a Sheppard family confidant and received Bailey's name when he asked the Keeler Institute for a recommendation as to a suitable attorney for the job.

68. "The Boston Prodigy," *Time*, December 9, 1966, 52, 57.

69. "The Defender: Bailey for the Defense," *Newsweek*, April 17, 1967, 41–43.

70. Bailey to Sherman, November 8, 1962, 2. Available at the EngagedScholarship @ Cleveland State University website, https://engagedscholarship.csuohio.edu/sheppard_habeas/34. Bailey had used a similar enticement when writing to Sherman to recruit him to the cause: "if you get mixed up in this thing you will tend to develop a strong personal feeling about the sickening failure of the law to maintain the unbending strength we were told about in school. As to the Sheppard case, the state has been a government of men and not of laws. The reward in smashing to a smidgeon those responsible for this mess is something which can only be calculated in terms of the individual." Bailey to Sherman, August 1, 1962, 2–3. Available at the EngagedScholarship @ Cleveland State University website, https://engagedscholarship.csuohio .edu/sheppard_habeas/35.

71. F. Lee Bailey, *The Defense Never Rests* (New York: Stein & Day, 1971), 64.

72. Bailey to Holmes, October 31, 1962, 1–2. Available at the EngagedScholarship @ Cleveland State University website, https://engagedscholarship.csuohio.edu/sheppard_habeas/32.

73. Holmes to Bailey, November 4, 1962, 1. Available at the EngagedScholarship @ Cleveland State University website, https://engagedscholarship.csuohio.edu/sheppard_habeas/36.

74. Judge James Finley Bell's 1961 Ohio Supreme Court dissent read as follows: "Murder and mystery, society, sex and suspense were combined in this case in such a manner as to intrigue and captivate the public fancy to a degree perhaps unparalleled in recent annals. Throughout the pre-indictment investigation, the subsequent legal skirmishes, and the nine-week trial, circulation-conscious editors catered to the insatiable interest of the American public in the bizarre. . . . In this atmosphere of a 'Roman holiday' for the news media, Sam Sheppard stood trial for his life." 165 Ohio St. at 294, 135 N.E.2d, 342.

75. Samuel H. Sheppard, F. Lee Bailey, Russell A. Sherman, and Alexander H. Martin, "Petition for a Writ of Habeas Corpus." Available at the EngagedScholarship @ Cleveland State University website, http://engagedscholarship.csuohio.edu/sheppard_habeas/4, 13. Note that Bailey also used the state's denial of a polygraph test as grounds for appeal: "He has sought to produce evidence favorable to him by submitting to a polygraph test, participated in by the state, and to hypnosis, to restore a memory which the record clearly shows to be vague with respect to events occurring on the night of the murder. Polygraph tests are not without judicial approval, . . . and the right to the aid of a hypnotist has been established by sound authority (Cornell v Superior Ct, 52 Cal 2d, 1959). Ohio administrative officials refused to let petitioner use these methods." F. Lee Bailey, Alexander H. Martin, and Russell A. Sherman, "Memorandum of Petitioner." Available at the EngagedScholarship @ Cleveland State University website, http://engagedscholarship.csuohio.edu/sheppard_habeas/12, 5.

76. Justice Tom C. Clark, "Sheppard v. Maxwell (Supreme Court, 1966) [Conviction Reversed with Instructions]." Available at the EngagedScholarship @ Cleveland State University website, http://engagedscholarship.csuohio.edu/sheppard_habeas/14, 3.

77. Clark, "Sheppard v. Maxwell," 4.

78. Clark, "Sheppard v. Maxwell," 29.

79. Bailey, *The Defense Never Rests*, 82.

80. Bailey, *The Defense Never Rests*, 88.

81. "Volume 04, 1966 Trial Transcript (Retrial of Samuel H. Sheppard): Testimony; State Rests; Motions for Directed Verdict and Time." Available at the EngagedScholarship @ Cleveland State University website, http://engagedscholarship.csuohio.edu/sheppard_transcripts _1966/1, 764.

82. "Volume 04, 1966 Trial Transcript," 768.

83. "Volume 05, 1966 Trial Transcript (Retrial of Samuel H. Sheppard): Testimony; Defense Rests; Closing Arguments." Available at the EngagedScholarship @ Cleveland State University website, http://engagedscholarship.csuohio.edu/sheppard_transcripts_1966/5, 1687, 1688.

84. Alder, *Lie Detectors*, 230.

# PART TWO

# PRACTICES OF POWER
# AND POLICING

# CHAPTER FIVE

# Death and Empire

## Medicolegal Investigations and Practice across the British Empire

JEFFREY JENTZEN

As the British acquired political control over vast overseas territories, they transported both colonists and the English law in their conquest and settlement of distant lands. The expansion of the British Empire extended the reach of the British Parliament and legitimized British colonialism and political ideologies.[1] The period from the early seventeenth to the late nineteenth century marked Britain's increasing presence in the world, its expansion of colonial trade, and the further consolidation of the United Kingdom. Along with English law that covered both civil and criminal matters, the British transported their legal institutions that not only included judges and barristers, but also coroners, who investigated and presided over inquests into the causes and circumstances of sudden and unexpected deaths.[2] England would eventually transport its coronial system to its global colonies across the globe, including: Ireland (1607), the Virginia colony (1607), Canada (1763), Australia (1788), Singapore (1824), New Zealand (1840), and Hong Kong (1842), to name a few.

For more than three hundred years, the British maintained a system of death investigation implementing a unique form of colonial medical jurisprudence. Military physicians acting as civil surgeons assisted the coroners and magistrates with the medical investigations of sudden and suspicious deaths. In the distant colonies, these British civil surgeons encountered harsh physical and cultural conditions unsuited for their European-based medical theory and practice, which required them to adapt to a more empirical type of colonial medicine. This empirical approach resulted in the creation of professional networks of medicolegal innovation that evolved from distant colonies and not in the metropole, where dogma and a rigid medical theory reigned. These civil surgeons were tools of empire, at times both complicit in state-controlled violence and the civilizing efforts of native populations.[3]

In the colonies, the British encountered a variety of indigenous geography, racial groups, cultures, poisons, religions, local law, and traditions that challenged the establishment of a uniform system of British colonial death investigation. In those settler colonies where the native population was decimated or otherwise marginalized through colonial conquest in such regions as in Australia, Caribbean, the North America colonies, and South Africa, the British colonial government continued to function under statutory laws developed for centuries under the foundation of English law. In these territories, the political institutions were derived directly from their British counterparts, and as a result the emigrant institutions and government were "ultra-English" by race, religion, and interest. In the other trading territories such as British India and Burma and the extrepot colonies of Hong Kong and Singapore, the complexity of the indigenous culture required the amalgamation of English and indigenous legal systems. In these lands, pluralistic systems of law and death investigation emerged to both control the local peoples and provide protection to British and European subjects.[4]

Essential to maintaining the rule of law was a judicial system capable of investigating sudden, unexpected, and suspicious deaths. In their developing colonies, the British created a legal system based on extraterritorial rights, insisting on their own courts, laws, and medical experts for the protection of British and foreign citizens and their property. The British feared manipulation of the judicial proceedings and over time increasingly relied on demonstrative scientific evidence to provide necessary objectivity rather than accept suspect oral testimony of native witnesses alone.

Violence was a common feature of colonization throughout the Empire. Racial violence was not just practiced but at times condoned by a colonial system that institutionalized legal inequality.[5] With huge expanses of territory and a general lack of sufficient administrative resources, British colonial power was inherently weak and relied on violence to maintain control of the indigent population and to protect foreign colonists. The demarcation between sanctioned violence and the legal methods to control it evolved constantly over the colonial period. Colonial rule saw a racially based system of unequal justice in which medical experts and court officials provided European colonists with the legal alibis for mistreatment and violence.[6] Initially, the British colonial administration turned a blind eye to sanctioned violence inflicted upon the native population by colonists or as a result of indigenous archaic religious or cultural beliefs.

The British transported their judicial apparatus—the lay jury, public oral testimony, and coroner's inquest—to their colonies in order establish a rule of law. But British colonial law differed from that in the metropole in that it did not allow for cross-examination of medical witnesses or questioning of their reports. Initially developed for economic purposes, by the early twentieth century, colonial law evolved to further Britain's expanding humanitarian goals.[7] Medicolegal death investigation provided protection of British colonists and control of the indigenous peoples through the investigation of crime and surveillance and control of epidemics. Death investigation was the convergence of not only medicine and law but also the influences of politics, local cultural and ethnic practices, and geography of the colonial sphere.

By the late nineteenth and early twentieth centuries, the new field of forensic medicine became increasingly instrumental in British efforts to provide humanitarian support to native peoples, resulting in the elevated status of the medicolegal expert and production of more reliable evidence. Medicolegal experts became a crucial component in the humanitarian efforts for social justice. Under the British administration, the system of death investigation created for colonized subjects an imperial system of justice and provided humanitarian protection for both native and foreign peoples.[8]

The British civil surgeon, who provided medicolegal dissections, toxicology analyses, jail inspections, and other "multifarious duties," was the backbone of colonial legal medicine.[9] The physicians who responded to cases of suspicious and sudden death functioned as arms of the state. Scottish-trained physicians stationed in isolated colonial posts provided the bulwark of medicolegal expertise for the British Empire. Products of Scottish universities in Edinburgh or Glasgow, these physicians were generally from the lower middle class, often Scots and Irish, known as the "Celtic Fringe." They received a medical training by professors who preached an empirical and innovative approach to medicine.[10] The role of the medicolegal expert changed over time as British political and cultural values evolved. In Britain's colonial outposts, the rule of law and those institutions and their personnel who provided it were tools of empire that served practical, economic, and ideological functions.

In recent years, the postcolonial influence in history has led to a surge of scholarly interest in the historical role of legal medicine in the colonial state. Previous studies, predominantly by legal and medical historians, were restricted almost exclusively to the ancient roots of Anglo-American coroners.

A new body of historians from a variety of scholarly fields are now producing specialized global studies concentrating on questions of cultural, gender, race, environmental, and political influences, gradually clarifying the role that death investigators played in colonial imperialism. Despite the growing interest in colonial legal medicine, the primary and secondary sources remain scant.

However great a role the advances that DNA technologies have played in the contemporary criminal justice system, a number of recent reports have criticized the methodology, literature, and practice of a number of fields in criminalistics such as fingerprints, firearm analysis, patterned evidence, and dental identification. The 2009 National Academies of Science report in the United States has gone so far as to argue that many criminal fields "lack a scientific basis."[11] In addition, challenges to the criminal justice system around the globe continue to focus on racial bias, political influences, police misconduct, the use of excessive force, and professional incompetence. These challenges are as apparent today as they were in the British colonial state in past centuries.

In my own field of forensic pathology, critics in highly contested trials have charged medical experts with cognitive bias, human error, and incompetence. Forensic pathology is a field of medicine relying primarily on observation, description, training, and experience rather than comparative or quantitative methods. In the courtroom, forensic pathologists are constantly implicated to be under the direction and control of aggressive prosecutors and overzealous police.[12] Despite the modern popular media portrayals, the tools of the forensic pathologist remain "low-tech" and have changed little over the centuries. The challenges faced by today's practicing forensic pathologists are in many instances the same as they were for the British district police surgeon isolated in a rural colony in centuries past. For this reason, an exploration of the forensic practice of colonial medical practitioners helps to illuminate the allegations, mistakes, and misconceptions of past, present, and future legal institutions, forensic experts, and courtroom justice.

British colonial administration was not a monolithic global institute in dealing with the various cultures, geographic, ethnic, and religious cultures it encountered around the globe. In this chapter, I focus on death investigative practices of select British colonies, dominions, and protectorates of Egypt, India, North America, and South Africa to illustrate the colonial practices of the race-based extraterritorial law, the colonial legal administration, and the role of medicolegal death investigation. I also examine the role they played in ascertaining the cause and manner of death while at the same time facilitating the imperialistic and humanitarian motives of the British.

## The English Coroner

The modern English coroner system emerged as a direct consequence of the Norman Conquest and later colonization of the ancient Anglo-Saxon lands.[13] As the Normans sought ways and means to protect themselves, they instituted an inquiry procedure, the *lex murderum*, which assumed every murder victim to be a Norman and forced the local Anglo-Saxon inhabitants to prove "Englishry" of the decedent. This practice of inquiry evolved into what would become known as the English coroner's inquest, which would later form the foundation for the investigation of unexpected death and rule of law, which Britain transported with its colonizing armies around the globe.

The coroner became an officer of the state in England in 1194 when King Richard established the requirement that coroners be elected in each county.[14] Coroners became increasingly important over the thirteenth century as many duties of the sheriff shifted to office of coroner, including the investigation of sudden deaths. The coroner, or "crowner," was also required to collect various fees for the king, including the property of the deceased depending upon the circumstances. In any death the coroner determined unnatural, the personal property or the implement that caused the death, called the deodand, was forfeited and distributed to the king or the church.

In order to determine the cause of death, the coroner was required to convene a public inquest to view the bodies of all those who had died unnaturally, suddenly, suspiciously or in prison at their places of death.[15] In contrast to European continental medicolegal practices based on Roman law, which relied on university scholars to inform the courts, English coroners impaneled a lay jury and heard oral testimony, only occasionally calling upon medical experts. Over the centuries, with the practice of public inquiry of judicial decisions through an elected coroner, the English came to appreciate the office of coroner as a "bulwark of English democracy."[16]

## English Coroners in American Colonies

The British transported the system of the coroner's inquest to its North American colonies beginning in the early seventeenth century. The English judicial system, based on trial by lay jury, oral testimony, and public coroner inquests, was easily transported to and implemented in colonial lands, where there was a paucity of administrative personnel, flexible methods of legal proof, and little regard for medical expertise.[17] There were few written guides for colonial coroners or physicians in the conduct of the inquest, and they depended upon English works such as *Lex Coronatoria; Or, Office and Duty of Coroners*, by Edward Umfreville, and, after independence, the *Conductor Generalist*, published in New York.[18]

In their examination of homicide cases in Maryland from 1633 to 1683, Helen Brock and Catherine Crawford observed that, even though "the new colony had few men capable of even the most rudimentary system of criminal justice," the American colonies practiced a highly organized system of English colonial justice through the coroner's inquest and the occasional use of medical experts and postmortem dissection in their investigation of sudden deaths.[19] Colonial coroners in America adopted verbatim the statutory language for oaths and inquest verdicts from English law. They were charged to "Doe all and everything done by any coroner of any county in England."[20]

English colonial juries were reluctant to convict colonists for murder of a native Indian, servant, or slave. Servants were often beaten for lassitude, possibly the result of debilitating effects of malaria and other endemic diseases common to the English colonies. Native Americans were subject to the laws of the English colony and punished severely when they were believed to have committed a crime. Coroner's juries in the English colonies, however, were more reluctant to convict colonists of capital crimes, a relatively rare outcome considering the cruelty inherent in English law.

With increasing migration to the America colonies in the seventeenth and eighteenth centuries, the demand grew for medical expertise in coroner's inquests, and we begin to see growing numbers of autopsy dissections ordered by coroners. Those American colonial physicians who traveled overseas did so predominately in medical schools and hospitals, in Edinburgh, Leyden, and London where they came under the influence of Scottish and enlightened "continental" system of medical jurisprudence and its emphasis on the concepts of medical policing and public health. The majority of American physicians, however, received their medical training through local apprenticeships in the colonies.[21] Colonial physicians commonly participated in coroner's inquests as witnesses or as member of the jury.

The office of English coroner had remarkable staying power on American soil. Coroners, unlike the continental system of death investigation, had investigative powers to determine both the cause and the manner of death. This allowed the coroner to determine the extent of the investigation and whether a crime had been committed. Similar to the British, Americans coveted their democratic right to elect coroners from the citizenry as a protection against governmental intrusion and prosecution.[22] In the twenty-first century, coroners still cover 40 percent of the American population and are elected in a majority of counties.

## The East India Company and British Justice in India, 1757–1857

Britain's industrial revolution of the late eighteenth century initially encouraged the country to establish an economic as well as a civilized empire. The British used the power of English law and political force to protect as well as punish its colonial subjects. By the end of the nineteenth century, the British were motivated to pursue a more humanitarian policy to aid in legitimatizing imperial rule.

Historians have recently begun to investigate the practice of British colonial legal medicine from more diverse institutional, social, political, and racial perspectives. Recent scholarship on the role of colonial legal medicine in such works as those of historians Jordanna Bailkin, Elizabeth Kolsky, Ishita Pande, and Mitra Sharafi has revealed that physicians within the judicial machinery of Indian colonial death investigation and in British law itself were tools of the evolving imperial state. At times, these tools enabled and protected certain acts of violence largely by placing European British subjects above the law. Other times, medical experts provided the native population a level field to challenge the unjust acts of the British. More specifically, scholarly depictions of unequal death investigation practices, corrupt medical experts, and race-based justice, although limited in number, are beginning to emerge from the view of both English colonists and the colonized within the imperial enterprise in India. British medical experts also provided testimony for the benefit of the innocent from harsh and demeaning traditional religious and cultural practices.

The English established the East India Company in 1600, principally for trade purposes, and through acquisition and conquest brought nearly the entire Indian subcontinent under English rule. India was a complex country of many existing religions, cultures, and ethnicities. The British administered India for a century through the indirect rule of the East India Company until 1857 through a pluralistic legal system based on extraterritorial law, which provided English law for the British and foreign colonists and local religious-based laws for the Hindu and Muslim populations in rural areas. This pluralistic legal arrangement protected British citizens and property and at the same time allowed the native Indians to enjoy their own laws. It later served as a model for establishing pluralistic legal systems in other British colonial regimes. Legal historian Jörg Fisch has outlined the methods by which the early colonists attempted to live within these two separate legal systems.[23] This included retaining existing indigenous laws where possible and installing indigenous Muslim and Hindu lawyers to interpret the law to British magistrates.

The Charter Act of 1793 empowered the governor general of the Indian presidencies to appoint coroners with the same powers as in England. The Bengal Regulation XX of 1817 later formalized the legal procedures for all police investigations to require the consultation of civil or police surgeons.[24] British police surgeons responded to the requests of the local chiefs of police or village constable, the *darogah*, and British magistrates or coroners were charged with investigating sudden, unexpected deaths. In cases of unnatural or suspicious deaths, the darogah was instructed to examine the body of the wounded or dead person to assert the number, size, and locations of wounds and injuries, and to note what the weapons had been used. After making this initial inquiry, the body was to be handed over to the family, except in cases of murder or poisoning.[25] The cause of death was more difficult to determine in India than in Europe owing to countless factors unique to the Indian landscape, such as rapid decomposition of bodies, religious practices of prompt cremation and burial, the myriad poisons available, and the unwillingness of native Hindu officials to properly examine dead bodies for religious reasons.[26]

A system of medicolegal death investigation would not be clearly formalized by law until the Indian Penal Act of 1862, but it existed out of administrative traditions of the English legal system. J. C. Marshman's *The Darogah's Manual* (1850) provided practical advice to the British colonial authorities for the handling of dead bodies.[27] The general practice in cases of murder and mortal wounding was to have the body initially inspected by local civil surgeons, although this was not a formal mandated duty.

British administrators considered Islamic and Hindu laws of homicide inferior to English law in that the former treated murder as a violation against the victim and extended family or tribe and not as an offense against the state. The Crown Courts were tribunals of English law with jurisdiction over all British subjects and natives living in the presidencies of Bengal, Bombay, and Madras. Indian magistrates were not allowed to punish British subjects. In the interior of the country, the *mofussil,* or company courts ruled by British judges with little or no prior legal experience, governed the native Indians and non-British Europeans.[28] Under Islamic law, the state was subordinated to the wishes of the family, who were commonly reimbursed through the custom of blood money.[29]

Kolsky observed that medical expertise was not commonly involved in determining court rulings in early nineteenth-century India, and both laypersons and trained medical personnel provided expert testimony. In addition, physical or demonstrative evidence did not receive any special status in Islamic courts over the oral testimony of witnesses. Judges were routinely untrained and unable to admit special medical evidence into the trial.

Initially, Kolsky recognized, the East India Company authorities were not tolerant of white violence against either colonists or natives. British subjects met with the severe punishments of English law as a result of the strict discipline imposed the company.[30] Reflecting an "other" race-based attitude toward of their Indian colonial subjects, the British took an ambivalent attitude toward culturally sanctioned or "hidden" crimes attributed to the indigenous religious cultures, including rape, infanticide, assault, and culturally sanctioned crimes. It was considered the dark side of native culture, hidden from view, to which the British justice turned a blind eye.

Medical evidence during the early nineteenth century often played an ambiguous and uncertain role. Postmortem dissection, although performed, lacked evidentiary veracity. Indian courts valued oral testimony, which was often subjective and tainted with lies and untruths, and considered medical evidence from physicians to be equivalent to that of laymen. The use of medical evidence in India took a dramatic shift with the Charter Act of 1833. In an attempt to check the power of the East India Company, the British removed its administrative powers and eliminated the existing pluralistic laws, creating a "systematic body of law and centralized system of command" for all of India. This gave the governor general absolute control over British territories.[31]

One of the major differences between the English and indigenous Muslim and Hindu legal methods was the reliance on postmortem dissection. British police surgeons argued that "the only certainty we have in this country of the actual cause of death is from the postmortem examination of the surgeon."[32] Police and medical doctors began to provide more detailed medicolegal reports and examinations of crime scenes, weapons, and corpses and relied on demonstrative evidence rather than simple testimony.[33] Reports of postmortem examinations and other medical evidence began to appear in all murder cases. Poisoning cases, however, did not always attract the attention of legal authorities owing to the large number of indigenous poisons the difficulty of their detection. Unlike in Britain, where corpses at the time were scarce and there was black market in bodies, in the colonies, the bodies of the natives and British soldiers in India were plentiful for dissection and scientific inquiry. Motivated by the French advancements in morbid pathology, British surgeons in the colonies commonly dissected victims of disease and sudden death.

Coupled with Charter Act legal reforms of 1833, the British ended their experiment to teach an indigenous form of medicine in the vernacular by translating Western textbooks and lectures to natives in the Native Medical Institution in Calcutta. The absence of Western anatomy lectures was cited as one of the major reasons for the change. The British instead created a medical school in its place to "propagate the Western system and train Indians in its use."[34] Lectures were taught using Western methods in anatomy, materia medica, medicine, and surgery. With the creation of the school in 1835, British chemists were appointed to Indian medical schools to investigate the myriad organic and mineral poisons used in sinister ways.[35] In 1836, the first public autopsy by a Hindu physician was performed at the Calcutta Medical College. This widely celebrated event reflected the desire of Indian physicians to establish a Western scientific veracity that would eventually find its way into the courtroom.[36]

## After the Rebellion: The Indian Reform Period, 1857–98

The successful repression of the Indian Mutiny (1857–59) resulted in the complete removal of East India Company rule and the establishment of British state power in India and new efforts to justify the legitimacy of the Raj. The British enacted civil and criminal reforms to provide for enhanced surveillance and control of their Indian colonial subjects. According to Kolsky, colonial medical jurisprudence depended upon British medicolegal experts, along with the enactment of racially unequal laws and the development of other forms of "truth technologies" to provide British legal officials with reliable, scientific evidence.[37] The British feared Indian manipulation of the law and distrusted the oral testimony of native witnesses whom they saw as untruthful and sinister. The "truth technologies"—such as fingerprints, anthropometry autopsy pathology, ballistics, and chemical testing—assisted colonial officials in maintaining control of the native population, especially in areas of identification and the determination of the cause of death.

A universal and centralized British "death investigation machinery" and the acceptance of expert medical testimony in India were not formalized until the passage of the Coroner's Act in 1871 and the later Indian Evidence Act in 1872.[38] The former prescribed methods for performing medicolegal work, while the latter attached special value to expert evidence and testimony. The codification of the Indian Penal Code in 1862 had failed to distinguish between murder and culpable homicide, as in Britain, and allowed for exculpatory grounds such as the mental state and intent of the assailant as well as the health of the victim to influence the decisions of the judge and jury. The code also placed a higher burden of proof on the prosecution to demonstrate that the accused had intent to kill. In England, the law assumed any homicide to be an act of murder. The code explicitly differed from English law in expressing the right of self-defense and eliminated the adversarial method of the cross-examination.[39]

Over the course of the late nineteenth century, medical experts and forensic evidence played an increasingly prominent role in judicial inquires and trials in India. Although expert testimony had been used in colonial trials prior to the enactment of the Indian Evidence Act, it had no unique or special legal standing. The goal of making criminal investigations a new science afforded physicians and scientific evidence a privileged status in juridical inquires and trials.[40] Scientific facts were increasingly brought to bear in colonial courtrooms, not only in new forms of physical evidence, such as toxicological analysis for drugs and poisons, fingerprints and photographs, but also in the person of the medicolegal expert. The body and its latent evidence now provided demonstrative evidence that could be used in court to disprove the falsified testimony of assailants and victims.[41]

The prevalence of endemic diseases, such as malaria, rendered colonial bodies more susceptible to minor trauma.[42] The penal code, in allowing for the use of self-defense and consideration of an existing condition of a vulnerable victim, sanctioned a system of unequal justice for Britons and native Indians.[43] Jordanna Bailkin has described how British police surgeons and coroners became complicit in validating the "ruptured spleen" defense by providing testimony of the common occurrence and known frailty of the spleen when Indians died from a soldier's kick. Under these circumstances, the death was more often than not considered an accident in that the assailant lacked the intent to kill, and Indians tended to have enlarged, delicate spleens as the result of endemic malaria.[44]

The ruptured spleen defense provided the judicial rationale with which Britons could inflict mortal wounds on native Indians and not be charged with murder. Such was not the case in Britain, where common law supported the plaintiff in cases concerning the "eggshell defense," where the assailant retains culpability and "takes their victim as he finds him."[45] As a result, it was much harder to obtain a conviction for murder or assault in India than in Britain. As civil surgeons and judges looked askance at deaths at the hands of their cruel masters, they perpetuated the exploitation of the native laborers.[46] With reference to colonial India and the British occupation, historian David Arnold observed that "there was an undoubted growth in the British emphasis upon racial difference, and especially the perceived physical characteristics of race, over the course of the nineteenth century."[47]

The published work of English medicolegal expert Alfred Swayne Taylor, *The Elements of Medical Jurisprudence* (1836), provided a standard guide for a British-based practice of legal medicine for police surgeons in both England and the colonies. Multiple editions were published up to the mid-nineteenth century. Taylor provided medicolegal advice and guidance along the same model as in English adversarial courts. His work focused on scientific investigations of poisons and trauma. He warned unwary physicians of the difficulty of presenting scientific evidence in the courtroom for the medical expert who considered practicing legal medicine.[48]

*A Manual of Medical Jurisprudence for Bengal and North-Western Provinces* (1856), by the British physician Norman Chevers, was the first medicolegal work dedicated solely to Indian medical jurisprudence. In this "new field of forensic medicine," Chevers emphasized the cultural "differences" of the native perpetrators and British victims of colonial violence and claimed that the civil surgeon should "possess an intimate acquaintance with the dispositions, customs, prejudices, and crimes" of the indigenous peoples.[49] For Chevers, death investigation was a cultural, ethnographic, and anthropological process. Chever's anthropological theory of criminal behavior was an attempt to justify the British goal of "advancing the progress of civilization."[50] His textbook validated the defense that even "trifling" violence from the kick of a boot or blow from a fist would cause death from ruptured diseased spleens infected with malaria. Chevers also freely admitted that accusations of rape were often spurious, reflecting the British laissez-faire attitude that openly sanctioned rape in India as in Britain.[51]

For legal historian Elizabeth Kolsky, legal medicine in the British colonies constituted a unique type of "colonial legal medicine."[52] The primary motivation for this distinctive form of medicine as observed by Chevers was the influence of local culture, law, medicine, natural environment, and political forces on matters of life and death. Colonial legal medicine, unlike colonial medicine itself, differed in many ways from the practice of legal medicine at home in Britain.

## Age of Humanitarianism, 1892–1947

By the late nineteenth century, medicolegal investigations performed by British colonial police surgeons had made way for a new age of humanitarianism. According to historian Ishita Pande, the medicolegal expert gained renewed importance and validity to the political project of colonialism by the publication of the objective facts derived during the conduct of autopsy and inquest testimony to counteract subjective and abusive traditional cultural and religious practices of indigenous peoples.[53]

In 1889 inquest testimony in the case of *Queen Empress v. Hurry Mythee*, the police surgeon publicized the plight of the horrible rape and death of Phulmoni Dasi, an eleven-year-old Hindu "child-wife," by her adult husband.[54] The case drew widespread attention of the British public when published in the 1904 edition of Isadore B. Lyons's *Medical Jurisprudence for India with Illustrative Case*, the most important medicolegal textbook of the period. In earlier times, police surgeons, and authors of medicolegal publications such as Chevers, emphasized an ethnological approval of child rape. The courts, too, were generally accepting of and ambivalent about rape and considered it simply one of the culturally race-sanctioned "colonial pathologies." Authorities casually mentioned that rape may cause death, and "such a cause of death was not uncommon amongst the child-wives in Bengal." At the time, British medical jurisprudence had a tradition of treating allegations of rape charges both at home and the colonies with skepticism.[55] By providing actual cases and illustrations of child rape, Lyons's textbook intended to instruct and educate medical experts in testifying in rape cases in colonial courts. The autopsy report and courtroom testimony of the medical expert proved so powerful and crucial to the case that Ishita Pande refers to the police surgeon as "taking a scalpel to the Indian society."[56]

The British colonial ambivalence to the murder of natives depended upon a systemic and mutual agreement between legal officials and medical experts who colluded to practice a racially biased system of death investigation. Murder charges against British subjects eventually returned in the early twentieth century with a growing sense of justice and equality. According to

Bailkin, the renewed prosecution of murder as a result of the production of a "new kind of medical jurisprudence" emerged by the 1920s and 1930s, introduced by indigenous and British medical researchers and practitioners. By the early twentieth century, the ruptured spleen defense had been replaced by rigid murder charges for Britons guilty of interracial violence and abuse during the interwar years.[57]

## Egypt: Continental Law and British Occupation

Beginning with Napoleon's 1798 invasion, the French introduced the parquet system of law, an administrative branch of law that dealt with the investigation and prosecution of crime. Here, the police functioned both as the investigating authority and public prosecutor. As early as 1850, the close connection between medicine and law and a system of death investigation was already firmly established in Egypt, especially in criminal and homicide cases. Egyptian officials assimilated a death investigation system not too different from that found on the European continent.

By the time the British took control of the Egypt protectorate in 1882 as part of their attempts to protect their interest in the Suez Canal, they had formalized death investigation in the Department of the Ministry of Justice, a remnant of the French parquet administration, whose head was the procurator general. The parquet was responsible for the investigation of crimes, performed by the police. Its officers had the status of magistrates and could examine witnesses under oath and order arrests. The chief of the parquet determined to which court a particular case should be sent for trial, and when he did, he acted as prosecutor. This followed the French system of death investigation based on the Napoleonic Code rather than the British common law practices.[58]

Under the articles of capitulation, as was the practice in other British colonies, foreigners enjoyed certain extraterritorial rights in connection with both criminal and civil proceedings. Extraterritorial law was a concern in most British colonies where foreign residents were exempted from the jurisdiction of local courts and locals were excluded from foreign consular courts, and sought to remove the burden of extraterritorialism from existing and future foreign trade agreements. If a matter concerned foreign colonist alone, the British consular courts dealt with it. If both foreigner and Egyptian were involved, the proceeding took place in the mixed courts.[59] British medical experts were officials represented in the Egyptian government, and their evidence frequently affected foreigners, who represented both accuser and the accused.

British colonial administrators relied only on evidence from the forensic testimony of a European physician who had performed an autopsy to determine the cause of death and did not acknowledge inspections of deceased persons by local authorities without an autopsy.[60] Egyptians retained control over domestic administration of society, mainly family and property law, while the British exerted control in international and military law. The British protectorate was short-lived, lasting only until 1924, and they never intended to create a colony along the lines of the Raj in India.

Historian Khaled Fahmy has documented the central role that legal medicine and the autopsy played in the Egyptian criminal justice system, the progressive reform of Islamic law in the Ottoman Empire, and the establishment of a continental style of rule of law beginning in the early nineteenth century.[61] This eventually resulted in a system of mixed courts, Islamic and secular, by the late nineteenth century. Egypt had been a part of the prior Ottoman Empire under Islamic sharia law, which placed a higher value on oral witness testimony than on written case law. In cases with no fixed religious provisions provided in sharia law, the siyāsī allowed the ruler or state to intervene in the administration of justice and admitted evidence in court when it was in the public interest. Siyāsī law ultimately allowed for the use of demonstrable evidence, such as the autopsy, in courts of law.[62]

According to Fahmy, by the mid-nineteenth century, the Ottomans had firmly established a death investigative procedure. On receiving news of a death, the local barber-surgeon (*hallāq*) or midwife (*dāya*) would be dispatched by the police to conduct an external examination of the victim. If suspicious signs were found on the body, the provincial physician (*hakīm*) would be summoned to give a second opinion. If this second investigation proved inconclusive, the body would be sent to the provincial hospital, where a complete autopsy wound be conducted and an "autopsy committee" (*jam'iyya tashrīhiyya*) would write a report. In Cairo, the resident doctor conducted the preliminary investigation.[63]

For the Ottoman government, the education of medical practitioners was an important factor that contributed to the significant position of forensic medicine in nineteenth-century Egypt. Postmortem examinations, Islamic religious officials argued, were permissible because they were a necessary part of modern medical education and crucial to control of epidemics and crime. Arab doctors were not trained in Egypt until 1855, when a school for the training of Arab doctors opened in 1875. Until then, French-trained physicians assisted in the performance of autopsies and exhumations.[64]

Fahmy argued for the siyāsī law's gradual inclusion of the functions performed by medical experts along with Islamic sharia law, which enabled the modern Egyptian state developing under foreign control to exercise greater control over society, especially in the area of criminal and murder cases. Egyptian state authorities introduced the autopsy and forensic medical expertise to aid the legal system without overriding existing religious law. The autopsy thus played an increasingly important role in the detection of crime in Egyptian criminal investigation. Autopsy reports became a crucial part of "a body of evidence" introduced into courts.[65] The introduction of forensic medicine also allowed the state to monitor disease.

The local population welcomed autopsies, which they saw as both supporting sharia law and protecting against criminal violence and the state. Although police authorities ordered the majority of autopsies, families often sought independent autopsies to redress serious miscarriages of justice. People were not deterred by the sanctity of death if they thought that an autopsy was their only chance to establish justice.[66] Fahmy observed that the common people understood the important role that medical evidence, including the autopsy, played in the administration and enforcement of the law. They were willing to go to great lengths to use the death investigative system for their benefit despite the presence of strict religious law and the ignobility of the autopsy. Eventually, the government issued strict orders forbidding the burial of corpses without a physician's examination.

The Muslim culture that predominated in much of Egypt provided many challenges to British medical experts and played a major role in many investigations. The Islamic law that allowed for up to four wives and ease of divorce, coupled with rigid inheritance laws, created the potential for crimes resulting from jealousy and revenge. Child murders by competing wives were also seen as common. British forensic experts voiced their appreciation to their Egyptian colleagues who tutored them in the morass of local customs, which frequently played a large role in investigations.[67]

As in India, Egypt was home to a wide variety of poisons available to murderers, and death by poisoning was common. Visiting pathologists to Cairo's world-class forensic science institute were shocked to observe the number of Marsh tests for the detection of arsenic conducted by British chemical examiners in suspected murder cases.[68] Under the British, exhumations were not uncommon and came to be an accepted inconvenience of Egyptian crime investigations owing to the hastily arranged initial burials required by Islamic law, the torrid heat of the region, and attempts to conceal crimes. As in other equatorial colonies, decomposition was rapid, and for this reason bodies were not typically buried but placed in superficial tombs without coffins, which hastened the decomposition process, making both identification and the determination of the cause of death difficult, if not impossible, for colonial investigators.[69]

Forensic examinations were not restricted to the dead. Living persons frequently visited the forensic laboratory for examination of alleged disabilities; faked stabbings, shootings, and other injuries; and to rule out malingering. In Islamic societies, legal physicians were responsible for sexual assault examinations and the premarital examinations of women to ascertain virginity.

The race-based perception by the British of perjury by the native population was widespread throughout the empire and created a general lack of reliance by the British courts of eye-witness testimony. In Egypt, the courts thus relied more on medical expert testimony and laboratory evidence, where "laboratory evidence assumed an importance of the first magnitude, on the understanding that while men may lie, inanimate objects do not," a concept that helped to stimulate scientific investigation in the colony.[70]

Anticolonial resentment eventually resulted in increasing unrest and revolt. Between 1919 and 1922, there were thirty murders or attempted murders of British government officials. The British government formally ended the Egyptian protectorate and declared Egypt an independent sovereign state. The new state received a new king and constitution. British authority and medicolegal expertise were initially undiminished, and British judges continued to sit in courts alongside Egyptian judges. With time, during the interwar period, the new Egyptian government gradually removed British physicians from permanent government positions; however, the scientific practices and traditions instilled by British forensic experts remained strong in the forensic institute. The recognition of an unequal system of law created by the imposition of extraterritorial laws by foreign countries provided the crucial motivation for the Egyptian government to turn to new forms of medicolegal expertise in Western legal, medical, scientific, and policing practices in death investigation in order to create legislation and institutions to address the problems created by the compromised sovereignty of the state.

## South Africa: British Law in Continental Courts

The introduction of legal medicine in South Africa can be traced to the Dutch occupation of the South African Cape by the Dutch East India Company

(VOC) in 1652.[71] The Dutch introduced the Roman-Dutch legal system from Holland, which over time eclipsed the traditional law and customs of the indigenous nomadic peoples, the Khoikhoi. VOC law superseded any religious or cultural beliefs of the Khoikhoi, who were frequently subject to medical dissection. The Dutch performed some autopsies under the protection of the Roman-Dutch law to determine the cause of death, and others simply for the purpose of advancing colonial medicine and medical research.[72]

The imposition of postmortem dissection could also provide justice to the indigenous peoples. Indigenous victims of assault and sexual violence especially sought to consult Dutch district surgeons in cases of rape and murder. In Cape Town, for example, the body of a murdered Khoikhoi servant was exhumed in 1764 after his friend, who witnessed the murder, insisted that authorities perform an autopsy. The local Landdrost (magistrate) agreed, and the state physician performing the autopsy reported that the person in question had "died of multiple blows sustained from a blunt object. The assailant responsible for the murder was subsequently executed and his accomplices banished from the Cape."[73]

With the arrival of the British in 1795 following the Anglo-Dutch War, the British took control of the Cape as its colony. The legal system transformed into a mixed system based on English and Roman-Dutch law. Over time, the Roman-Dutch laws and judges were replaced in various ways by English law and later, after the Afrikaner republics, became British dominion following the Boer War 1899–1902.

In all cases of homicide, magistrate courts ordered postmortem examinations, which were performed by a local district surgeons, if possible.[74] The most important contributions of the British were in the area of South African criminal law procedure, especially laws of evidence and those governing inquest procedures as applied in English courts. The British were influential in other areas, including the statutes that allowed for postmortem examinations along with the establishment of the legal next-of-kin. Criminal trials in cases of murder, rape, and serious assaults were conducted by a judge and a special jury. Magistrate courts maintained jurisdiction over all criminal proceedings except murder, rape, and treason, which were referred to the supreme court.[75]

In 1910, the former British colonies created the Union of South African States. Under the Inquests Act of 1914, British magistrates continued to convene inquests and routinely requested assistance of physicians as medical experts. In accordance with the Inquests Act, magistrates convened inquests and routinely requested the assistance of physicians with medical expertise. State-employed district surgeons were mostly private medical practitioners with no special forensic training who functioned on a part-time basis, appointed by district magistrates through the Department of Health. In southwest Africa, inquest courts were under the control of the medical officer, whose principal duties were to investigate and examine complaints in criminal matters as well as to conduct postmortem examinations and exhumations. Toxicology examinations in the whole of the South African Union were performed in Johannesburg.

Under Public Health Acts, the public health officer could order compulsory autopsies in deaths from diseases suspected to be infectious or formidable epidemics. In special cases under the Silicosis Act of 1916, district surgeons were directed to determine compensation for suspected occupational silicosis deaths of native gold miners. Medical practitioners were ordered to "open the body of a deceased person whom he believes to have been a miner or native labourer, he shall preserve the respiratory organs of the deceased," and then sent the sample to the chemical laboratory for inspection.[76]

With the rise of the National Party by 1948, apartheid segregation statutes were instituted through parliamental decree. Statutes authorized the performance of postmortem examinations in cases of suspected homicide or unnatural deaths. "The wishes of the deceased, the executor, or next-of-kin are clearly [to be] ignored," district surgeons were advised, "and must be ignored."[77] British governmental pathologists were under the control of the South African Police Service (SAPS) and were obligated to use their facilities and personnel. The SAPS attended the death scene and was responsible for the control of evidence. Government pathologists performed official postmortem examinations on homicide cases in Cape Town, Durban, and Pretoria. In Johannesburg, trained medicolegal pathologists, assisted by district surgeons, carried out these examinations.[78]

In the past, the SAPS was implicated repeatedly for political killings common during apartheid. Most district surgeons were forced to conduct autopsies under auspicious control of the SAPS, and complained that the police hampered or attempted to influence their examinations in these cases.[79] During this period, as was the case in other British colonies, the racially biased medicolegal death investigation system in British South Africa failed to provide for adequate and consistent democratic justice for all British subjects.

Steven Biko, a South African reformer and leader of the Black Consciousness Movement, died in 1977 at the age of thirty from head injuries following an assault while shackled in a South African jail cell. Biko's death in custody brought widespread international scrutiny to the system of apartheid and medical collusion at the hands of the South African police in covering up deaths of individuals in custody.

The police had arrested Biko under the 1967 Terrorism Act, which allowed for indefinite detention and solitary confinement without trial or charges for the purposes of interrogation. Biko was detained for 110 days. While in jail, Biko was beaten by his interrogators and sustained brain injuries related to an assault in which he was punched, beaten with a hosepipe, and thrown into a wall—where he collapsed.[80]

News of Biko's death was followed by universal outrage. The minister of police initially offered that Biko had died by suicide, similar to their explanation for the deaths of many other prisoners in custody of the South African police, or possibly from a hunger strike. Up until this time, South African courts had not substantiated police culpability in any case of torture or death of a police detainee.[81]

The autopsy was performed by the state forensic pathologist, and the family employed a private independent pathologist. The pathologists' postmortem reports provided clear and convincing evidence during the inquest that Biko had suffered impacts to the brain. The autopsy report was essential in establishing a violent cause of death that contradicted the events alleged by police personnel.[82]

The end of apartheid in 1994 resulted in an overhaul of forensic science services in South Africa. After decolonization, district surgeons were abolished, and their role transferred to the junior-level resident medical officers. The result was a dramatic decline in the clinical expertise and experience of those medical officers investigating injuries and testifying in courts. The SAPS continued to provide the overall physical structure of the forensic pathology services until 2006, when control was transferred to provincial divisions of forensic pathology. The SAPS, however, continued to secure and control crime scenes, as well as perform the identification and recovery of evidence.[83]

The British Empire consisted of a collection of various religious, legal, and cultural institutions in diverse geographic locales. The British initially attempted to create pluralistic institutions by introducing English traditions and laws into existing native society, until they could no longer to control the native subjects. Struggling to control a legal system within such a diverse environment of religions, ethnology, cultures, and physical environments,

British colonial legal medicine required specialized medical techniques, "truth technologies," an expanding colonial medicolegal literature, and local laws to function in variable colonial settings. In doing so, it transformed into a death investigation system that departed from the English model. The scientific practices of colonial medicine developed forensic techniques that preceded and exceeded those in the metropole.

Many of the former and current British colonies—Australia, Canada, India, and the United States—have abolished the previous English coroner system or made major reforms of death investigation legislation. But remnants of the British death investigation system remain in current and former colonies and provide a fitting memorial to the British civilizing attempts and humanitarian efforts to establish a modern system of death investigation and the foundation of English law and democracy.

## NOTES

1. Elizabeth Kolsky, *Colonial Justice in British India* (Cambridge: Cambridge University Press, 2010), 2.

2. Bernard S. Cohen, *Colonialism and Its Forms of Knowledge: The British in India* (Princeton, NJ: Princeton University Press, 1996), 57.

3. Jordanna Bailkin, "The Boot and the Spleen: When Was Murder Possible in British India?," *Comparative Studies in Society and History* 48, no. 2 (2006): 462.

4. Karuna Mantena, *Alibis of Empire: Henry Maine and the Ends of Liberal Imperialism* (Princeton, NJ: Princeton University Press, 2010), 53–54.

5. Kolsky, *Colonial Justice in British India*, 233.

6. See Ishita Pande, "Phulmoni's Body: The Autopsy, the Inquest, and the Humanitarian Narrative on Child Rape in India," *South Asian History and Culture* 4, no. 3 (2013): 9–30; Kolsky, *Colonial Justice in British India*; and Bailkin, "Boot and the Spleen."

7. Mitra Sharafi, "Corruption and Forensic Experts in Colonial India" (presented at the Center for South Asian Studies, University of Michigan, Ann Arbor, February 17, 2017).

8. Kolsky, *Colonial Justice in British India*, 233.

9. D. G. Crawford, *A History of the Indian Medical Service 1600–1913* (London: W. Thacker, 1914), 293–94.

10. Catherine Kelly, *War and the Militarization of the British Army Medicine, 1793–1830* (London: Pickering & Chatto, 2011), 4. According to medical historian Brenda White in her review of Scottish forensic medicine, "One of the many distinctive features of nineteenth-century Scottish medical education was the unique system of teaching forensic medicine and public health together under the heading of medical jurisprudence and medical police." Brenda White, "Training Medical Policemen: Forensic Medicine and Public Health in Nineteenth-Century Scotland," in *Legal Medicine in History*, ed. Michael Clark and Catherine Crawford (Cambridge: Cambridge University Press, 1994), 145. Alfred Crosby recognized the importance of Scottish medical schools in providing colonial physician: "the sun never sets on the empire of the dandelion." See Lisa Rosner, "Thistle on the Delaware: Edinburgh Medical Education and Philadelphia Practice: 1800–1825," *Journal of the History of Allied Science and Medicine* 5, no. 1 (1992): 19–42. For a description of the French system of naval and

colonial medicine, see Michael A. Osborne, *The Emergence of Tropical Medicine in France* (Chicago: University of Chicago Press, 2014), 11–46. In the seventeenth century, the Dutch East India Company required that outgoing ships employ a ship's surgeon and sufficient medical supplies. See Russel Viljoen, "Medicine, Medical Knowledge and Healing at the Cape of Good Hope: Khoikhol, Slaves and Colonists," in *Medicine and Colonialism*, ed. Poonam Bala (London: Pickering & Chatto, 2014), 42.

11. National Research Council, *Strengthening Forensic Science in America: A Path Forward* (Washington, DC: National Academies Press, 2009).

12. Andy Baker, "History, Cognitive Bias, Incompetence and Corruption Are Not the Same Things" (presented at the plenary session of Human Factors in Forensic Science, Annual Meeting of the American Academy of Forensic Science, Orlando, Florida, February 18, 2015).

13. Marc Morris, *The Norman Conquest: The Battle of Hastings and the Fall of Anglo-Saxon England* (New York: Pegasus Books, 2012), 366.

14. R. F. Hunnisett, *The Medieval Coroner* (Cambridge: Cambridge University Press, 1961), 1–9.

15. Hunnisett, *Medieval Coroner*, 9.

16. Jaroslav Nemec, *Highlights in Medicolegal Relations* (Bethesda, MD: US Department of Health Education and Welfare, Public Health Service, National Library of Medicine, 1976), 15. Added duties included, most notably, the holding of the inquests, participating in the witnessing of wounds of felonies (wounding, rape, and housebreaking), and making inquiries concerning treasure trove and wrecks at sea. Hunnisett, *Medieval Coroner*, 10.

17. Catherine Crawford, "The Emergence of English Forensic Medicine," in *Legal Medicine in History*, 187–88.

18. See Edward Unfreville, *Lex Coronatoria; Or, the Office and Duty of the Coroner*, 2 vols. (London, 1761), and James Parker, *Conductor Generalis* (New York: John Patterson, 1788).

19. Helen Brock and Catherine Crawford, "Forensic Medicine in Early Colonial Maryland," in *Legal Medicine in History*, 25–43. The first recorded autopsy in colonial America was performed in 1639, on an "ill disposed boy" who had sustained a skull fracture as the result of a beating by his master. Brock and Crawford note the first autopsy in Maryland as occurring in 1643.

20. W. Keith Kavenaugh, ed., *Foundations of Colonial America: A Documentary History*, 3 vols. (New York: Chelsea House, 1973), 2:1014.

21. The first medical school in America was founded in 1765 by John Morgan, an Edinburgh graduate.

22. For a history of the American system of death investigation, see James C. Mohr, *Doctors and the Law: Medical Jurisprudence in Nineteenth-Century America* (New York: Oxford University Press, 1993); Jeffrey Jentzen, *Death Investigation in America: Coroners, Medical Examiners and the Pursuit of Medical Certainty* (Cambridge, MA: Harvard University Press, 2009).

23. Jörg Fisch, *Cheap Lives and Dear Limbs: The British Transformation of the Bengal Criminal Law, 1769–1817* (Wiesbaden: Franz Steiner Verlag, 1983), 14–15.

24. Kolsky, *Colonial Justice in British India*, 123.

25. Norman Chevers, *A Manual of Medical Jurisprudence for Bengal and the North Western Provinces* (Calcutta: F. Carbery, Bengal Military Orphan Press, 1856), 16.

26. Chevers, *Manual of Medical Jurisprudence*, 18.

27. J. C. Marshman, *The Darogah's Manual* (Serampore, 1850), section XIII, in Kolsky, *Colonial Justice in British India*, 123.

28. Kolsky, *Colonial Justice in British India*, 31.

29. Bailkin, "Boot and the Spleen," 474. Murder in England, however, was treated as a crime against the state and severely punished.

30. Bailkin, "Boot and the Spleen," 43–48. There were far more executions of Europeans in the India before 1860 than there were after the Indian Mutiny of 1857 and the promulgation of India Penal Code in 1861. After 1860, there were only four recoded executions of nonmilitary Europeans in India. See Kolsky, *Colonial Justice in British India*, 120.

31. John Wilson, *The Chaos of Empire: The British Raj and the Conquest of India* (New York: Public Affairs, 2016), 199, 203.

32. Chevers, *Manual of Medical Jurisprudence*, 15.

33. Kolsky, *Colonial Justice in British India*, 127.

34. David Arnold, *Toxic Histories: Poison and Pollution in Modern India* (Cambridge: Cambridge University Press, 2016), 51.

35. Arnold, *Toxic Histories*, 51.

36. Ishita Pande, *Medicine, Race and Liberalism in British Bengal: Symptoms of Empire* (New York: Routledge, 2010), 73–77.

37. Kolsky, *Colonial Justice in British India*, 141. Here I appropriate Kolsky's "truth technologies" to designate a wide range of medicolegal methods that developed in the colonial sphere. In the case of India, Kolsky points to the "particular circumstances of the colony . . . deemed to demand special legislation at departed from contemporary legal rules and practices in England," where colonial legal medicine in the face of unreliable testimony, and colonists "produced a medico-legal machinery that was grounded in a distinctly subjective ethnographic approach. Indian medical jurisprudence and medical experts played a critical function in mitigating Europeans criminal capability in cases of violence and murder by emphasizing the vulnerability of the Indian body and hidden causes of death."

38. Kolsky, *Colonial Justice in British India*, 124.

39. See Sharafi, "Corruption of Forensic Experts in Colonial India," 18.

40. Kolsky, *Colonial Justice in British India*, 110, 120. Refer also to Binyamin Blum, "Forensic Technology in the Age of Empire" (presented at the Law and Society Association, Minneapolis, Minnesota, 2014).

41. Bailkin, "Boot and the Spleen," 477.

42. Kolsky, *Colonial Justice in British India*, 110.

43. Bailkin, "Boot and the Spleen," 475–76. See also Martin J. Wiener, *An Empire on Trial: Race, Murder, and Justice under British Rule, 1870–1935* (Cambridge: Cambridge University Press, 2009), 157.

44. Wiener, *Empire on Trial*, 164. In 1893, a soldier had given a laborer who had fallen asleep at his job several kicks, which proved to be fatal. But "the medical evidence showed that the cause of death was rupture of the spleen, which was in such a state that the slightest blow might have broken it; and there were no external marks of violence."

45. The eggshell skull rule, or what's known as "you take your victim as you find them," is based on English common law and basically represents that the frailty of the injured person is not a defense in a tort case. See 701 F.2d 1217 in William L. Prosser, *Handbook of the Law of Torts*, 4th ed. (St. Paul, MN: West, 1971), 261.

46. Kolsky, *Colonial Justice in British India*, 140. In 1902, a civil surgeon in Hooghy studied three hundred cases of death from ruptured spleen and dismissed the claim that the pathology of the spleen was a convenient judicial alibi for murder offered by police surgeons. By contrast, the Indian frustration with the "ruptured spleen defense" cast suspicion on the work of police surgeons: "It often happens that whenever a Native is killed by a European, the Civil Surgeons, after examining the remains, are of the opinion that the deceased was in such a weak state of health that he might have died at any time without violence. We are always suspicious when this is the result of an examination."

47. Quoted from Russel Vilojoen, "Medicine and Medical Knowledge and Healing at the Cape of Good Hope: Khoikhol, Slaves and Colonists," in *Medicine and Colonialism*, 54. The

concept of race, presented as biological fact, became one of the governing ideas of the high imperial era, and sustained attempts were made in the name of science to give race an anatomical, even mathematical, precision.

48. Alfred Swaine Taylor, *The Principles and Practice of Medical Jurisprudence* (London: Churchill & Sons, 1865), 2. See also Noel. G. Coley, "Alfred Swaine Taylor (1806–1880): Forensic Toxicologist," *Medical History* 35, no. 4 (1991): 409–27.

49. Chevers, *Manual of Medical Jurisprudence*, 2. See also Isadore B. Lyons, *Medical Jurisprudence for India with Illustrative Cases*, 3rd ed. (Calcutta: Thacker, Spink, 1904), 143.

50. Kolsky, *Colonial Justice in British India*, 130.

51. Chevers, *Manual of Medical Jurisprudence*.

52. Kolsky, *Colonial Justice in British India*, 129.

53. Ishita Pande, "Phulmoni's Body," 10.

54. Pande, "Phulmoni's Body," 1. Lyons summarized the case in the third edition of *Medical Jurisprudence for India*, the most popular medicolegal publication in India at the time. "The court held that when a girl is a wife and above the age of consent although it therefore not rape, still the husband has not he absolute right to enjoy the person of his wife without regard for her safety. Found that the prisoner caused the death of the girls by rash and malignant act." The court transcript and detailed autopsy report were widely circulated after the trial by reformers seeking to end the centuries-old practice of the abuse of the "child-wife" as a part of the Hindu culture. "The age of consent refers to the age at which the law recognized an individual's eligibility to consent to sexual intercourse, and was set at 10 years by the Indian Penal Code (1860)" (23).

55. Pande, "Phulmoni's Body," 11. The medical evidence produced in the *Queen Empress v. Hurry Mythee* so incensed the judicial members of the Legislative Council that in 1891, the Age of Consent Act was enacted, raising the age of consent from ten to twelve for a Hindu child-wife.

56. Pande, "Phulmoni's Body," 14.

57. Pande, "Phulmoni's Body," 490. Bailkin goes so far as to suggest that the reform on interracial violence was a result of a change in the British public perception of killing in light of the horrific English losses during World War I and a reaction to the atrocities of other colonial regimes such as Germany and the Netherlands in the wars southwest Africa as well as their own.

58. Sydney Smith, *Mostly Murder* (New York: Dorset Press, 1988), 54. The system was more like that of Scotland, with an office of the procurator-fiscal. Smith delighted to present cases in which he and other forensic experts testified in support of the accused native peoples and against prosecution witnesses, all for the cause of impartiality of forensic science.

59. The Egyptians were eager to end the capitulations system that allowed the European consulates in Egypt to have their own courts and to abuse the extraterritorial rights granted to them by the capitulations. Mixed courts were independent of the Khedive and British occupation authorities. The creation of mixed courts in Egypt was generally viewed as a watershed, essentially establishing the rule of law. According to Nadav Safran, historians credit the function of mixed courts with a leading role in the rising Nationalist movement, that the mixed courts "instructed Egyptians in the constitutional foundations of Western law . . . and civil liberties which became the basis of the nationalist movement." See Khaled Fahmy, "The Anatomy of Justice: Forensic Medicine and Criminal law in Nineteenth-Century Egypt," *Islamic Law Society* 6, no. 2 (1999): 230.

60. Quentin Pearson, "From 'Inauspicious' to 'Suspicious' Death: Inquest in Turn of the Twentieth-Century Bangkok" (presented at the History of Science Society Meeting, Philadelphia, July 12, 2012). Pearson argues that colonies were obliged to hire Western-trained physicians to perform autopsies in order to liberalize and end extraterritorial obligations.

61. Fahmy, "Anatomy of Justice," 224–71.

62. Fisch, *Cheap Lives and Dear Limbs*, 20.

63. Fahmy, "Anatomy of Justice," 226.

64. Fahmy, "Anatomy of Justice," 237.

65. Fahmy, "Anatomy of Justice," 226.

66. Fahmy, "Anatomy of Justice," 271.

67. Smith, *Mostly Murder*, 63.

68. Smith, *Mostly Murder*, 74.

69. Smith, *Mostly Murder*, 77.

70. Smith, *Mostly Murder*, 56.

71. Russel Vilojoen, "Medicine and Medical Knowledge and Healing at the Cape of Good Hope: Khoikhol, Slaves and Colonists," in *Medicine and Colonialism*, 41–59.

72. Vilojoen, "Medicine and Medical Knowledge," 57.

73. Vilojoen, "Medicine and Medical Knowledge," 57.

74. Ian Gordon, R. Turner, and Tom William Price, *Medical Jurisprudence* (Edinburgh: E & S Livingstone, 1952), 239.

75. Gordon et al., *Medical Jurisprudence*, 3–4.

76. Gordon et al., *Medical Jurisprudence*, 243.

77. Gordon et al., *Medical Jurisprudence*, 242.

78. Gordon et al., *Medical Jurisprudence*, 231. All criminal proceeding are carried out under the Criminal Procedures and Evidence Act No. 31 of 1917. For any person who appears to have died from violence, criminal neglect, or otherwise non-natural causes, report of the death must be made to the magistrate of the district.

79. Mahomed Dada and Joanne Clarke, "Courting Disaster? A Survey of the Autopsy Service Provided by District Surgeons in Kwazulu-Natal," *Med Law* 19 (2000): 763–77. See also Gert Saayman, "Forensic Medicine in South Africa—Time for Change?," *Med Law* 13 (1994): 129–32.

80. In the 1987 film *Cry Freedom*, produced by Richard Attenborough, South African journalist David Woods is forced to leave the country after his attempt to investigate the death of his friend and African activist Steve Biko, who was a major organizer for the Soweto Uprising in 1976. The film was adapted from Wood's book of the same title.

81. Hillel David Braude, "Colonialism, Biko and AIDS: Reflections on the Principle of Beneficence in South African Medical Ethics," *Social Science Medicine* 68, no. 11 (2009): 2057–58.

82. Derrick Silove, "Doctors and the State: Lessons From the Biko Case," *Social Science and Medicine* 30, no. 4 (1990): 428.

83. Following Steve Biko's trial, the police continued to work invisibly to investigate the killing, kidnapping, and torture of anti-apartheid activists with counterinsurgency units. Antijie Krog, "The Transformation of an Evil Man," *New York Times*, March 15, 2015, 7.

# CHAPTER SIX

# Fingerprints and the Politics of Scientific Policing in Early Twentieth-Century Spain

JOSÉ RAMÓN BERTOMEU-SÁNCHEZ

The introduction of fingerprinting has been described as a major development in the new methods of identification in the twentieth century. These new methods were usually portrayed as the answer to the challenges posed by nineteenth-century migrations, the new "society of strangers," and the advent of novel forms of crime and violence. According to these master narratives, the changes rendered inefficient the traditional methods of identification based on interpersonal recognizance and encouraged the arrival of new technologies based on medical and scientific knowledge. This picture has been largely revised during the last decades, thanks to historical studies on the technologies of identification from the late Middle Ages to the present. These studies have challenged the previous narrative while introducing new scenarios, protagonists, and problems. Many changes in practices of identification have been tracked from the early modern period in different contexts, not only concerning the development of state bureaucracy but also in other aspects of social and economic life (e.g., bank transactions, commercial contracts, or electoral registration processes). Even in the nineteenth century, most of the novelties did not appear as the result of dramatic scientific breakthroughs, but rather as a consequence of the refinement and enlargement of tedious practices of registration and classification made by clerks, police officers, or prison employees. Historians have recently become interested in the changing uses of the new technologies, either as a tool of control or as a way of granting citizenship rights. The focus has moved from "identifiers" to "identified people," and their reaction to the new technologies, ranging from enthusiastic or passive acceptance to riots and resistances and new unexpected forms of identity forgery.[1]

Many of these topics have been studied by following the development of fingerprinting as an identification technology since the second half of the nineteenth century. In traditional narratives, the main protagonists are authors such as Henry Faulds (1843–1930) or William James Herschel (1833–1917). These pioneers emphasized the main features of fingerprints as identification technology: uniqueness and invariability. The method soon attracted the interest of many European scholars, including Francis Galton. He investigated many other features of fingerprints—for instance, the relationship between ridge patterns and race or inheritance—and offered the first attempt at classification. By the end of the nineteenth century, fingerprinting started to compete with other forms of identification (as Bertillon anthropometry) in the control of recidivism in prisons while making its first steps in criminal investigations. These developments took place at the turn of the twentieth century, particularly in Argentina and England, where two important police departments emerged thanks to the activities of Juan Vucetich (1858–1925) and Edward Henry (1850–1931), respectively.

Geographically, studies have tended to focus on Argentina, England, and India, the places where the new method was allegedly "discovered." Apart from a detailed review of the twentieth-century developments in the United States, studies on the uses of the new technology in other countries are less frequent but include interesting case studies on France, Italy, or South Africa.[2]

This chapter introduces a new scenario for discussing a new range of problems. Fingerprinting practices emerged in early twentieth-century Spain from the confluence of interests of a group of physicians, police officers, and politicians. The protagonists played different but sometimes overlapping roles in this process. On the one hand, university professors of medicine such as Federico Olóriz Aguilera (physical anthropology) and Antonio Lecha-Marzo (forensic medicine) pursued new research on anthropometry, providing new technologies of identification. On the other hand, the police and prison employees introduced new forms of organization while refining their practices of registration, classification, and identification. A new generation of politicians implemented regulations using the new technologies for a broad range of purposes, from the identification of recidivists to the prosecution of crime and the control of political dissidence. In this chapter, I pay attention to the interactions, tensions, and exchanges in the making of Spanish fingerprinting technology.

To begin, two main historical features have to be borne in mind. First is the political turmoil, cultural anxiety, and social malaise after 1898, the critical year in which Spain lost its last overseas colonies. This period was also characterized by new forms of violence that prompted the reform of the Spanish

Police. The second feature is the encounter of a new generation of Spanish scientists and physicians with the "regenerationist" movement, a loose group of politicians, journalists, engineers, and intellectuals concerned with the "decadence" of Spain and the means to "regenerate" the nation. They frequently adopted the rhetoric of modernity and progress by science, which resonated with the interests of an emerging generation of Spanish scientists and physicians (the so-called silver age of Spanish science). The confluence of these features created a propitious scenario for exchanges between different academic and professional groups interested in fingerprinting: physicians, politicians, lawyers, prison officers, and police. In this chapter, I focus on the hybrid spaces created by these academic and professional groups. This hybrid area, placed at the crossroads of medicine and policing, took shape around 1910 by means of courses, meetings, networks, trials, textbooks, and journals. First, I describe the methods of identification in Spanish prisons during the late nineteenth century, including a brief summary of the rise and fall of anthropometry. I then summarize the research developed by Federico Olóriz Aguilera, starting from his early work on medical anthropology to his first activities at the head of the identification services in Madrid. I follow his path from his first works on fingerprinting (aimed at the control of recidivism in prisons) to his later proposals of national identification cards (intended for the whole Spanish population). The next part of the chapter is focused on the activities developed by police departments in Madrid and Barcelona and the early work of the young forensic physician Antonio Lecha-Marzo. His role was crucial in transforming Olóriz's work into "Spanish dactyloscopy." Finally, I discuss the confluence of these different groups in an editorial project: the journal *La Policía Científica* (*The Scientific Police*), which was published by the lawyer Gerardo Doval between 1913 and 1914.

## Anthropometry, Recidivism, and Prisons

As with many other countries, challenges to traditional practices of identification emerged in nineteenth-century Spain. While interpersonal recognizance remained important in most aspects of social life, new practices based on official papers and archives were developed along with the enlargement of the state bureaucracy. Different categories of police were established for confronting new forms of delinquency in both rural and urban areas. Additional problems were the control of recidivism in prisons and the management of a growing population of prisoners. New methods of identification were introduced, including those related to photography and anthropometry at the end of the century. One of the most successful methods was introduced by Alphonse Bertillon (1853–1914), combining bureaucratic practices of registration, classification, and archiving with quantitative measurements, disciplined photography, and a sophisticated vocabulary for describing body singularities (*portrait parlé*). Adopting different versions of these techniques, new anthropometric offices were established in many European countries (England, France, Italy, etc.) and in the Americas (Argentina, Brazil, Peru, the United States, etc.) by the final decades of the nineteenth century. The International Conferences on Terrorism, first held in Rome in 1898, encouraged this process.[3]

In Spain, after the turbulent years of the First Republic, the 1880s was the decade of the final codification of the criminal law, including new and more lasting regulations concerning criminal trials, the role of experts, and methods of identification. According to the law of September 14, 1882, personal identity had to be established by close acquaintances, relatives, and, in cases where this was impossible, by any other means. The need for "other means" emerged first from the problem of recidivism and then control of the increasing number of prisoners held for political activities and terrorist attacks. The first anthropometric and photographic offices were established in the prison of Barcelona in 1895. It was one of the outcomes that followed the "terrorist wave" of these years and the ensuing criticism of the Barcelona Police's perceived incompetence. The following year (on September 10, 1896), a law was passed to create a new Spanish office of "anthropometric identification" "based on the Bertillon system" under the direction of a "forensic doctor." The head of the office was the Spanish doctor Enrique Simancas Larsé. Until then, he had only published books on hydrotherapy and other practical medical topics, but he quickly learned Bertillon anthropometry, started the organization of cabinets and registers of identification, and trained a large group of employees in the new techniques.[4]

Anthropometric offices were established in many Spanish prisons during the 1890s.[5] But many important cities, such as Cádiz, Málaga, Sevilla, Valladolid, and València, with crowded prisons lacked such an office at the end of the nineteenth century. Most of the budget was paid by the local councils, which were reluctant to spend large quantities of money on instruments and employees. Fernando Cadalso Manzano (1859–1939), the director of one the main prisons in Madrid, also argued that anthropometric measurements should be applied to recidivists, sentenced prisoners, and remand prisoners in pretrial detention. It was against the Spanish law, Cadalso claimed, to submit

"murders, thieves" and people who enjoyed "all their rights" to the same practices. He also spoke of the main problem of anthropometric measurements:
"If anthropometric identification is performed with exactitude, it becomes
very expensive . . . If it lacks exactitude, the service becomes useless."[6]

As in other countries, critics of anthropometry like Cadalso remarked on
the problems of replication, the difficulties in training officers, and the com-
plications for applying this technology to children and women. A minor
mistake, or just some lack of precision caused by inexperience, rendered the
obtained data "completely useless, making the retrieval of the record very difficult."[7] The forensic doctor Antonio Navarro Fernández also remarked that
anthropometric measurements were embarrassing when applied to women
("impossible from a moral point of view") and useless for the growing group
of young delinquents (under the age of twenty-one) whose bodies were still
changing.[8] The new regulations published on February 18, 1901, included a
critical review of other problems related to the anthropometry offices: lack of
skilled people, limited budgets, and absence of instruments. In order to deal
with these problems, a new school of anthropometry was created in Madrid
during the first years of the twentieth century. Two courses were held per year,
in which fifteen to twenty students were trained in anthropometric techniques.[9]
Anthropometry faced another unexpected problem in early twentieth-
century Spain: its main supporter, Enrique Simancas, suddenly passed away
in 1900 and was replaced by Federico Olóriz Aguilera at the head of the Identification Office.

## Federico Olóriz and the Making of the Spanish Fingerprints

Federico Olóriz Aguilera (1855–1912) studied medicine at the University of
Granada between 1870 and 1875. He was appointed chair of anatomy at the
University of Madrid in 1883. His first publications were on anatomy, but he
soon specialized in the emerging domain of medical anthropology. His first
substantial work was a large anthropometrical survey of the Spanish population. He collected many skulls from Madrid hospitals, which became part
of a new Museum of Anthropology at the Faculty of Medicine in 1885. His
purpose was "to collect and study all the items which could be employed in
knowing the material part of the Spanish people, aiming to creating a fragment of national science which could be named the physical anthropology of
Spain." In 1889, the collection included a large archive with more than fifteen
thousand anthropometric measurements. Olóriz obtained about two-thirds
of the data, but the rest came from "public institutions such as the Madrid

Anthropometric Cabinet for Identification and several military schools, as well as private institutions such as insurance companies, schools and others." Olóriz faced for the first time the problem of handling a large amount of anthropometric data. He conceived a card index and forms of classification that allowed him to perform "arithmetical analysis and comparative and statistical analysis."[10]

With this anthropometric data and the collection of skulls, Olóriz started to take craniometrical measurements, aiming to obtain the "cephalic index" of Spanish population (the ratio of the maximum width of the head divided by its maximum length). This ratio was connected to contemporary attempts at mapping the European races. Olóriz soon realized that his project faced many technical problems. Measurements were affected by problems of replicability and inexactitude, particularly when several observers were involved and data were shared. Moreover, his collection was far from being a representative sample of the Spanish population from the point of view of both geography (most of them were from Madrid) and social origins (many of them came from hospitals). Olóriz acknowledged that it did not represent the "whole Spanish people but the most disadvantaged classes, maybe the most degenerated ones."[11] To obtain a more representative sample, Olóriz visited many military establishments and prisons during 1892 and performed many head measurements. When possible, he also included data obtained from schools or provided by learned societies whose members were willing to collaborate in his research. After solving problems of classification, he presented his first results at the Hispano-Portuguese Conference on Geography, held in Madrid in October 1892. Two years later, he published his study on the Spanish cephalic index based on more than eight thousand measurements. He confirmed salient differences according to different Spanish regions. His study was positively received at international meetings and in publications, and he even obtained the prestigious Godard Prize from the Paris Society of Anthropology.[12]

In these first anthropometric studies, Olóriz learned how to deal with problems related to his work on identification: gathering anthropometric data coming from hospitals, prisons, and military establishments; replicating and ensuring exactitude in the measurements; and classifying, retrieving, and analyzing a large amount of information. Even so, during the late nineteenth century, Olóriz managed to establish a sort of Latourian "center of calculation," gathering and processing a growing influx of anthropometric data. His position was reinforced when he was named "inspector of the service of identification" in 1891, and after the death of Simancas, he was appointed professor and head of the new service of identification in Madrid.[13] At this point, he got privileged access to the central archive of prison records, so he could request any anthropometric data available from Spanish prisons. At the same time, and in contrast with Simancas, Olóriz kept the academic support provided by his chair as professor of anatomy at the University of Madrid. This bridge between university and prisons was a key issue in Olóriz's further research and work on fingerprinting, a topic he became interested in around 1902. His first paper was presented at the International Medical Meeting held in Madrid in April 1903. The topic was the identification of young criminals, a persistent problem for anthropometry. Assuming that papillary ridges were invariable, and relying on his previous craniometrical studies on children, students, and young criminals, Olóriz concluded that fingerprinting associated with the cephalic index was the "most reliable method for identifying young people."[14]

During the next several years, Olóriz's work changed dramatically after discovering the burgeoning international literature on fingerprinting, particularly the work of Juan Vucetich, the Argentinian police inspector who developed one of the most influential methods for organizing fingerprint archives.[15] Olóriz introduced minor changes to Vucetich's classification. He renamed the groups by using medical-like Greek terminology and a single criterion: the number and position of the "deltas," one of the most important features of papillary ridges.[16] He then focused on refining subclassifications and extending the uses of fingerprinting technology. Both issues were closely connected. Groups of fingerprints were not evenly distributed in the population: while some formulas could be applied to a large number of individuals, other formulas were rare and hardly found in everyday practice. If fingerprinting were to be applied to large groups of the population, as Olóriz projected, the archive would include a large number of cards for certain groups, which would be difficult to handle. He thought the issue could be solved by gathering statistical data about the unequal distribution of the patterns. He focused on these problems and progressively abandoned his previous attempts of using the cephalic index as complementary data for identification. In fact, apart from his first studies around 1903, he published no studies connecting fingerprinting and race or inheritance.[17]

Rather than serving as a source of data, statistical inferences, or refined classifications, Olóriz's former studies on anthropometry were a research model for his work on fingerprinting. Like in previous years, he employed the network provided by his position as head of the service of identification in prisons and member of the Consejo Penitenciario (Prisons Council). For instance, in early 1907, Olóriz asked his fellow members of the Prisons Council to start a gradual change from anthropometry to fingerprinting as the main identification method for prisoners. During the next several months, he sent detailed instructions to the directors of prisons, in which he ordered the fingerprinting of all the prisoners. The data were sent to a central register, then transformed into formulas, and classified by Olóriz and his reduced team of skilled employees. With this division of labor, he managed to collect more than ten thousand cards in just a few months. He detected the most frequent formulas (making the largest groups of prisoners) and conceived means of subclassification, so that the large groups could be split into smaller ones that were easier to handle. He employed different strategies for this purpose, from ridge counting to topographic divisions of the fingerprint.[18]

Over the next several years, Olóriz organized the transition from anthropometry to dactyloscopy in the identification of prisoners.[19] In his writings addressed to politicians and decision makers, Olóriz offered arguments for

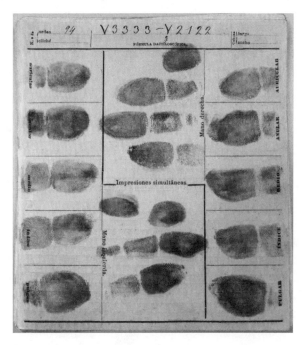

Fingerprints collected by Olóriz from Burgos Prison, June 1907. His new classification is written in the top part of the card (V3333–V2122). José Ramón Bertomeu-Sánchez

this change, pointing to the problems of anthropometry as method of iden-tification (technical difficulties, limitation to adult males, degrading nature, and high costs) and praising the advantages of fingerprinting (reliable, gen-eral, simple, cheap, and mildly annoying for the identified person).[20] By the end of the 1900s, the Spanish service of identification was a mixture of an-thropometry and dactyloscopy. The center of the network was the Central Register in Madrid, with the archive of criminal records and a school in which students were trained in anthropometry, fingerprinting, and photog-raphy. Some of the graduates were appointed to one of the thirty-eight pro-vincial cabinets of identification in the main Spanish cities.[21]

Relying on this practical experience, Olóriz imagined further uses of fin-gerprinting, including not only the identification of prisoners and recidivists in prisons, but also forensic investigations (the detection of latent fingerprints in crime scenes) and the confirmation of the authenticity of documents in banking and other economic activities (replacing manuscript signatures). He also suggested a new National Register of Identity, in which fingerprinting would become the main identification technology in social life, from recently born children to anonymous corpses.[22] To make these new uses possible, Olóriz worked on two related issues: new monodactylar classifications and the detailed descriptions of single fingerprints. The formula of ten fingers was efficient for the identification of recidivists, when all the fingerprints of the prisoner could be compared with the data available in the archive. But it was useless in criminal investigations when one "latent" fingerprint (sometimes just a fragment) was found at the crime scene. The fingerprint had to be identi-fied by its minor details ("minutiae") and compared with cards in a monodac-tylar archive. That was a real challenge for Olóriz. He developed a language for the detailed description of fingerprints, what he named the "speaking portrait" (*retrato hablado*) of the fingerprint in reference to Bertillon's *portrait parlé* of criminals.[23]

Olóriz performed many "experiments" on monodactylar identification with the help of his students and in front of political and academic authori-ties. These experiments served to endow fingerprinting technology with the epistemological virtues of the natural sciences, while creating an influential network of direct witnesses who could confirm the credibility of Olóriz's claims in crucial decision-making spaces. In June 1909, Olóriz presented his results to the Spanish minister, who was delighted to see how several police-men, carefully trained by Olóriz, could solve several problems of identifica-tion by retrieving the due information from an archive organized by means of the formula and the classifications suggested by Olóriz. The session was fully reported in Madrid newspapers.[24] Another public demonstration was organized in April 1910 with the help of around one hundred students from the Madrid Faculty of Law. In front of the dean and other professors, Olóriz obtained an impressive 80 percent of successful identifications. The remaining cases were always false negatives, so Olóriz could argue that fingerprinting "might let some criminals go unpunished in certain cases, but it never leads to the conviction of innocent people." He pointed out that the method could be employed by the new scientific police, while providing the basis of a new National Register of Identification.[25]

Like Vucetich in Argentina, Olóriz was pioneer in suggesting a national register of identification based on fingerprinting of the entire population. He announced his project in a meeting of the Spanish Society for the Advancement of Science in October 1908: he regarded fingerprinting as the best technology for "granting the identity of citizens in every act of social life."[26] Two years later, in June 1911, Olóriz presented his work to both political authorities and to scientists who gathered in Granada at the annual meeting of the Spanish Society for the Advancement of Science. He convinced the meeting participants to submit to the government his proposal for a "National Register of Identity," in which all citizens would be included.[27]

While expanding its uses, Olóriz was keen to present fingerprinting as proof of the social benefits of Spanish contributions to science. In doing so, he adopted a rhetoric that was common among the Spanish "regeneracionists," particularly those who regarded science as the crucial tool for modernizing the Spanish society in early twentieth century. At the Madrid Academy of Medicine, which seated many members of this group, Olóriz described in 1911 the new "scientific identification" methods as a sort of lubricant for smoothing the social tensions, thus fixing the "great social machine" and preventing its quick wear and tear.[28] Presenting fingerprinting as a form of science, he argued that resistance could only arise from ignorance and "prejudices against scientific identification." Because it had been hitherto applied just to "thieves and killers," many people regarded being summited to fingerprinting as a source of social stigma (*nota de infamia*) and "incompatible with the honorability of the citizens."[29]

To address resistance to his projects, Olóriz developed a large and varied campaign in different spaces and media around 1910. He kept lecturing at the School of Police in Madrid, so training a group of devoted disciples in fingerprinting. He also organized further public "experiments" of identification, whose purpose was less to obtain additional empirical data than to provide convincing proofs to his audience made up of lawyers, police officers, prison employees, and politicians.[30] In addition, he wrote detailed reports in both medical and professional journals and presented them to academic societies, scientific congresses, and professional meetings for prison officers. He also performed self-experiments to determine whether papillary ridges could be purposely modified by criminals, thus making difficult or impossible the recognition of fingerprints in crime scenes.[31] One of his last and most popular publications, which was translated into French and enjoyed broad circulation, was a portable register of common delinquents, in which he showed that police agents could easily recognize the criminals in the street by combining the methods of anthropometry and fingerprinting. He took care to send his publications to his colleagues in Europe and America, while keeping a frequent exchange of letters concerning technical developments and organizational problems.[32]

## Disciplining Spanish Police

With his lectures, talks, papers, public experiments, and international networks, Olóriz aimed to obtain both the recognition of his colleagues and medical fellows and the support of politicians and police. He connected his work with the attempts to modernize the Spanish Police at the beginning of twentieth century. It was at this time that the conservative regime dominating Spanish political life entered its last years, marked by delegitimation, corruption, and social violence. The crisis of 1898, in which Spain lost the last colonies in Cuba and Filipinas, is usually regarded as a turning point in this long-term process, which was also caused by the economic crisis, the poor conditions of the working classes, and the increasing power of trade unions as well as other socialist and anarchist groups. Many terrorist attacks were carried out against prominent political figures, including several first ministers and the king. Riots and strikes were frequent as well, as ruthless repression against activists by both the police and armed groups leaded by Spanish employers. By the end of July 1909, a general strike turned into a widespread insurrection followed by repression of political dissidence and trade unions.

The press criticized the police for their inefficiency in dealing with these revolutionary episodes and the increasing crime in cities. New ways of organizing the police forces were discussed, particularly the unification of different police units, from the militarized "Guardia Civil" (predominant in rural areas) to the local and provincial forces of security. The existence of these different units acting in urban and rural spaces, along with the excessive militarization of some units and the chronic low budgets for material and human resources, was regarded as a major problem for making a unified model of police. Many authors demanded a more rational and specialized police in tune with the new times and the developments of other parts of the administration. For one famous Catalan politician, Enric Prat de la Riba, the Spanish Police was "an apparatus of a primitive type, an unusable fossil." To deal with modern problems with such old-fashioned police was like "using flint spears and stone hatches to fight against multitudes armed with Mausers and Krupps." He regarded this terrible situation as further evidence about the "impotence" of the Spanish state for dealing with the new problems resulting from "intense civilization."[33]

This sense of decadence and urge for reform, combined with the challenges created by social violence and terrorism, paved the way for Olóriz's proposals, which echoed crucial ideas of the regeneracionist movement. The new "scientific identification" (i.e., fingerprinting technology) was conceived as one of main instruments for modernizing the Spanish Police. In supporting Olóriz's ideas, one of the crucial protagonists was the minister Juan de la Cierva y Peñafiel (1864–1938), who undertook a major program of police reform between 1907 and 1909. He introduced new regulations against nepotism in appointments, a new meritocratic system of selection based on experience and regulated examinations, centers for professional training (schools of police in Madrid and Barcelona), and further material resources for the bureaus of information and identification of criminals, in which fingerprinting played an increasing role. When Juan de la Cierva was dismissed in October 1909 (after the so-called Tragic Week in Barcelona), Olóriz angrily wrote to Vucetich informing him that the change of government had ruined his project of a general register of identification "covering all the aspects of social life, from recidivism to official documents, as well as recruitment [in the army], emigration, passports, banking transactions, etc."[34]

In spite of these frustrating results, the years between 1907 and 1909, when Juan de la Cierva was minister, were decisive in Olóriz's projects on fingerprinting. He could create a group of disciples while nullifying the attempts to implement alternative methods of identification. The crucial spaces for disciplining fingerprinting technology in Spain were the police schools created in Madrid and Barcelona.[35] When lecturing at the School of Police in Madrid, he organized practical activities intended to both transmit the required know-how and convince his students on the reliability and efficiency of fingerprinting, so making them publicists of the new scientific methods. In some sessions, students were requested to solve "practical problems," for instance, the identification of a particular person from a group of two hundred agents (either using the identification cards or the portable register conceived by Olóriz). These problems reproduced in detail the activities of vigilance and detection realized by the police: the cards provided anthropometric information (*portrait parlé*) for the visual identification of suspects, and their identities had to be finally confirmed by fingerprinting. Olóriz devised several variations of these practical activities but always with similar consequences: students could confirm how Olóriz's methods provided positive results in an idealized situation, which was portrayed as identical to regular criminal investigations.[36]

The courses yielded a group of disciples who played a crucial role in the development of fingerprinting after Olóriz's sudden death in 1912.[37] This group included policemen such as José Pastor Ferrer, Victoriano Mora Ruiz, and Jesús Lasuén Urrea; prison employees such as Vicente Rodríguez Ferrer; and lawyers such as José Jiménez Jerez, who became a police inspector. Being a well-known university professor, Olóriz also promoted fingerprinting at the Faculty of Medicine of Madrid. Some of his students wrote medical theses on this topic, for instance, Ramón Lobo Goya, who was a doctor at one of the Madrid prisons and became director of the anthropometric and identification service in this same institution.[38] After Olóriz's death, many of these students published manuals on fingerprinting or papers supporting this technology in different uses. They also actively participated in the publication of the journal *La Policía Científica*, to be discussed in the last section of this chapter.[39]

Another school of police in which fingerprinting was taught was established in Barcelona in 1907, during the years of increasing social unrest and political violence that culminated in the Tragic Week during the summer of 1909.[40] The director of the Barcelona School of Police was Francesc Molins, who had spent several months at Scotland Yard learning fingerprinting techniques under the supervision of Edward Henry. Many other police officers spent several months in England thanks to the support of the Spanish government. Moreover, Charles Arrow, a recently retired metropolitan detective chief inspector, was appointed the head of a new Criminal Investigation Department in Barcelona against anarchist terrorism. A police force of around thirty people was put at his disposal, and he started to compile registers of suspected anarchists whose movements where noted in the Scotland Yard system.[41]

Being trained in London, the new Catalan detectives, supported by Arrow, introduced Henry's classification of fingerprint patterns in the Barcelona School. When the news reached Madrid, Olóriz was alarmed and redoubled his effort for having his own classification accepted. Having two different systems, he insisted, would create errors and inefficiency, so just one of them could be adopted. In March 1909, at the headquarters of the Madrid Registration Office, he organized a public comparative analysis of the two methods. He invited Francesc Molins and other police officers from Barcelona, and together they performed several identification essays with registration cards. Finally, they wrote a brief report that was sent to the Spanish government and later published by Olóriz. Unsurprisingly, the main conclusion was that "the dactyloscopic system to be applied in Spain should be the so-called Vucetich or Argentinian system with the modifications already introduced" in the Madrid Office (i.e., those suggested by Olóriz).[42] As in this text, Olóriz employed still in 1909 in both public and private writings the expressions "Argentinian" or "Vucetich" systems for referring to the methods employed in Madrid, even if highlighting the "modifications" introduced by himself.[43] The transformation of these minor changes into an original "Spanish dactyloscopy" was mostly the work of Olóriz's disciples, particularly of a young forensic doctor: Antonio Lecha-Marzo (1888–1919).

## Fingerprints, National Pride, and Forensics

Even if also trained as physician, Antonio Lecha-Marzo's path to fingerprinting was quite different than Olóriz's. In contrast with the anthropometric studies of Olóriz, Lecha-Marzo's first publications were on legal medicine, particularly on the detection of blood and seminal stains, under the supervision of his uncle, a professor of legal medicine at the University of Valladolid. His further work is a good example of how legal medicine evolved in the early twentieth century thanks to several entangled issues: the shifting focus from the body to the crime scene, the new role of the technologies of trace analysis, and, even more important, the creative, even if sometimes contentious, encounters with the emergent culture the scientific police that developed in identification offices and police departments.[44]

Adopting this approach, Lecha-Marzo reaffirmed the change of focus in the late work of Olóriz, moving fingerprinting technology from prisons to crime scenes. Lecha-Marzo became interested in fingerprints by the mid-1900s. He was part of a new generation of young scientists that enjoyed the favorable atmosphere for science in early twentieth-century Spain. The situation was reinforced by the organization of new scientific institutions, the building of laboratories, and more investment in training and permanent positions for scientists.[45] Scientific travels were encouraged by the Junta de Ampliación de Estudios (JAE, the Board for Advanced Studies and Scientific Research). This support was crucial for promising students such as Lecha-Marzo, who established fruitful connections with important European scholars. In Liège (Belgium), Lecha-Marzo met Eugène Stockis (1875–1939), who lectured on legal medicine at the university, and whose research was focus on fingerprinting and palm ridges. He also worked with Henri Welsch, a young member of the laboratory of legal medicine in Liège, and together they published a French textbook on dactyloscopy, with a focus on recent advances in photography, coloration, and transport of latent fingerprints. When back in Spain, Lecha-Marzo finished his medical thesis on palm ridges that he presented in October 1912, just some months after the death of Olóriz.[46]

Antonio Lecha-Marzo was also in touch with Italian scholars such as Salvatore Ottolenghi, professor of legal medicine at the University of Roma, whose works he translated into Spanish. Like Ottolenghi, Lecha-Marzo regarded the new methods for detecting different forms of trace evidence as the basis for a new "scientific judicial police," in which "the characters imagined by the mind of [Émile] Gaboriau and [Arthur] Conan-Doyle will be made real." According to Lecha-Marzo, scientific police consisted in "the application of scientific knowledge to criminal investigations" with the double aim of detecting the traces of crime (by the methods provided by legal medicine) and preventing the action of criminals (thanks to the new criminal anthropology).[47] Lecha-Marzo wrote some texts on this last issue (including a book on "the criminal brain") in line with early twentieth-century degeneration theory. One of his papers was on Mateo Morral, the famous anarchist who organized the bomb attack during the wedding of the Spanish Royal Family in Madrid. Lecha-Marzo reviewed the data from the autopsy and recognized some characteristic features of the "Lombroso criminal type." And yet he strongly opposed the associated idea of irresponsibility in terrorists and murderers. In more general terms, he rejected the common criminological view of connecting subversion, crime, and madness. Accordingly, he opposed sentence reductions for criminals affected by mental troubles. The purpose of the trials was to establish whether the defendant was dangerous for society and the measures to be taken to prevent further crimes.[48]

In spite of these works, Lecha-Marzo's main research was on the forensic detection of trace evidence. With the help of his uncle, and during his years as a medical student, he developed innovative experimental techniques on blood and seminal stains, which were followed by an increasing interest in the new methods of identification (fingerprinting), which became the focus of his research in his last years. In these publications, Lecha-Marzo always highlighted the Spanish contributions to fingerprinting. In 1910, he published a comparison between the systems of Vucetich and Olóriz. Lecha-Marzo not only judged both systems equal, but also went further in drawing attention to the unique and valuable Spanish contributions: a more accurate and affordable nomenclature, subdivisions of the most repeated formulas (in order to organize the registers), the monodactylar classification, and the *portrait parlé* of fingerprints (both of them intended for the identification of the latent fingerprint in crime scenes). He regarded these advancements as essential tools for the development of the new scientific policing in Spain.[49]

The tone of Lecha-Marzo's texts was so unctuous that Olóriz was compelled to present his apologies to Vucetich for the "excessive enthusiasm" of his "young friend Lecha-Marzo," whom he described as the most "active propagandist" of his work.[50] In the following year, Lecha-Marzo published another long paper on the "value of Spanish contributions to the study of the means of identification," in which he praised the work of Olóriz and the "Spanish dactyloscopy." Lecha-Marzo proudly included more than ten pages of excerpts from publications and personal letters written by famous European and American authors praising the advantages of the Spanish dactyloscopy. The list of authors included celebrities such as Edmond Locard, Eugène Stockis, and Juan Vucetich.[51]

Enlarging the network previously established by Olóriz, the travels of Lecha-Marzo served to popularize Spanish studies on fingerprints in the international context. He presented the "Spanish dactyloscopy" as both a tool for the emergence of the new scientific police and a source of national pride. In a more marked way than Olóriz, Lecha-Marzo adopted the rhetoric of many authors of his generation who defended science as a way of "regenerating" Spanish society after the disaster of 1898. Exaggerating the originality of Spanish contributions to "scientific identification" methods was perfectly in tune with this trend. He was also keen to defend Olóriz's views of extending the uses of fingerprinting in the regeneration of many aspects of Spanish society, including the making of a new "scientific police" and new personal identity cards for the whole population. He created powerful heroic narratives around Olóriz and the Spanish dactyloscopy, which persist in many current historical accounts. This narrative was also important in overcoming reluctances concerning the extension of fingerprinting to the whole population.[52]

## La Policía Científica

National pride and the reform of policing are common topics in the textbooks published by the police and prison officers who studied with Olóriz. These ingredients are also found in the publication that best represented the hybridization of academic and professional cultures of scientific policing: *La Policía Científica*. The journal was created at the beginning of 1913 by Gerardo Doval (1863–1940), a lawyer specializing in criminal trials who had also studied medicine at the University of Santiago de Compostela. When he was at the peak of the fame, he recalled in an interview his medical training and how it had been crucial in his successful career as lawyer in high-profile trials.[53] One of his first famous affairs started when a corpse with a bullet in the chest was found in November 1902 on a road near the village of Mazarete (in central Spain). Doval was the defense lawyer of the two villagers who were found guilty and sentenced to death. In a desperate move, Doval contacted Tomas Maestre (1857–1936), a forensic physician, who had been recently appointed to the chair on legal medicine in the Madrid Medical Faculty. After reviewing the autopsy, Maestre became convinced that the victim died by suicide. He wrote a book showing how forensic science could solve this judicial mistake. The issue was widely discussed in general press, and the controversy turned into a more general debate on the "regeneration" of Spanish justice administration by means of science.[54] Doval participated in other famous judicial affairs involving science, such as that involving the mysterious bomb-maker Joan Rull Queraltó (1881–1908) in Barcelona. Doval was also a member of the Liberal Party and was elected at different times to the Spanish Parliament. Around 1913, when he started his editorial project *La Policía Científica*, he wrote on hygienism and degeneration theory, opposing marriage between people affected by tuberculosis, an issue hotly debated at the time in the Spanish Parliament.[55]

This brief biographical account confirms that Doval was in tune with the other protagonists discussed in this chapter, particularly regarding the idea of regeneration of Spanish society by means of medicine and science. This idea applied to the police was the main motto of his editorial project *La Policía Científica*, whose first issue appeared on March 5, 1913. Three issues (around ten pages each) per month were published until the end of 1914. Both its main intended audience and its general aim were clearly expressed in the subtitle,

"Journal for Identification," addressed to policemen, "Guardia Civil," and prison employees. *La Policía Científica* included both academic papers and issues related to the organization of the police and professional interests. This approach was one of the most original features of the new publication, even in the European context. Raoul Ruttiens, who reviewed similar European journals in 1913, affirmed that these publications lacked "unity, ensemble, and cohesion." He urged the creation of a journal "especially devoted to the scientific police" in order to support both the advancement of research and the professional identity of the new policing.[56]

This was the main challenge facing Doval, who acknowledged that combining academic and professional interests was complex. He defined the "scientific police" (*policía científica*) as the rational application to all police activities of the knowledge produced by "anthropology, biology, psychology, legal medicine, physics and chemistry." In fact, anthropology and criminology were almost absent from the publication, and legal medicine, toxicology, and chemical analysis played a minor role. Fingerprinting and identification problems were the main issues discussed in the journal, as Doval remarked in the title and in the first issue. Doval announced that the journal would include a small number of articles ("mostly on practical matters and addressed to those in charge of chasing the villains"), including contributions from academicians, police officers, and prison employees. He also announced the publication of textbooks (distributed in chapters with the issues of the journal) and of problems of identification, which would allow the readers to construct their own archive of identification cards and use them in practical cases. He wanted to offer tools "to place the Spanish Police, in its different activities, at the top of the best ones in Europe."[57]

The hybrid nature of the journal announced by Doval is reflected in the profiles of the authors. The group included a limited number of professors of legal medicine and foreign scholars, but many papers were written by police officers, lawyers, and prison employees. From the first group, the main author was Antonio Lecha-Marzo, who published new papers on latent fingerprints, palm ridges, and other forms of trace evidence, always highlighting Spanish contributions and their advantages to scientific police. The role of other Spanish professors was rather limited, but some meaningful examples can be quoted. Take, for instance, Tomás Maestre, the abovementioned professor of history of legal medicine in Madrid. Another paper was published by a professor of the Faculty of Law, José Valdés Rubio (1853–1914), who had strongly supported Olóriz in his research on fingerprinting.[58] Domingo Sánchez y

Sánchez (1860–1947), a student of the Nobel Prize winner Santiago Ramón y Cajal, reviewed several laboratory techniques for coloring fingerprinting, research that was closely connected with his former work on histology. Confirming its academic nature, the paper was also published in the journal of the Spanish Society of Biology.[59]

A substantial number of publications were translations of papers published by Belgian authors with connections to Olóriz and Lecha-Marzo. This group also included authors with both academic and professional profiles. Among members of the police corps, the most prolific was Eugène Goddefroy, a member of the judicial police and founder of the municipal police school in Ostende (Belgium), who corresponded with Olóriz during his last years. Another frequent contributor was Henri Welsch, member of the Institute for Legal Medicine in Liège, who collaborated with Lecha-Marzo in many papers in this journal. Some papers by the professor of legal medicine Eugène Stockis (on the role of forensic physicians in the making of the new scientific police) were also translated into Spanish.[60]

Many other papers published in *La Policía Científica* were written by police officers, prison employees, and members of the Spanish cabinets of identification. The main group were disciples of Olóriz, particularly Vicente Rodríguez Ferrer and José Jiménez Jerez, whose textbooks were published by the journal. Some of these papers dealt with professional problems, but many others were focused on fingerprinting, sometimes describing new techniques (photography, crime scene investigation, etc.), problems of classification, or the new social uses to which it could be put (e.g., supporting Olóriz's ideas about the national card of personal identification). They also published about professional issues: unification of the police, appointments, new regulations, and the like. The sections on professional matters usually appeared at the end of the journal and included factual data about the available positions, laws, and examinations.

Apart from the information about professional life and academic papers, *La Policía Científica* included reports of famous criminal cases (reporting mistakes of old methods of identification or advantages of the new "scientific" ones), practical problems of identification (using fingerprinting records and classifications), and reviews of both academic and professional meetings. Adopting a utilitarian approach, these issues conveniently reinforced the hybridization of academic research and professional interests. The approach is described by one of the supporters of the journal, Manuel de Castro Alonso (1864–1944), an influential Catholic authority who had previously studied law at the university of Valladolid. He affirmed that "abstruse and purely theoretical matters" had to be "carefully avoided" as well as "problems which divide opinions and rouse emotions." Manuel de Castro claimed that *La Policía Científica* had to be helpful for those in charge of prosecuting crime against the "social order" and preserving "our properties and lives."[61]

This utilitarian, apodictic, and noncontroversial tone is clear in the academic papers published in the first pages of the journal. The contents were in tune with current international research on classification and analysis of fingerprints, but authors paid particular attention to practical matters appealing to the intended audience of policemen and prisons employees. Many of the papers encouraged training in fingerprinting techniques and offered practical guidelines about the use of the new identification technology in different activities related to criminal investigations and the management of prisons. For instance, one of the papers by Welsch and Lecha-Marzo was a discussion about the best way to present fingerprinting in courts, either projecting them onto a wall (so the coincidences could be appreciated by judges), or presenting an expert report with detailed information about the coincidences. They supported the second option because they thought that fingerprints required the trained judgment of experts: "The probatory value of fingerprints arose from the demonstrated and certified competence" of experts, they claimed, referring to Bertillon. The reliability and usefulness of fingerprinting in criminal investigations were taken for granted.[62]

Other contents reinforcing these ideas included *La Policía Científica's* problem-solving section. It was usually a collection of "problems of identification" concerning fingerprints. Readers were asked to identify ridge patterns and place them in the general classification, so finally one could establish the formula based on Olóriz's nomenclature. The contest was popular: the names of the winners were published in the next issue of the journal, and some prizes were even offered. The problems concerned real prisoners who had been identified and fingerprinted in Spanish prisons, so their identification cards were available at the central register of identification, including both fingerprint formulas and anthropometric information. Many of the prisoners had been arrested for robbery or false identity, whereas others were famous for their criminal activities or for having been involved in terrorist attacks or other violent acts of political dissidence. One of the most famous prisoners whose data were employed in a problem of identification was the anarchist Francisco Jordan Gallegos (1886–1921), who had been arrested after actively participating in the general strike of Barcelona in 1910. Just after being released, he became president of one the most important Spanish anarchist trade unions in 1916. The problems of identification highlighted the usefulness of fingerprinting while conveying the idea that a trained prison employee or police agent could perform identification and retrieve the correct record from an archive of carefully classified cards.

In that way, fingerprinting was portrayed as a way to create a new, more rational police, and so to replace the old, inefficient practices and forms of organization that were conspicuous sources of inefficiency and errors. Reports of famous criminal cases also served this general purpose. The failures and mistakes were represented as problems emerging from traditional methods and ineffective organization, which were so damaging for the public image of the Spanish Police. The authors frequently suggested that those failures could have been solved by using new technologies such as fingerprinting. For instance, the journal devoted a long section to the problems of identification of the anarchist Manuel Pardiñas (1886–1912), who killed the president of the Spanish government in November 1912.[63] Famous crimes were also reported, for instance, a robbery case that took place in several jewelry stores in Barcelona and was fully described in newspapers. The report remarked that police could confirm the identity of one of the thieves thanks to the two fingerprints that he left in a safe. The author of the report highlighted that it was "a real success of the investigation service whose glory belongs entirely to the police."[64] Another paper praised the success of the Madrid Police in discovering a famous thief who adopted different names and managed to rob some famous singers and popular actors throughout Spain.[65] The case was fully described in the Spanish newspapers and even in Argentina, where the thief was also wanted, so his arrest was described as an international success of the Spanish Police. With pictures of the main protagonists and policemen, the report conveyed the idea that a new and more efficient police was emerging in Spain. Rather than laboratory practice or mathematical apparatus, the meanings associated with the word "scientific" when applied in policing were efficiency and modernity.

In this sense, fingerprinting was also presented both as science and as a source of national pride, since the scientific identification had been developed by Olóriz, a Spanish doctor. In many papers published in *La Policía Científica*, Olóriz's contributions to fingerprinting were highly praised and positively commented on, and indeed were overstated. With Spain as the cradle of such impressive advancements, wondered one of the authors, how on earth could the nation's police resist the advantages of fingerprinting? The mixture of national pride, police reform, and fingerprinting was clearly expressed in the long description of Juan Vucetich's visit to Spain in October 1913. During a tour of Europe, Vucetich stopped in Madrid to visit the new headquarters of the Spanish Police (Dirección General de Seguridad). He was also invited to the Institute for Social Reform (with members of the regeneracionist movement), the Institute for Criminology, and the Laboratory for Legal Medicine, under the direction of Tomás Maestre at the Faculty of Medicine, to which he delivered a lecture on fingerprinting.[66]

This special issue on Vucetich in *La Policía Científica* included a long paper by Jiménez Jerez on "the two great masters of dactyloscopy," putting on the same footing the contributions of Vucetich and Olóriz. He remarked on the coincidences between the two approaches but always highlighted the originality of Olóriz's contributions.[67] Another paper was written by Lecha-Marzo, who had accompanied Vucetich on his visits in Madrid. He highlighted Vucetich's laudatory comments on the organization of the Spanish Police and the service of identification, particularly concerning the handling of criminal records of "anarchists, socialists and trade unionists." The positive comments were contrasted with excerpts from the critical report written by Vucetich on other European police centers. The final picture was a positive image of the Spanish Police, which was clearly connected to recent reforms in organization and methods. Lecha-Marzo concluded: "I think we can congratulate each other for having deserved from Mr. Vucetich so flatting comments, which are a high honor for our homeland."[68] This last example, which was so in tune with the editor's aims, confirms that national pride, fingerprinting, and social regeneration were intermingled in *La Policía Científica* as well as in the other publications reviewed in this chapter.

## Conclusions

*La Policía Científica* is a good example of the hybrid spaces that encouraged the advent of fingerprinting in early twentieth-century Spain and largely shaped how this new technology was conceived, legitimized, and employed through the years. Fingerprinting likewise had a dynamic interaction with previous methods of identification, such as photography or Bertillon anthropometry. In these hybrid spaces, a broad range of protagonists were involved. They played different, but sometimes overlapping, roles, so their interaction was complex and cannot be pictured in terms of linear narratives, even less using old-fashioned difussionist models. The crucial issue was the emergence of contact zones (such as *La Policía Científica*) that encouraged exchanges of epistemic virtues, practical knowledge, and sources of legitimation. These hybrid areas emerged as a result of a partial and instable overlap of academic and professional interests. Physicians such as Olóriz and Lecha-Marzo selectively appropriated international studies on fingerprinting and performed new research, providing new technologies for identification and classification of criminal records. They followed different academic paths to fingerprinting: Olóriz started from anatomy, studies of races, and Bertillon anthropometry, while Lecha-Marzo's research was on the new areas on legal medicine related to trace analysis. Relying on his early work, Olóriz imagined around 1903 a method of identification based on the "cephalic index," which was employed to map European races at the end of the nineteenth century. This approach was abandoned when Olóriz discovered the work of Argentinian police inspector Juan Vucetich. After 1907, Olóriz attempted to move fingerprinting to new areas outside prisons and established links with a broad range of new collaborators. Forensic physician Lecha-Marzo developed this trend by including fingerprints as part of the trace evidence managed in crime scenes.

Crucial aspects were the unequal exchanges between academic and professional worlds. Both Lecha-Marzo and Olóriz developed activities outside the academic world, the first one as forensic physician, the second as head of services of identification in prisons. These activities produced an invaluable influx of empirical data (e.g., on the distribution of fingerprint patterns) that proved crucial to their research. At the same time, while working at the identification service or teaching at the Madrid School of Police, Olóriz established a network of disciples (police officers and prison employees) who played a major role in the legitimation of fingerprinting as reliable identification technology. During his last years, Olóriz pursued a crusade for the extension of fingerprints to new aspects of social life, including the making of a national identity card. The crusade involved publications in medical and professional journals, talks in both academic and professional meetings, and public "experiments." While providing data for refining the methods, these experiments offered convincing proof to police, prison employees, politicians, and other decision makers about the usefulness and reliability of his methods. Moreover, these public experiments, along with the academic papers and the other activities of the two university professors, contributed a great deal to invest fingerprinting technology with the cultural authority of the sciences. In this way, fingerprinting technology, which was initially created by colonial servants, clerks, and police, played a crucial role in discourses on modernity, progress, and science, which were so popular among early twentieth-century Spanish "regenerationists." This issue also helps to explain why fingerprinting technology became such a crucial ingredient in the social imaginary of the new "scientific police" in Spain.

The other group of protagonists in this chapter were politicians, lawyers, police officers, and prison officers. On the one hand, influential lawyers such as Doval allowed more room for medicine and science in the administration of justice. His work as editor of *La Policía Científica* was decisive for placing fingerprinting (understood as a scientific identification method) at the center of the renewal of Spanish policing. Politicians such as de la Cierva created new spaces, promoted practical training, and created new regulations to accommodate the new technologies of scientific identification, from the control of recidivists to the prosecution of crime and the control of political dissidence. Using their practical knowledge in these matters, police and prison employees suggested improvements regarding problems of identification, classification, and retrieval of the information. Many of them were students of Olóriz or close collaborators in his studies on fingerprinting. After Olóriz's death, they became the most important group of supporters of fingerprinting thanks to both their role in prisons and police departments and the publications addressed to their colleagues, such as textbooks, guidelines, and journals like *La Policía Científica*. The contents of these publication were similar to the papers described above: they featured an apodictic tone, utilitarian approach, and a highly selective range of noncontroversial topics, with the focus always placed on the new fingerprinting technologies. With the support of academic authorities such as Lecha-Marzo and his European colleagues, fingerprinting was therefore presented as a technology invested with the alleged virtues of science: exactness, neutrality, and reliability. From the point of view of regenerationists, fingerprinting was a perfect tool for the "modernization" of the Spanish Police. In addition, the "Spanish dactyloscopy" became a source of national pride, a convincing proof of the contributions of Spanish scientists (such as Olóriz and Lecha-Marzo) to the "regeneration" of Spanish society.

This chapter offers clues of the cultural anxieties and social processes that prompted the exceptional visibility of fingerprints in modern Spain. National pride and the urge for regeneration, along with ideals of modernity and progress by science, produced a favorable cultural ferment for fingerprinting in early twentieth-century Spain. Professional and academic groups in connection with reformist politicians created hybrid spaces for exchange, discipline, and legitimation. All these issues largely contributed to the acceptance of fingerprinting as a reliable identification method during the 1910s in Spain, while paving the way for more panoptic uses during the years of Franco dictatorship during the 1940s. And yet there was some reluctance and resistance. One of the most important controversies was about the range of people who could be fingerprinted: prisoners, migrants, political dissidents, or the whole population. Many further debates were related to the perceived uses of fingerprinting technology, not only in policing crime and political dissidence, but also in banking, controlling borders, and granting (or limiting) citizenship rights. There was no room in this chapter for the study of these resistances, but exchanges between police officers, politicians, and physicians were marked by tensions, discontinuities, misunderstandings, and failures, reflecting the array of disparate interests and agendas involved. Resistances remained for several decades and, from that point of view, the campaigns of Olóriz proved to be less effective than military discipline and political punishment. His national identification card project was viewed with suspicion by many social groups during the next three decades, and it was never implemented before the Franco dictatorship. It was not until 1944, amid the brutal repression pursued by the new totalitarian regime installed after the so-called civil war, that a universal identification card—including fingerprints—was introduced for the whole Spanish population. Fingerprints would remain part of Spanish identity cards until recently.[69]

### NOTES

1. For a general overview, see Jane Caplan and John Torpey, eds., *Documenting Individual Identity: The Development of State Practices in the Modern World* (Princeton, NJ: Princeton University Press, 2001); James Brown, Ilsen About, and Gayle Lonergan, eds., *Identification and Registration Practices in Transnational Perspective: People, Papers and Practices* (New York: Palgrave MacMillan, 2013); Gérard Noiriel, ed., *L'identification: Genèse d'un travail d'État* (Paris: Belin, 2007); Ilsen About and Vincent Denis, *Histoire de l'identification des personnes* (Paris: La Découverte, 2010). On changes taking place before the nineteenth century, see Valentin Groebner, *Who Are You? Identification, Deception, and Surveillance in Early Modern Europe* (Brooklyn, NY: Zone Books, 2007); Vincent Denis, *Une histoire de l'identité, France, 1715–1815* (Champ Vallon: Seyssel, 2008).

2. See Allan Sekula, "The Body and the Archive," *October* 39 (1986): 3–64; Simon A. Cole, *Suspect Identities: A History of Fingerprinting and Criminal Identification* (Cambridge, MA: Harvard University Press, 2002); Chandak Sengoopta, *Imprint of the Raj: How Fingerprinting was Born in Colonial India* (London: Pan Books, 2004). On Argentina, see Kristin Ruggiero, *Modernity in the Flesh: Medicine, Law, and Society in Turn-of-the-Century Argentina* (Stanford, CA: Stanford University Press, 2004); Julia Rodríguez, *Civilizing Argentina: Science, Medicine and the Modern State* (Chapel Hill, NC: University of North Carolina Press, 2006); Mercedes García

Ferrari, *Marcas de identidad: Juan Vucetich y el surgimiento transnacional de la dactiloscopia (1888–1913)* (Rosario: Prohistoria, 2015). On England, see Edward Higgs, *Identifying the English: A History of Personal Identification from 1500 to the Present* (New York: Continuum International, 2011). On France, see Ilsen About, "Les fondations d'un système national d'identification policière en France (1893–1914)," *Gèneses* 54 (2004): 28–52; Ilsen About, "La police scientifique en quête de modèles: Institutions et controverses en France et en Italie (1900–1930)," in *L'Enquête judiciarie en Europe au XIXè siècle*, ed. J. C. Farcy et al. (Paris: Creaphis, 2007), 257–69. On Italy, see Massimiliano Pagani, "Fingerprinting at the Bar: Criminal Identification in Liberal and Fascist Italy" (PhD diss., University of Exeter, 2009). On South Africa, see Keith Breckenridge, *Biometric State: The Global Politics of Identification and Surveillance in South Africa, 1850 to the Present* (Cambridge: Cambridge University Press, 2014). See also P. Knepper, "The Empire, the Police, and the Introduction of Fingerprint Technology in Malta," *Criminology and Criminal Justice* 9 (2009): 73–92. For more examples, see the books cited in note 1 above.

3. See Richard Bach Jensen, *The Battle against Anarchist Terrorism: An International History, 1878–1934* (Cambridge: Cambridge University Press, 2014). On Bertillon, see Pierre Piazza, ed., *Aux origines de la police scientifique: Alphonse Bertillon, précurseur de la science du crime* (Paris: Karthala, 2011). See also Pagani, "Fingerprinting at the Bar," 48–50; Mercedes García Ferrari and Diego Galeano, "Cartografía del bertillonage: Circuitos de difusión, usos y resistencias del sistema antropométrico en América Latina," in *Delincuentes, policías y justicias en América Latina, siglos XIX y XX*, ed. Daniel Palma Alvarado (Santiago: Universidad Alberto Hurtado, 2015), 279–311. For a review of literature on Bertillon, see Pierre Piazza, "Alphonse Bertillon et l'identification des personnes (1880–1914)," *Criminocorpus*, December 4, 2018, https://criminocorpus.org/en/expositions/suspects-accuses-coupables/alphonse-bertillon -et-lidentification-des-personnes-1880-1914/.

4. Royal Order, September 10, 1896, in Antonio Navarro Fernández, *Estado actual de la dactiloscopia en España* (Madrid: Rojas, 1912), 10–11.

5. Royal Order, March 15, 1897.

6. *Revista de prisiones y de policía* 6, no. 8 (January 16, 1898): 5–31, quoted on 27–29. See Jorge Alberto Nuñez, "Fernando Cadalso y Manzano: Medio siglo de reforma penitenciaria en España (1859–1939)" (PhD thesis, Universidad de Valladolid, 2013).

7. Antonio Lecha-Marzo, *Los últimos progresos de la identificación de los reincidentes: Dactiloscopia Vucetich y dactiloscopia Olóriz* (Granada: Guevara, 1910), 7–8. See also Fernández, *Estado actual de la dactiloscopia en España*, 66.

8. Fernández, *Estado actual de la dactiloscopia en España*. See similar arguments in Lecha-Marzo, *Los últimos progresos*, 7. The difficulties in obtaining reliable measurements was a common criticism against anthropometry. On problems regarding honor, see Ruggiero, *Modernity in the Flesh*, 102–6, 184–96, and García Ferrari, *Marcas de identidad*, 70.

9. Federic Olóriz offers a review of the situation and challenges in his unpublished manuscript notes. Cf. *Notas de dactiloscopia, Plan de una memoria sobre la reforma de la identificación judicial en España*, Archivo de la Universidad de Granda (AUG), Granada.

10. Federico Olóriz, *El laboratorio de antropología de la facultad de medicina de Madrid* (Madrid: Idamor Moreno, 1899), 5–6. On Olóriz, see Miguel Guirao, *Discurso de apertura por el catedrático de la facultad de medicina* (Granada: Universidad de Granada, 1955); Miguel Guirao Pérez and Miguel Guirao, *Federico Olóriz Aguilera: Biografía íntima del Profesor* (Granada: Comares, 2008). The main historical sources are at the AUG. Olóriz's letters have been published in Rafael Sánchez Martín, "El epistolario (1886–1912) de Federico Olóriz (1855–1912)" (PhD thesis, Universidad de Granada, 1979). See, e.g., pp. 165–68 for letters on the making of the Museum of Anthropology. For more details about Olóriz's early research on anthropology, see Elena Arquiola, "Anatomía y antropología en la obra de Olóriz," *Dynamis* 1 (1981): 165–77.

11. Olóriz, *El laboratorio de antropología*, 17. See also his letters in Sánchez Martín, "El epistolario de Federico Olóriz," 131–32.

12. Federico Olóriz Aguilera, *Distribución geográfica del índice cefálico en España deducida del examen de 8.368 varones adultos* (Madrid: Memorial de Ingenieros, 1894). See also his letters from different European scholars who praised the publication, including Cesare Lombroso. Cf. letter dated December 25, 1894, in Sánchez Martín, "El epistolario de Federico Olóriz," 159. On Godard prize, see letter dated November 14, 1894, in Sánchez Martín, "El epistolario de Federico Olóriz," 181. See J. Deniker, "Rapport sur le concours pour le prix Godard, en 1895," *Bulletins de la Société d'Anthropologie de Paris* 6 (1895): 717–22; and Jean-Claude Wartelle, "La Société d'Anthropologie de Paris de 1859 à 1920," *Revue d'Histoire des Sciences Humaines* 1, no. 10 (2004): 125–71.

13. Olóriz wrote an interesting diary during his first days working in prison. See Fondo Olóriz, *Diario de prisiones*, 1898–99, AUG.

14. Federico Olóriz Aguilera, "Identificación personal en los jóvenes," in *Comptes Rendus du XIVe Congrès International de Médecine, Madrid, avril 23–30 1908*, ed. A. Fernández Caro (Madrid: Sastre, 1904), 2:109–12 (quoted on p. 112).

15. Juan Vucetich, *Dactiloscopía comparada, el nuevo sistema argentino* (La Plata: Pruser, 1904). On Vucetich, see Mercedes García Ferrari, *Ladrones conocidos/Sospechosos reservados: Identificación policial en Buenos Aires, 1880–1905* (Buenos Aires: Prometeo, 2010), and García Ferrari, *Marcas de identidad*.

16. E.g., he changed the initial "B" in "Bideltos" to a "V" to keep in tune with Vucetich's last group. On the differences between Vucetich and Olóriz on primary classification of fingerprints, see Federico Olóriz Aguilera, "Procedimiento de identificación. Cual es preferible. Importancia de su generalización," *Revista general de legislación y jurisprudencia* 116 (1910): 50–72; Lecha-Marzo, *Los últimos progresos*, 21–23.

17. Many decades after, these issues were discussed in colonial settings in the 1940s and 1950s. See Rosa Medina-Doménech, "Scientific Technologies of National Identity as Colonial Legacies: Extracting the Spanish Nation from Equatorial Guinea," *Social Studies of Science* 39 (2009): 81–112.

18. Fondo Olóriz, "Identificación dactiloscópica según lo prevenido en la Real Orden de 31 de mayo de 1907," AUG. The archive gathers data from prisons of Cartagena, Burgos, Chinchilla, Ceuta, Tarragona, Santoña, Granada, Ocaña, Valencia, Palencia, Madrid, Barcelona, etc., AUG. Olóriz's proposal was formally approved in the March 23, 1907, session of the Consejo Penintenciario. See *Revista Penitenciaria* 4 (1907): 604. See Fernando José Bunillo Albacete, *La cuestión penitenciaria: Del sexenio a la Restauración (1868–1913)* (Zaragoza: PUZ, 2011), 134–45.

19. Fondo Olóriz, "Circular dirección general de prisiones, Madrid, 31 de julio de 1909," AUG.

20. Fondo Olóriz, *Ante-proyecto de Decreto sobre Identificación*, c. 1909–10, AUG: "dificultades técnicas, su limitación a varones adultos, su carácter vejatorio, y su coste"; "medio de reconocimiento seguro, general, sencillo, poco molesto y económico."

21. Fondo Olóriz, "Consideraciones y anteproyecto, ca. 1909–1910," AUG.

22. Federico Olóriz, "Las firmas dactilar y escrita en las operaciones de Previsión,"*Anales del Instituto Nacional de Previsión* 2–3 (1909–10): 66–71, 5–13. See also letter dated February 12, 1910, in Sánchez Martín, "El epistolario de Federico Olóriz," 316. See also Federico Olóriz, *Morfología socialística* (Madrid, 1911).

23. Antonio Lecha-Marzo, *Sobre el valor de la contribución española al estudio de los medios de identificación* (Madrid: Tordesillas, 1911).

24. *La Época*, June 22, 1909, 3. Olóriz, "Procedimiento de identificación," 58. Fondo Olóriz, *Registro manual para la identificación de delincuentes por el Dr. F. Olóriz Aguilera* (Madrid, 1910), 2–3.

25. [Fingerprinting] "podrá todavía en algún caso dejar impune a un criminal, pero jamás conduce a la condena de un inocente." Cf. Federico Olóriz Aguilera, *Experimentos de identificación monodactilar* (Madrid: Reus, 1910), 13–14. The final report was published in *Revista de Legislación y Jurisprudencia*. The report was signed in Madrid on May 11, 1910.

26. Federico Olóriz, "Conferencia sobre Dactiloscopia presentada en el Congreso de Zaragoza el 24 de octubre de 1908," in *Actas del Congreso de la Asociación Española para el Progreso de las Ciencias* (Madrid: Eduardo Arias, 1909) 7:215–248 (quoted on p. 248).

27. "Congreso de las Ciencias," *La Alhambra, Revista quincenal de Artes y Letras* 14, no. 319 (June 30, 1911): 377–79, at 378: "Considerando las razones expuestas y el registro manual de identidad, inventado por el Dr. Olóriz Aguilera, propone al Congreso que éste eleve al Gobierno la siguiente conclusión: Dada la perfección actual de la Dactiloscopia, es ya posible y muy conveniente para el mejor cumplimiento de los actos sociales, que requieren la determinación rigurosa de las personas, que se cree un Archivo nacional de identidad en que puedan figurar todos los ciudadanos."

28. Olóriz, *Morfología socialística*, 43: "lo mismo para proteger al bueno que para perseguir al malo y para facilitar a todos el ordenado cumplimiento del papel que como partes de la gran máquina social les corresponda, vendría a ser la identificación científica algo semejante a la grasa que en las máquinas industriales lubrifica las piezas, suavizando sus roces, manteniendo su ajuste y evitando su rápida alteración o su desgaste."

29. "El único obstáculo capaz de retrasar el cumplimiento de la misión utilísima a que la Dactiloscopia se halla destinada, es de orden moral, consiste en el prejuicio que contra la identificación científica existe hoy, por ser exclusivamente judicial y ser tenida por muchos como nota de infamia, propia de ladrones y asesinos e incompatible con la honorabilidad del ciudadano." Cf. Olóriz, *Morfología socialística*, 44. Olóriz expressed similar concerns in his letter to Vucetich, dated November 24, 1909 (Museo Policial de la provincia de Buenos Aires, Caja 31), courtesy of Mercedes García Ferrari. On the problem of the criminal stigma associated with fingerprinting, see Edward Higgs, "Fingerprints and Citizenship: The British State and the Identification of Pensioners in the Interwar Period," *History Workshop Journal* 69, no. 1 (2010): 52–67; Breckenridge, *Biometric State*, 96–114 (on South Africa and Gandhi). On other examples of resistances against policial identification, see Mercedes García Ferrari, "Una marca peor que el fuego: Los cocheros de la ciudad de Buenos Aires y la resistencia al retrato de identificación," in *La ley de los profanes: Delito, justicia y cultura en Buenos Aires (1870–1940)*, ed. Lila Caimari (Buenos Aires: Fondo de Cultura Económica, 2007), 99–13; and Eduardo Andrés Godoy Sepúlveda, *La Huelga del Mono: Los anarquistas y las movilizaciones contra el retrato obligatorio (Valparaíso, 1913)* (Santiago: Quimantú, 2014). On British reluctance to identification cards, see Jon Agar, "Modern Horrors: British Identity and Identity Cards," in *Documenting Individual Identity: The Development of State Practices in the Modern World*, ed. Jane Caplan and John Torpey (Princeton, NJ: Princeton University Press, 2001), 101–20.

30. Olóriz wrote detailed reports of his "experiments" that are preserved in his personal archive. Cf. Fondo Olóriz, "Cinco problemas de identificación monodactilar," July 22, 1911, AUG.

31. On Olóriz self-experiments, see the document reproduced by Fernando Girón Irueste and Miguel Guirao Piñeyro, "'Influencia del roce y desgaste epidérmico en el dibujo papilar': Experiencias inéditas sobre dactiloscopia de Federico Olóriz Aguilera (1855–1912)," *Dynamis* 35 (2015): 177–91. Similar studies had been previously developed by other European authors, such as Locard or Stockis. See Lecha-Marzo, *Los últimos progresos*, 12–14, for a contemporary review of these studies.

32. Fondo Olóriz, "Registro manual para la identificación de delincuentes en Madrid por el Dr. Federico Olóriz Aguilera, Ms., Madrid, October, 1910," AUG. It was translated into French: Federico Olóriz Aguilera, *Manuel pour l'identification des délinquants de Ma-*

drid: *Traduction revue par Th. Borgerhoff, attaché au Ministère de la Justice, Bruxelles* (Brussels: Larcier, 1911). Copies were sent to several police departments in America, Australia, and Europe, and to authors such as Henry, Goddefroy, Bertillon, and the like. See letters included in Sánchez Martín, "El epistolario de Federico Olóriz," 363–80.

33. Enric Prat de la Riba, "Les Bombes," *La Veu de Catalunya*, December 27, 1906, 1. Translation by Jensen, *Battle against Anarchist Terrorism*, 322. See Eduardo González Calleja, *La razón de la fuerza: Orden público, subversión y violencia pública en España de la Restauración (1875–1917)* (Madrid: CSIC, 1998); Eduardo González Calleja, *En nombre de la autoridad: La defensa del orden público durante la Segunda República Española (1931–1936)* (Granada: Comares, 2014). On similar developments in Britain, see Haia Shpayer-Markov, *The Ascent of the Detective: Police Sleuths in Victorian and Edwardian England* (Oxford: Oxford University Press, 2011).

34. Olóriz to Vucetich, November 24, 1909, Museo Policial de la provincia de Buenos Aires, Caja 31. Courtesy of Mercedes García Ferrari. On the exchanges between Olóriz and Vucetich, see José Ramón Bertomeu-Sánchez and Mercedes García Ferrari, "Huellas dactilares a través del mundo transatlántico: Las vidas paralelas de Juan Vucetich y Federico Olóriz," *Dynamis* 38 (2018): 131–62.

35. For other examples of similar schools, see Ilsen Abot, "La Police scientifique en quête de modèles: Institutions et controverses en France et en Italie (1900–1930)," in *L'Enquête judiciarie en Europe au XIXè Siècle*, ed. J. C. Farcy et al. (Paris: Creaphis, 2007), 257–69.

36. Fondo Olóriz, *Escuela de Policía de Madrid, 22 de Junio de 1909*, AUG (quoted on p. 2). Some of the lectures were taken by one of his students and revised by Olóriz before being published in Federico Olóriz, *Guía para extender la Tarjeta de Identidad* (Madrid: Imprenta Hernández, 1909).

37. Some examples are provided in Fondo Olóriz, "Relación del personal afecto a la comisaría de la Universidad y grado de instrucción en que se encuentran para el manejo del Manual de Identidad," Madrid, April 7, 1911, AUG.

38. Ramón Lobo Goya, "Identificación judicial. Estudio sobre dactiloscopia: Memoria presentada para aspirar al grado de doctor" (PhD thesis, Universidad Central de Madrid, 1912). Another thesis on fingerprinting was written by Jesús Losón Dalama, "Aplicaciones de la dactiloscopia en la vida civil" (PhD thesis, Universidad Central de Madrid, 1913).

39. Some examples: José Jiménez Jerez, *Sistema dactiloscópico de Olóriz y retrato hablado de Bertillón* (Madrid: Alvarez, 1914); idem, *La dactiloscopia al alcance de todos: Catecismo de la identificación personal* (Madrid: La editora, 1915); José Pastor Rodríguez, *Nociones elementales de dactiloscopia* (Madrid: Guardia Civil, 1914); Jesús Lasuén Urrea, *Dactiloscopia* (Madrid: Pontones, 1913).

40. The school is described by the forensic doctor Fernando Bravo Montero, in Eugenio Stockis, *La identificación judicial y la filiación internacional . . . Traducido por D. Fernando Bravo y Moreno* (Barcelona: La Académica, 1909), 53–54.

41. On the reform of the Spanish Police, see Eduardo González Calleja, *La razón de la fuerza*, 392–95, 402–9; Eduardo González Calleja, *En nombre de la autoridad*, 32–37. The main regulations were published in *La Gaceta de Madrid*, February 3 and 29, 1908. See also the memoirs by Juan de la Cierva, *Notas de mi vida* (Madrid: Reus, 1955), 92–100. On Arrow and Scotland Yard, see Jensen, *Battle against Anarchist Terrorism*, 322, and Shpayer-Markov, *Ascent of the Detective*, 119. His months in Barcelona are colorfully described in his memoirs. Cf. Charles Arrow, *Rogues and Others* (London: Duckworth, 1926), 193–209 (esp. 194–99).

42. The report was signed on March 9, 1909, by Francesc Molins, Luis Bachiller, and Federico Olóriz. Printed in Olóriz, "Procedimiento de identificación," 55.

43. See Olóriz to Vucetich, September 20, 1909, Museo Policial de la provincia de Buenos Aires, Caja 31. Courtesy of Mercedes García Ferrari. "[Mi contribución] se reduce, como verá,

si se digna hojear mis folletos, a trabajos estadísticos aun no terminados, a tentativas de reseña dactilar y a ligerísimas modificaciones de instalación, ordenamiento de tarjetas y otros detalles, muy secundarios, que en nada afectan a la esencia del Vucetichismo, del que me declaro sencillamente admirador sin pretensiones de reformador."

44. On these changes, see Ian Burney and Neil Pemberton, "Making Space for Criminalistics: Hans Gross and Fin-de-Siècle CSI," *Studies in History and Philosophy of Biological and Biomedical Sciences* 44, no. 1 (2013): 16–25; Ian Burney and Neil Pemberton, *Murder and the Making of English CSI* (Baltimore: Johns Hopkins University Press, 2016). I am grateful to the authors for allowing me to read the draft version of this book.

45. For an introduction, see L. Lopez Ocón, *Breve Historia de la ciencia española* (Madrid: Alianza Editorial, 2003); *La Junta para Ampliación de Estudios e Investigaciones Científicas: Historia de sus centros y protagonistas (1907–1939)*, ed. E. Caballero (Gijón: Trea, 2010).

46. See Archive of the Junta de Ampliación de Estudios, Madrid, File 84-106. H. Welsch and A. Lecha-Marzo, *Manuel pratique de Dactyloscopie* (Liège: Vaillant-Carmanne, 1912). On Lecha-Marzo, see Carmen Meer, "Antonio Lecha-Marzo y la Junta para Ampliación de Estudios," in *La Junta para Ampliación de Estudios e Investigaciones Científicas*, ed. José Manuel Sánchez Ron (Madrid: CSIC, 2007), 647–60. See also José Martínez Pérez, "Sobre la incorporación del método experimental a la medicina legal española: El estudio de las manchas de la sangre en la obra de Lecha-Marzo," in *Estudios sobre Historia de la Ciencia y de la Técnica* (Valladolid: Junta de Castilla y León, 1988), 833–44; idem, "La contribución de Lecha-Marzo a la tanatología médico-forense," in *Actas del IX Congreso Nacional de Historia de la Medicina* (Zaragoza: PUZ, 1990), 4:1429–42. Lecha-Marzo's letters have been preserved and edited by Carmen de Meer Lecha-Marzo, "Antonio Lecha-Marzo (1888–1919): Contribución al estudio de la Historia de la Medicina Legal contemporánea" (PhD thesis, Universidad de Valladolid, 1985). I am grateful to Carmen de Meer for her kind help concerning Lecha-Marzo's documents.

47. Antonio Lecha-Marzo, *Estado actual de nuestros conocimientos sobre policía judicial científica por . . . Interno de la Facultad de Medicina de Valladolid: Con una carta abierta del profesor S. Ottolenghi* (Granada: López Guevara, 1907).

48. Ricardo Campos Marín, "La construcción del sujeto peligroso en España (1880–1936): El papel de la psiquiatría y la criminología," *Asclepio* 65, no. 2 (2013): 1–7, on 6–8.

49. Antonio Lecha-Marzo, *Los últimos progresos de la identificación de los reincidentes: Dactiloscopia Vucetich y dactiloscopia Olóriz* (Granada: Guevara, 1910).

50. See Olóriz to Vucetich, May 16, 1910, Museo Policial de la provincia de Buenos Aires, Caja 31. Courtesy of Mercedes García Ferrari.

51. Antonio Lecha-Marzo, *Sobre el valor de la contribución española al estudio de los medios de identificación* (Madrid: Tordesillas, 1911).

52. Lecha-Marzo, *Sobre el valor*. The heroic narrative of Spanish fingerprinting is dominant in the publications related to the centenary of the emergence of Spanish scientific police in 1910 and the death of Federico Olóriz in 1912. See, e.g., José Miguel Otero Soriano, ed., *Policía científica: 100 años de Ciencia al Servicio de la Justicia* (Madrid: Ministerio del Interior, 2011); Miguel Guirao, *Federico Olóriz Aguilera: Biografía íntima del Profesor* (Granada: Comares, 2008). The celebrations, which took place between 2010 and 2013, have also provided exhibitions, publications, and digital editions of important sources. Yet many important archives (including Olóriz's personal one) still remain to be inventoried and analyzed from fresh perspectives.

53. "Los Intervius de 'El Fígaro': El Criminalista Gerardo Doval," *El Fígaro*, December 17, 1918.

54. See José Martínez Pérez, "Restableciendo la salud del Estado: Medicina y regeneración nacional en torno a un proceso judicial en la encrucijada de los siglos XIX al XX," *Dynamis*

18 (1998): 127–56. The issue was fully discussed in the general press and even compared with the Dreyfuss Affair in France.

55. Quoted by Ricardo Campos Marín, "La teoría de la degeneración y la medicina social en España en el cambio de siglo," *Llull* 21, no. 41 (1998): 333–56, at 345. More biographical details on Doval appear in L. Barrio y Moraita, "Doval," *El Foro Español*, November 20, 1911.

56. See *Revue Critique de Police Scientifique*, October 1913, and Raoul Ruttiens, "Une revue de police scientifique," *Revue de droit pénal et de criminologie* 7 (1913): 659–61, quoted on p. 661. On the British *Police Journal*, a publication that played a similar role in the development of forensic science in the United Kingdom, see Alison Adam, *A History of Forensic Science: British Beginnings in the Twentieth Century* (London: Routledge, 2015), 144–47; Burney and Pemberton, *Murder and the Making of English CSI*, 107–18, On German journals, see Peter Becker, "Les étranges chemins de la perfection: L'innovation criminologique en Allemagne et en Autriche au XIXè siècle," in *L'identification: Genèse d'un travail d'Etat* (Paris: Belin, 2007), 97–122.

57. *La Policía Científica*, March 5, 1913, 1–2.

58. *La Policía Científica*, March 15, 1913, 6–7.

59. "Sobre revelación y fijación de huellas dactilares invisibles," *La Policía Científica* 2, no. 37 (1914): 1–6.

60. See letters to Olóriz by Goddefroy reproduced in Sánchez Martín, "El epistolario de Federico Olóriz," most of them between 1910 and 1911. On fingerprinting in Belgium, see David Somer, "The Criminology and Forensic Police School: The Twofold Project to Humanize Judicial Practice and to Implement Technical Police in Belgium," in *Policing New Risk in Modern European History*, ed. Jonas Campion and Xavier Rousseaux (Basingstoke: Palgrave, 2015), 36–57. I am thankful to the author for providing a copy of his work.

61. *La Policía Científica*, May 5, 1913.

62. Henri Welsch and Antonio Lecha-Marzo, "Demostración de la identidad de dos impresiones," *La Policía Científica* 1, no. 9 (1913): 1–4. On the debates about who can speak for the fingerprints, see Cole, *Suspect Identities*.

63. *La Policía Científica*, April 5, 1913.

64. *La Policía Científica*, December 25, 1914.

65. *La Policía Científica*, December 15, 1913.

66. *La correspondencia de España*, November 19, 1913, 5.

67. José Jiménez Jerez, "Los dos grandes maestros de la dactiloscopia," *La Policía Cientím fica* 1, no. 25 (November 5, 1913): 10–12. This was a special issue on Vucetich.

68. *La Policía Científica*, October 25, 1913, 6: "Vucetich is today our guest." The special issue was published on November 5, 1913 (quoted on p. 10: "Creo podemos congratularnos de haber merecido al Sr. Vucetich tan halagadores conceptos que constituyen un alto honor para nuestra patria"). A similar report was published in medical journals. See "Vucetich En Madrid," *España Médica* 3, no. 100 (1913): 15. On Vucetich's travels to Europe, see García Ferrari, *Marcas de identidad*, 249–53. On other similar travelers who reviewed police departments in Europe, see Ilsen About, "Qu'est-ce qu'un système policier? Le voyage de Raymond B. Fosdick à travers les polices d'Europe, 1913–1915," in *Circulations policières, 1750–1914*, ed. Catherine Denys (Villeneuve d'Ascq: Presses Universitaires de Septentrion, 2012), 63–83.

69. "Decreto de 2 de marzo de 1944 por el que se crea el Documento Nacional de Identidad," *Boletín Oficial del Estado* 81 (March 21, 1944): 2346–47. Similar attempts were made in France during the Vichy regime, but they did not last after World War II. See Pierre Piazza, *Histoire de la carte nationale d'identité* (Paris: Odile Jacob, 2004), 179 and *passim*.

# CHAPTER SEVEN

# From Bedouin Trackers to Doberman Pinschers

## The Rise of Dog Tracking as Forensic Evidence in Palestine

BINYAMIN BLUM

## Introduction

On a moonless Friday night in June 1933, Dr. Haim Arlosoroff, head of the Political Department of the Jewish Agency, was assassinated.[1] The victim and his wife were strolling along the Tel Aviv beach when two men who had been stalking them approached and asked for the time. One of the men then shined his torch on Arlosoroff's face while the other drew a weapon and shot him. The murder investigation began that night and was overseen directly by Harry Rice, head of Palestine's Criminal Investigation Department (CID). Large spotlights illuminated the dark coastal crime scene, allowing Bedouin trackers to follow the shooters' prints in the sand before they were washed away by the tide. Speculation concerning the assassins' identities and motives was rife: some believed they were Arab nationalists, while others thought them to be anticolonial Communists. Some speculated they were Nazi operatives, while others believed the murder to be a British conspiracy to sow disunity between Zionist factions in an attempt to divide and rule. Ultimately, Avraham Stavsky and Zvi Rosenblatt, two members of the right-wing Revisionist Party, the political rival of the Labour Party, were charged with the murder.

At trial, the prosecution relied heavily on the Bedouin trackers' testimony to corroborate the widow's identification of the two defendants. The trackers had followed prints found at the scene to help reconstruct the crime. Additionally, at a lineup held after Stavsky and Rosenblatt's arrest, the trackers confirmed that the two men's shoeprints matched those found at the scene. Yet cross-examination of the trackers laid bare the apparent fallibility of their methods. First, rather than following the footprints of the alleged murderers, the trackers conceded that they may have mistakenly tracked spoors left by policemen who had trodden through the crime scene. Second, they had breached protocol by inspecting the soles of the defendants' shoes prior to the lineup.[2] In his decision to acquit, Chief Justice McDonnell concluded that the trackers' evidence should be given absolutely no credit.[3] The case's high profile meant that the acquittals were a painful setback for the newly restructured Palestine Police's CID. Despite considerable resources invested in the investigation, a conviction had not been secured. The trial's celebrity also placed Bedouin tracking skills under considerable public scrutiny, raising profound doubts concerning the reliability of their methods.[4]

Inspector General Roy Spicer took Stavsky and Rosenblatt's acquittal as an opportunity to replace the force's Bedouins with what he believed to be superior trackers: Doberman Pinschers. Weeks after the trial, Sergeant John Kenyon Parker and Constable Alexander R. Pringle were dispatched to South Africa for a six-month dog-master training course.[5] They returned to Palestine in December 1934 with three dogs named Gift, Mayer, and Ria.[6] Initially employed to investigate agricultural crime, the Dobermans came to play a prominent role in policing Palestine during the Arab Revolt (1936–39). Though imported strictly for investigative purposes—not to furnish evidence—expediency led Palestine's authorities to rely on dog tracking to justify extrajudicial action during the revolt. Meanwhile, prosecutors began offering canine evidence in court, with judges proving reluctant to exclude it. Once legitimized as a form of proof, dog tracking remained admissible evidence even after the revolt.

Replacing Bedouin trackers with Doberman Pinschers was significant for a number of reasons. First, the scientific foundation for canine identification was shaky, making it no more reliable than Bedouin tracking: dog tracking relied on olfaction, to this day the least understood of the five senses.[7] Second, unlike other trace evidence, scent by its nature could not be preserved for fact-finders to evaluate at trial, forcing judges to defer entirely to the opinion and expertise of others.[8] Third, that an animal had been employed to detect and match these traces deprived defendants of the chance to challenge such evidence and denied fact finders an opportunity to critically evaluate the technique. Unlike human trackers, canines could not be cross-examined.

Beyond its evidentiary aspects, dog tracking possessed significant performative power. Replacing human with canine trackers bore symbolic, moral, psychological, and religious significance that went far beyond the method's probative sway. Symbolically, substituting canines for native trackers suggested an equivalence—or even hierarchy—between animals and natives, marking the former as superior trackers. Beyond their ability to track more reliably than natives, dogs were believed to produce a moral effect upon the local population. The dogs' incomprehensible ability to follow even invisible tracks further contributed to the canine mystique, sowing the fear and panic that often led suspects to confess. In the Middle East, dogs also possessed cultural significance: in both Islam and Judaism, dogs were traditionally considered savage and impure, which was believed to further contribute to their effectiveness in investigations.

The rise of dog tracking as a method of detection and proof in Palestine offers several more general insights regarding the conditions under which forensic innovations thrive. First, the rise of dog tracking during the Arab Revolt demonstrates how crisis—real or perceived—may lead questionable methods of detection to evolve into "proof," which may serve to justify both judicial and extrajudicial action. Moreover, it illustrates how under such circumstances gatekeepers—namely, the judiciary—may prove unable or unwilling to disregard such disputable evidence. Second, the employment of dogs in the colonies at a time when they were rejected for policing uses in England demonstrates the greater willingness to experiment with novel policing and evidentiary techniques on "others," highlighting the colonies' distinctive role in shaping forensic culture. Third, the irreducible and incommunicable expertise of the dog complicates the conventional wisdom regarding forensic science's progression from the instinctive, ineffable proficiency of the specialist to the refutable expertise of the scientist. While Bedouin trackers in Palestine were required to make the indescribable comprehensible, dogs were exempt from meeting similar criteria. Arguably, one of the canine's greatest advantages over the human tracker was the inability to effectively scrutinize its evidence in court. The dogs' inability to testify was what rendered their evidence so powerful. In this, dog-tracking evidence presaged a whole category of "machine-produced" proof: difficult to classify, such evidence often continues to evade the judicial scrutiny reserved for direct and expert testimony in common-law jurisdictions.[9]

This chapter is divided into five parts. Analyzing structural reforms in the Palestine Police following the 1929 "disturbances," the first part provides the broader context for the force's adoption of new detection methods, including canine tracking. The second part then explores the proximate causes for the founding of Palestine's Dog Section: the publicly exposed fallibility of Bedouin trackers coupled with newly appointed Inspector General Spicer's enthusiasm about the Dobermans' tracking skills. The focus then shifts in the third part to the Dog Section's rise as a tool for solving political crimes at a time of waning public cooperation: the Arab Revolt (1936–39). During this time, dog tracking was transformed from a purely investigative method, aimed primarily at producing independently admissible evidence (such as stolen goods or confessions), into autonomous grounds for judicial and extrajudicial action. The fourth part examines the cultural aspects of dog tracking, analyzing canines' attributed psychological effects and their religious and cultural meanings, as well as the symbolic significance of replacing human, native trackers with canines. The final part of this chapter analyzes the legal aspects of classifying dog-tracking evidence, namely, its admissibility in court, a feature that distinguished Palestine from other parts of the British Empire, even those that employed tracking dogs.

## From Rifles to Notebooks: The Transformation of the Palestine Police

The three founding members of the Palestine Police's Dog Section arrived in Jerusalem on Christmas Eve in 1934. The policing transformation that led to their import began years earlier, however. The Dog Section's establishment was part of broader reforms following the 1929 "disturbances."[10] Violence erupted in Jerusalem's Old City on August 23, 1929, following the conclusion of Friday prayers at the Haram a-Sharif. The immediate cause for violence was the alleged Jewish breach of the status quo concerning prayer rights at the adjacent Western Wall, which struck a deeper chord in the relationship between Jews and Muslims in Palestine. The perceived increase in Jewish control over one of Islam's holiest sites fed broader fears of rising Zionist influence in Palestine.[11] Looming in the background were ongoing debates over Jewish immigration and land-purchase quotas, as well as breached British promises to convene a representative legislative council in Palestine.

Following the tensions in Jerusalem, violence spread throughout Palestine. On Saturday, August 24, rioters overwhelmed a Hebron police force of thirty-nine men, with multiple incidents of looting, rape, and murder throughout the city. By the end the day, the Jewish death toll reached sixty-seven. The British police force began shooting to kill, resulting in nine Arab casualties. Though the violence in Hebron subsided by Sunday, clashes continued in Jerusalem and its environs. Violence also began spreading northward to Safed. At the end of the violent week, the death toll reached 133 Jews and 116 Arabs. Over 500 were injured.

Critics cited several factors that led to the police's failure to contain the violence: the force's inadequate size (especially its British component), its deficient intelligence service, and its wanting leadership. Under High Commissioner Lord Plumer, in 1926, the British contingent of the force had been pared down to 175 members. The remaining 1,900 members of the force came from the local population: 1,600 Arabs and 300 Jews.[12] Plumer's plan to rely on military reserves from Egypt and Malta in case of emergency proved impracticable: by the time these reserves arrived in Palestine, matters had spiraled out of control.

Palestine's government had not only failed to prevent the violence but also was unsuccessful in prosecuting the perpetrators. In 420 indictable offenses, the accused remained "unknown." Moreover, even the cases that were prosecuted resulted in a 40 percent acquittal rate.[13] Zionist leaders accused Arab witnesses—including policemen—of perjuring themselves and thwarting justice.[14]

Following the disturbances, the inspector general of the Ceylon Police, Herbert Dowbiggin, was dispatched to Palestine.[15] Dowbiggin had gained a reputation as an imperial policing expert, sent to advise in "trouble spots" throughout the empire.[16] Perhaps surprisingly, Dowbiggin did not attribute the police's failure to its inadequate British component. His report instead scrutinized other features such as the force's overly militarized character and segregation, which contributed to the police being viewed as separate from the population it served.[17] These were, in Dowbiggin's eyes, features that characterized many police forces throughout the empire, which drew too heavily—in both their model and personnel—from the militarized Royal Irish Constabulary ("Black and Tans") rather than the unarmed Metropolitan Police.[18] In Ceylon itself, Dowbiggin promoted unarmed policing along the English model.[19]

According to Dowbiggin's vision, Palestine's policemen would be better educated and more integrated among themselves and within the communities they served.[20] British, Arab, and Jewish policemen would no longer be segregated into separate units but would serve alongside each other in a way that would promote coexistence in the population as well. In their service, the police would model a Palestinian civic identity that would transcend other affiliations and reflect impartiality in law enforcement. The new force would also display greater professionalism: new recruits would have to pass literacy exams, while existing members unable to "make the grade" would be dismissed. The police would be equipped with notebooks rather than armed with rifles.[21] Dowbiggin's reforms were part of his broader vision for bridging the gap between state and society, making the colonial police at least appear to be less coercive or external to the population they policed. Critics, however, sarcastically commented that such reforms replaced "old mounted warriors" with "pimply-faced youths from the training school."[22]

Still, the police's ability to better integrate and gain the trust of communities could not simply be assumed. If the police wished to demilitarize, law enforcement would have to guarantee alternative sources of authority. Therefore part of the shift from rifles to notebooks was restructuring the CID to better prevent and detect crime in less visible, and therefore less threatening, ways: intelligence gathering and scientific analysis of trace evidence.[23]

Drawing on Ceylon's pioneering experience with forensic science, in 1932, a forensic laboratory was established at Palestine Police Headquarters at Mount Scopus.[24] The newly established laboratory conducted ballistic, blood, and semen analyses, as well as identification of fabrics and firearm markings.[25] New recruits were brought to headquarters and given a "smattering of forensic stuff, such as the theory of blood grouping and the test for human or animal blood." They were "taught a bit about the impact of various calibers of firearm bullets" and "an outline of the fingerprint classification system, and how to search for, identify and remove a fingerprint from the scene of a crime." They were also "given instruction in footprints, and in general what to look for and how to investigate various types of crime."[26] Crime scene investigators were equipped with "Medico-Legal and Post-Mortem Specimens" forms to help them collect and label samples and ensure their admissibility in court. By 1937, a medicolegal course was also offered at the Jerusalem Law Classes.[27] Establishing the Dog Section fit neatly within this brave new vision of policing: the dogs provided a sophisticated technique for pursuing even invisible traces that were inadvertently left at crime scenes.

Forensic sciences addressed many of the deficiencies that had impeded investigation and prosecution during the 1929 disturbances. They provided the colonial state with direct access to evidence, unmediated by the indigenous population, thus freeing the police from reliance upon the cooperation of native eyewitnesses. The universal language of science also provided a semblance of precision, objectivity, and evenhandedness, which proved especially important in Palestine, where British authorities were frequently accused (by both Arabs and Jews) of partiality. With fresh memories of the police's ineptitude during the 1929 disturbances, and with mounting British fear of recurring unrest in Palestine, such direct access to evidence was deemed critical.

In line with this vision, the Dobermans were designated to assist the police in investigating otherwise difficult to solve "political crime." In his appeal to the secretary of state for the colonies to support the establishment of a canine unit, Palestine's High Commissioner Arthur Wauchope explained, "I think that there is no question that the use of dogs would assist the Police considerably in their work, particularly where the investigation of political crime is concerned . . . information which might be given by members of the public leading to the arrest and conviction of the criminals is often withheld out of fear and a misguided sense of sympathy."[28] Palestine's dog handlers Parker and Pringle similarly explained that "in countries like this, where the population has no tradition of assisting the authorities in apprehending criminals, the dogs are most necessary since they find their man without asking questions."[29]

Yet despite Dowbiggin's vision of bridging state and society through a "bluer" force, the forensic turn—including the establishment of the Dog Section—was potentially double edged: though it allowed the police to rely less heavily on overt militarized power, it also rendered the public's cooperation less crucial. Through the effective use of trace evidence, the police could afford less reliance on eyewitnesses. The police's new scientific capabilities thus risked further alienating Palestine's population by rendering them passive objects of surveillance rather than active participants in policing their own communities. Similarly, the expansion of the British web of informants that Dowbiggin advocated bore the potential of unraveling Palestine's social fabric by spreading fear and distrust.[30] The turn to notebooks was therefore not entirely harmonious with efforts to bridge state and society. The government's plans to incorporate dogs into the force similarly walked a troublesome line between trust and intimidation.

## Going to the Dogs: Establishing Palestine's Dog Section

While the Dog Section's establishment in 1934 must be considered within the broader context of Dowbiggin's reforms, the precise impetus for the canines' import was the Bedouin trackers' failure to provide incriminating evidence in the Arlosoroff case.[31] Inspector General Roy Spicer (Dowbiggin's protégé, who was appointed to head the Palestine Police after the 1929 disturbances) seized this opportunity to implement his long-awaited plan to employ tracking dogs in policing, which predated his tenure in Palestine. In 1927, while serving in Kenya, Spicer visited the South African Police's kennels and was captivated by the Dobermans' miraculous tracking abilities years later.[32] He was mesmerized by the skill with which the gifted canines could track his prints "across dry veldt" even though he had walked for seventeen hours and had made every effort to cover his tracks.[33] Though Spicer sought to adopt dog tracking in Kenya while serving there, he could not justify the expenditure.[34] When appointed to head the Palestine Police, however, "finding [himself] in a country where the Treasury coffers are full," his plans could be more easily realized.[35] The failed prosecution in the Arlosoroff case provided precisely the justification Spicer needed to import the canines, making his "dog dreams . . . come true."[36]

Soon after being imported, the Dobermans quickly proved their worth. By May 1935, only few months after their arrival, the dogs had already assisted in solving twenty-four cases, including four murders, two attempted murders, eleven crimes against agricultural property, one case of stock theft, and three cases of forced entry.[37] In their first year of operation, the three dogs were put to work at ninety-nine crime scenes, leading some suspects to confess.[38] They were reputed to have followed seventy-two-hour-old tracks, and distances of up to six kilometers from the crime scene.[39] They were able to track criminals even through terrain bereft of any visible tracks.[40] The dogs' fame spread beyond Palestine's frontiers: French authorities in Lebanon soon requested their assistance as well.[41]

The press played a central role in enhancing the canines' public image by disseminating frequent accounts of their successes. Between 1935 and 1939, the *Palestine Post* alone ran nearly four hundred stories about the dogs.[42] To counter any doubts, journalists from Palestine's leading newspapers in all three official languages were summoned to observe and experience the three Dobermans' skills soon after their arrival.[43] The Dobermans also conducted public demonstrations for incredulous spectators at both real and simulated crime scenes, where the dog masters asked audience members to hide objects for the dogs to retrieve.[44] On March 1, 1935, the front page of the Zionist Labour Movement newspaper *Davar* reported the Dobermans' first investigative triumph. According to the article, Gift and Mayer were brought to sniff a jewelry box from which £P.15 was stolen. Each leading their handlers separately, the two dogs followed identical trails to Dir Yassin, a few kilometers away from the crime scene. They then both went on to select the same individual.[45]

Though ostensibly imported to address "political crimes" where native eyewitnesses' cooperation was difficult to procure, dogs were initially employed mainly to address common but otherwise difficult to investigate "agrarian crime": tree cutting, crop burning, and animal maiming.[46] The latter in particular was regarded by colonial authorities to be endemic, representing one of "the most loathsome habits in this country."[47] To British observers, animal maiming epitomized the cruelty and irrationality of the Arab delinquent and distinguished him from his European counterpart: such crimes provided no tangible gain, satisfying only base, vengeful instincts, achieved through wanton cruelty toward innocent creatures.[48] As the *Palestine Post* noted, animal maiming displayed "beastliness of conception with great difficulty of detection and proof," thus rendering it "outside the pale of ordinary police methods." Dogs provided a practical solution that also bore symbolic significance that was not lost on contemporaries: "Brutality in humankind is confronted by a super-human intelligence in brute creation. Better still, one dumb creature has it in its power to avenge another."[49] That the dogs were first employed to these ends, rather than to investigate political crimes, perhaps stemmed from the authorities' desire to first establish dog tracking's acceptability as an investigative method before employing them in more contentious cases.

British inability to comprehend native criminal motives rendered not only dog tracking, but forensic science more generally, particularly crucial in the colonies. In his 1925 book *Forensic Medicine*, Sir Sydney Smith, the principal medicolegal expert to the Egyptian government, explained that it was this inability to make sense of natives' motivations that deemed forensic science so crucial in the Orient: "Motive, which plays so prominent a part in connection with Western crime, is often difficult to understand in the East, for murders of an extremely revolting nature may have what appears to be a most insignificant motive."[50] Not only were motives difficult to ascertain, but also eyewitnesses could not be trusted even during times of relative political calm: "As a rule, the statements of the victim bear no relation whatever to the facts, and even when the case is a genuine one the person will endeavor to improve it by telling the most ridiculous lies."[51] Unable to comprehend motives or to evaluate truthfulness, authorities instead focused on analyzing physical "trace evidence" such as tracks and marks.[52] Through the prism of Orientalist criminology, dog tracking—like other trace evidence—was particularly well suited to meet distinctly colonial challenges.

## Let Slip the Dogs of War: Suppressing the Arab Revolt

Though employed widely since their 1934 import, Palestine's Dog Section reached its heyday during the Arab Revolt (1936–39), when the unit was deployed to address attacks on British targets. The revolt was triggered by the murder of two Jewish drivers on April 15, 1936, which was followed by the murder of two Arabs in Petah Tikva. Violence spiraled rapidly and spread throughout the country. The Arab Higher Committee (AHC), an ad hoc national stirring body, called for a general strike to place economic pressure on the British government. Particularly in northern Palestine, the strike was accompanied by violent attacks on Jewish and British targets. With Jewish emigration from Europe increasing rapidly after the ascent of the Nazi Party in Germany, Arab leadership posed an ultimatum to the Palestine government for resolving key issues such as immigration, land sales, and representative government.[53]

The appointment of the Commission of Inquiry chaired by Lord Peel fostered hopes that these issues might be addressed, and in October 1936, the AHC called to end the strike. But in July 1937, the commission recommended that Palestine be partitioned. Soon thereafter, hostilities resumed and intensified: the September 1937 assassination of Acting District Commissioner of Galilee Lewis Andrews marked the beginning of the second, bloodier phase of the revolt, prompting the proclamation of martial law. Many members of the AHC fled Palestine, while others were administratively detained or exiled. During 1938, the British effectively lost control of many parts of Palestine, most notably the mountainous northern region surrounding Nablus.[54] Civil authorities ultimately handed the command over Palestine to the military, under the command of Major-General Bernard Montgomery. With overwhelming and uninhibited force, the 8th Infantry Division brought the revolt to an end in May 1939. By then, Jewish casualties were in the hundreds, while Arab casualties were estimated to be between three and six thousand.[55]

During the revolt, Dowbiggin's plan of relying on informants for early warning proved impracticable, as the colonial state's intelligence-gathering networks were shut out of Arab communities.[56] This made the state's reliance on trace evidence far more crucial. To meet increasing demand for canine assistance during the revolt, "aeroplanes were used in order to transport them expeditiously from one part of the country to another."[57] With airlifting, the Dobermans could arrive anywhere in Palestine almost immediately. Despite its small size, "To the rural villager who decided to break the law, the new dog section now began to appear ubiquitous."[58] With the assistance of the Royal Air Force (RAF), from their kennels in Jerusalem, the Dobermans could track criminals in Palestine's most remote districts; that prospect alone—it was hoped—would keep the rebels at bay. In so doing, the British merged two powerful technologies of imperial control: dogs and aircrafts.[59] Of 172 otherwise unsolvable mysteries in which the three dogs were employed in 1936, they were reportedly successful in detecting the culprit in as many as 87 cases.[60]

One such "successful" tracking took place on August 29, 1936. A search party of the York and Lancaster Regiment discovered a dead body of an Arab in possession of a rifle and a clip of cartridges. The dogs were "brought to the scene by aeroplane" and given scent, leading the forces on an eight-kilometer trail to a house where they "found a large stock of explosives, powder and lead, together with a photograph of the dead man."[61] If the official report is to be believed, more convincing corroboration could hardly be imagined.

During the Arab Revolt's more violent phase, the Dog Section became increasingly central to British plans to restore law and order. In September 1937, the secretary of state for the colonies dispatched Charles Tegart (former commander of the Calcutta Police) and David Petrie (director of intelligence for India and subsequently head of MI5) to advise on how to address the emergency. Titling the fourth section of their report "Dogs," Tegart and Petrie deemed the Dobermans indispensable for combatting "Arab terror."[62] "The trained instinct of these animals achieves results which are quite beyond the highest human intelligence," they noted.[63] Tegart and Petrie's observations relied not only on second-hand reports, but also on an investigation they had closely followed: in January 1938, archaeologist J. L. Starkey was murdered outside Hebron by "Arab bandits." Tegart and Petrie saw exactly how the dogs led detectives from the murder scene on a twenty-two-kilometer trail through mountainous terrain, from Beit Jibrin to Kharass, ultimately leading the police to a weapon concealed in a wall.[64]

In their report, Tegart and Petrie lamented the short supply of Dobermans. Even with the RAF at their service, the Dobermans could not meet demand.[65] In Hebron, only ten of fifty calls for the Dobermans' help were met.[66] As a result, many crimes went unsolved as the Hebron Police despaired of making additional requests. "Consideration of time, economy in transport and escort, as well as the health of the dogs themselves" required significantly increasing the number of dogs and spreading units throughout the country. The report recommended that there "be a complete and self-contained establishment in Nazareth," with the aim of eventually establishing independent dog units in Nablus and Haifa as well.[67]

Tegart and Petrie explicitly linked the demand for more dogs to waning public assistance, underscoring the need for stronger trace evidence during times of colonial unrest. "In every case where tracks are left, there is, with dogs, a good chance of success and without them a virtual certainty of failure, assistance from the public being rarely forthcoming." The revolt also added complexity to the use of Bedouin trackers: beyond the unreliability of their technique, the revolt placed their loyalty in question.[68] With no political agenda of their own, the Dobermans became an even more attractive alternative to the Bedouin trackers. The Tegart and Petrie report determined "There is nothing more likely to put an end to acts of sabotage and violence than tracking by dogs." They concluded: "The question of dogs we regard as of the utmost importance and urgency in the matter of restoring law and order."[69]

To implement Tegart and Petrie's recommendations, in early 1938, the Palestine Police spent considerable resources on acquiring eight additional Dobermans and temporarily commissioning two South African dog masters.[70] A new kennel was erected in Affula, which reportedly "greatly increased the efficiency of the section. Dogs can now arrive at the scene of a crime in any part of the country within two hours of being requisitioned."[71] The investment seemed imperative: with anticolonial sentiment mounting, police dogs had become indispensable.

When South Africa could no longer meet increasing demand for Dobermans, the Palestine Police turned elsewhere. Despite strained Anglo-German relations after the annexation of Sudetenland, in September 1938, the Palestine government acquired three dogs from Nazi Germany.[72] Beyond the political tensions and the language barriers that German-trained dogs would have to overcome, one Colonial Office official noted the irony in the "rather curious position that we should be seeking the aid of the German Government in acquiring animals designed to track down Arabs who murder Jews."[73]

During the revolt, dog tracking not only aided police investigations and evidence discovery but also increasingly justified extrajudicial punitive action.[74] As one British soldier reported in a letter to his family, "You may follow the police dogs into one village and upon this vague clue you may smash the village and burn it down."[75] His account was not mere hyperbole. Following the November 5, 1937, killing of two "Black Watch" soldiers near Jerusalem's Jaffa Gate, Dobermans led military forces to Silwan, where soldiers beat twelve civilians to death with rifle butts.[76] Similarly, after rebels ambushed and killed an RAF officer in February 1938, a dog led British forces to a house in the village of Ijzim. But when put on the scent again, the dog led to a house opposite the one it had originally marked. Rather than giving the inhabitants of both homes the benefit of reasonable doubt, both houses were razed, and the entire village was fined.[77]

Though the Dobermans were used mostly to track Arab suspects during the revolt, during the 1940s, the Palestine Police increasingly began using dogs to track "Jewish terrorists" agitating against British rule.[78] Dogs helped investigate the August 1944 attempt on the life of High Commissioner Harold McMichael.[79] Following a 1946 attack on a police post in Kfar Vitkin, dogs tracked the culprits back to Ramat HaKovesh.[80] That same year, dogs also helped uncover an ammunition stash outside of Birya, leading to a wave of arrests.[81] Dobermans would continue hounding Zionists exiled from Palestine for anticolonial activity. In 1947, the Kenyan government requested that the South African Police provide them with Doberman Pinschers to track down detainees who had escaped a British detainment camp at Gilgil.[82]

## Primitive or Modern? The Psychological, Religious, and Symbolic Aspects of Dog Tracking

Dog tracking's allure stemmed from more than merely the canine's ability to reliably follow a scent or their capacity to excuse arbitrary British actions. One of the dogs' primary appeals was their perceived psychological effect on the local population, who purportedly could not quite fathom how the dogs were able to detect them. After his 1927 visit to South Africa, Spicer noted that "half the effect" of the Dobermans was in the "influence they produce on the native mind."[83] In Palestine, Spicer similarly noted with satisfaction the "moral effect" that the dogs possessed over the "criminal classes," both primitive and sophisticated: "They are regarded with superstitious fear and dread by less educated offenders while the really intelligent criminal realises that they are possibly the most dangerous servants of public security that has yet been encountered."[84] In its annual report to the League of Nations for 1935, His Majesty's Government reported, "The Police dogs have . . . created a very useful deterrent impression in the minds of villagers."[85] Once apprehended, the dogs also proved useful in prompting suspects to confess.[86] Britain's conscious enlistment of superstition to enhance investigative techniques highlights the Dobermans' dualism: they were simultaneously considered modern and primitive, appealing to both the scientific (or pseudo-scientific) sensibilities of the European while also tapping into the superstitious fears—real or attributed—of natives.

Criminals' alleged inability to comprehend precisely how the dogs detected them was a significant component in what deemed canine tracking effective. Unable to fully grasp the dogs' mysterious ways, "primitive" Palestinian thieves reportedly tried every trick, including the wrapping of rags around their feet "in the vain hope of deterring the inevitable trail of Mayer or Gift."[87] Unlike eyewitnesses, the Dobermans' talents were constrained by neither temporality nor space—they could arrive at a crime scene hours or even days after the offense and identify the culprits even without perceiving their actions. The fear of such surveillance capabilities was thought to instill an anxiety that reduced criminality. Though the loss of informants during the revolt meant that the government did not possess eyes and ears everywhere, it hoped instead to rely on the dread of its enhanced nose.

## Harnessing Religion? Dogs in Judaism and Islam

Though less blatant in official reports, British authorities displayed an awareness of dogs' religious significance and seemed willing to tap into it to reduce crime. Spicer noted with gratification Palestinians' "superstitious fear" of dogs as a significant factor in their success. Both Islam and Judaism maintained a complex relationship with canines. According to some Islamic traditions, dogs were considered impure (*najis*). Some *hadith* described how angels refrained from visiting dwellings in which hounds resided, good deeds were believed to be discounted by dog possession, and the Prophet himself allegedly ordered the killing of canines.[88] As a result, dog ownership in the Middle East in the twentieth century was limited mostly to Bedouins, who used dogs for protection and herding, and Westernized elites, for whom dogs symbolized "Europeanization."[89] Judaism displayed a similar aversion toward dogs, dating back as far as biblical and Talmudic times.[90] In Europe, Jews were widely believed to suffer from an irrational fear of dogs.[91] As a Yiddish adage suggested, "if a Jew has a dog, either the dog is no dog or the Jew is no Jew."[92]

Other British officials serving in Palestine also displayed awareness of dogs' cultural significance for Palestine's inhabitants. Citing dogs' impurity in Islam, Palestine's first High Commissioner Herbert Samuel (1920–25) forbade the entry of dogs into his official residence out of consideration for his guests.[93] Similarly, when discussing the Dog Section of the Palestine Police in a letter home, one soldier noted that "Muslim Arabs considered dogs to be unclean,"[94] which was what rendered them effective in investigation. Dogs were not the only way in which the British harnessed religious taboo in criminal investigation: the CID reportedly used forced intoxication and blasphemy to coerce confessions.[95]

Though British authorities may have sought to harness religious taboo in employing the canines, their success remains questionable. Despite reports by the English press that Arab villagers referred to the police's Dobermans as "sons of the devil,"[96] Arabic newspapers such as *Filastin* and the more critically anticolonial *al-Difa'a* did not seem repulsed by the dogs. They published celebratory accounts of the canines' capabilities and successes.[97] The same was true of the Jewish press.[98] These papers expressed amazement with the dogs' skills and hopes that the dogs might be employed more broadly to curb crime, particularly agricultural crime that proved difficult to investigate.[99] So deep was the belief in the Dobermans' tracking abilities that victims—Muslims and Jews alike—soon began demanding that dogs be brought to crime scenes to investigate.[100]

Though one may dismiss celebratory press coverage as reflecting the views of Westernized elites, a more plausible explanation is that dogs did not truly tap into religious sentiment or possess the religious effects that the British had hoped.[101] As Alan Mikhail argues, dogs played a significant role in Egyptian society throughout the nineteenth century, their imputed impurity notwithstanding.[102] It was no longer clear that dogs were treated with fear or distaste among Jews, either. For Zionists, the fear of dogs was associated with the diasporic Jew's general cowardice and detachment from nature, which Zionism set out to obliterate. In an effort to forge a new Jewish identity, the Zionist movement therefore made a conscious effort to recast Jews' relationship with dogs, and with nature more generally.[103]

## Semi- or Superhuman? Canine versus Native Trackers and Britain's Civilizing Mission

Though the dogs' religious significance may have been exaggerated, replacing Bedouin trackers with Doberman Pinschers still bore considerable symbolic value. In occupying a liminal space between science and superstition, culture and nature, domesticity and wilderness,[104] British approaches to dog tracking paralleled a duality evident in European depictions of the African and Australian bushmen, the American Indian, and the Bedouin tracker.[105] Perhaps paradoxically, to European eyes, all fluctuated between quasi-human and superhuman, "Shaman" and "Sherlock."[106] What they supposedly lacked in intelligence they made up for in keener instincts, being more deeply in tune with their environment. Able to observe even the minutest interferences in nature, they possessed skills that European city dwellers had lost long ago. Yet for both native and canine trackers, "domestication" posed a threat, as it risked blunting these sharp instincts and—given their deficient intelligence—rendering them redundant.[107]

Spicer made the comparison between Bedouin trackers and Doberman Pinschers explicit. In his letter to Palestine's chief secretary, he noted: "The South African police dog is, with the possible exception of the Maori Aboriginal, the most wonderful tracker in the world."[108] Whereas some veterans of the Palestine Police believed that the Ottomans had no need for dogs because Bedouins trackers "are better than any dog,"[109] to Spicer, such "a tracker was better than nothing, but a highly trained dog was better still."[110]

The comparison between canine and native trackers carried to other locations and contexts in the British Empire.[111] Perhaps further effacing the boundary between nonhuman and native, British colonial authorities employed indigenous trackers alongside canines to reinforce one another. In 1950s Kenya, British forces commissioned Kikuyu and Sudanese trackers alongside canines to form "Tracker Combat Teams" (TCTs) charged with "deep penetration tactics."[112] Particularly challenging tracking was left, however, to canine rather than native trackers.[113]

Court decisions across the empire also echoed the analogy between indigenous and canine trackers: when addressing the admissibility of dog-tracking evidence (a topic discussed more fully below), judges pondered whether such proof could have been admitted had it been offered by a native who was unable to verbally articulate his findings. In South Africa, Judge Thomas Graham urged his readers to "Conceive a case where the police possessed the services of a native whose language no one in South Africa understood."[114] Yet there was nothing inherently significant about the tracker in Graham's analogy being indigenous; his legal point could have been pressed with equal force using the example of an uncommunicative, nonnative witness.

Such analogies between native and canine, as well as their joint employment, bore symbolic significance beyond hinting at their equivalence. For Europeans, one manifestation of native savagery was the wanton violence and cruelty that natives inflicted upon animals.[115] By indicating that dogs were equal or even superior trackers, British authorities signaled that dogs (and animals more generally) were worthy of the natives' respect and fear. Moreover, dog breeding and training—like colonialism itself—epitomized Europeans' ability to tame, master, and harness nature: it manifested asymmetrical power relations in which an unpredictable savage subject could be taught to realize the usefulness of his skills and trained to be loyal servant.[116]

## Trial by Canine: Dog Tracking as Judicial Evidence in Palestine

Though the Palestine Police began employing dogs as early as 1935, the admissibility of dog tracking as judicial evidence was not a forgone conclusion. From the fact that dogs could be used in investigation, it did not necessarily follow that their tracking would be admissible in court.

In South Africa, a pioneering country in dog tracking and the only common-law jurisdiction to have addressed the admissibility of such proof in depth in the early twentieth century, canine evidence had been excluded since 1920. In *R. v. Trupedo*, Chief Justice Innes ruled that dog tracking "testimony" constituted inadmissible hearsay, with all its concomitant dangers.[117] Beyond the dogs' unavailability for cross-examination, Innes ruled, the dogs' "assertions" when barking or laying their paws on a suspect at an identification parade were prone to "misunderstanding between the animal and his keeper" and therefore could not be presented at trial.

Remarkably, Innes did not deem the canines' lack of rationality or their reliance on instinct to be obstacles to admissibility; quite the contrary. He ruled that dogs—like other animals—might still provide reliable circumstantial evidence. In some cases, he ruled, "inferences may be quite properly drawn from the behaviour of animals." For example, where a dog had failed to bark upon the entry of an intruder, as in Conan Doyle's *Silver Blaze*, one could properly infer that the trespasser was someone familiar to the dog.[118] But this was not the case when prosecutors sought to introduce tracking and identification of an individual perpetrator previously unknown to the dog. In distinguishing permissible and impermissible inferences from canine behavior, Innes contrasted between behavior that was "instinctive and invariable" and skills for which animals "must be carefully trained before they can be relied upon." Whereas the former were founded upon the "instinct of self-preservation," the latter involved "processes closely akin to reasoning." As such—somewhat counterintuitively—whereas instinct was reliable enough to produce admissible evidence, animal skills honed through training could not. Reasoning and analogous processes were attended by risks of error and, more importantly, insincerity of either dog or handler.[119] Innes concluded by noting: "We have no scientific or accurate knowledge as to the faculty by which dogs of certain breeds are said to be able to follow the scent of one human being, rejecting the scent of all others." Properly analyzed, he ruled, dog tracking belonged in the "region of conjecture and uncertainty" rather than science.[120]

Consistent with South African law, when the Dobermans were first introduced into Palestine, the police declared publicly that their "function is not . . . to produce evidence . . . as naturally dog testimony is admissible in no court."[121] At most, the dogs could lead to other clues or evidence that was independently admissible, and that would in turn help the police identify the culprit or cause a suspect to confess. Even dog enthusiasts and experts such as Dr. Rudolfine Menzel, a world-renowned authority on dog training and tracking and founder of the *Palestine Canine Research Institute* in Kirjath Motzkin, expressed skepticism about dog tracking's "scientific foundation" (*wissenschaftliche fundierung*).[122] In a 1938 book, Menzel observed that dogs typically followed the most recent—and hence strongest—scent that they were able to detect. It was therefore not uncommon, she noted, for a dog to follow tracks for miles, finally leading to the home of a policeman who had investigated the crime scene. She opined that even the most talented dogs could not simply sniff an object and lead their handler to the criminal in a manner that laypeople often believed.[123]

Menzel's own experience with the Palestine Police Dog Section provided at least anecdotal grounds for skepticism: in at least two instances where Menzel arrived at murder scenes (often hours before the Palestine Police), her tracking dogs followed entirely different trails than the police's Dobermans. This suggested that a dog's training and handler significantly influenced the path followed and the suspect identified.[124] Other incidents in Palestine provided additional grounds for skepticism concerning their accuracy: in one robbery case, the Dobermans identified an old blind man, who clearly could not have committed the crime.[125] Critics further alleged that at lineups dogs invariably picked out someone, even if the perpetrator was not present.[126] Given their lack of verbal skills, the dogs' precise reasons for choosing a particular suspect could not be adequately explored. Moreover, critics alleged that, once selected, police would fabricate additional evidence of an individual's guilt.[127]

Though Dobermans may not have initially been expected to furnish evidence, the threat to civil order posed by the Arab Revolt led the government to renege on its pledge not to rely on such proof in judicial proceedings. As the Tegart report noted, during the revolt, the police were often unable to procure willing eyewitnesses; they could muster little more than the barking or pawing of a Doberman Pinscher to tie defendants to a crime. Yet neither the novelty of the method nor skepticism concerning its reliability led to its judicial exclusion. Moreover, two dogs could confirm each other's identifications and prove guilt beyond reasonable doubt, with criminal defendants sent to the gallows based on such identifications alone.[128] In extreme cases, even the testimony of one dog could suffice: in a 1938 shooting case, a dog led the police to Mustafa Mansour's village and house, where they discovered a few rounds of ammunition, some of which were spent.[129] The handler's testimony was sufficient to tie Mansour to the shooting, and he was sentenced to death.[130]

Perhaps the inability to scrutinize or effectively undermine the dogs' evidence was one of their greatest assets, especially during the Arab Revolt. The scent that the dogs followed could be neither seen nor preserved, and their "testimony" could not easily be challenged. Unlike Bedouin trackers, it would be difficult—if not impossible—to refute their evidence in court. This advantage of dog tracking is perhaps best illustrated by the case of four defendants of the Abdullah family from Kafr Qasm, who in 1938 were all convicted and sentenced to death for setting fire to a mill. In this case, the police employed both human and canine trackers to investigate the fire and its causes. The tracker's statement, however, was what provided grounds for appeal: "The tracker in his statement says that it was difficult to follow the trail along the rocky hills" and conceded that at one point he had "lost the trail." By contrast, the dogs that had pounced on two of the four defendants could not be impeached.[131]

Not only did Palestine's courts admit canine evidence, but they also often dispensed with the few available safeguards for evaluating such proof. Those testifying and interpreting the dogs' identifications in court were not always their handlers.[132] Parker and Pringle—Palestine's only certified dog masters for a number of years—were "virtually worked to death" during the Arab Revolt, and as a result were often unavailable to testify.[133] Instead, rank-and-file policemen offered evidence about the canines' behavior on the trail or during lineups: whether they pounced on the defendant or barked upon arriving at his home.[134] Given their lack of specialized expertise, these policemen could not speak to the dogs' pedigree, training, or reliability; nor could they testify concerning adherence to dog-handling protocols in the particular investigation. At most they could attest to compliance with general lineup guidelines, rendering their cross-examination of limited value.[135]

Yet dog identifications were admitted in Palestine even when their masters had clearly breached handling and tracking protocols.[136] In a 1936 murder trial, dog master Parker divulged during cross-examination that before holding a two-dog, ten-man lineup, he had permitted "a suspect person to come near the place in which there is a trace related to the crime." This raised fears of contamination and the possibility that the dogs—much like the Bedouin trackers in the Arlosoroff investigation—had followed tracks left at the scene *after* the crime. Though the judge deemed this breach "extreme negligence," he nevertheless admitted the evidence. Based on the dogs' tracking and the defendants' discernable motive, the two were convicted and sentenced to long prison terms, despite an eyewitness naming three other suspects.[137]

How did judges in Palestine justify the admission of dog-tracking evidence despite the legal challenges it posed? This question gains salience given dog tracking's exclusion as insufficiently reliable or scientific in South Africa. Because no judge addressed the question of canine tracking's admissibility directly, we may only surmise their reasons.[138] When dogs supplied the only proof linking alleged rebels to a crime, judges in Palestine may have been disinclined to insist on procedural niceties. This reticence may have also stemmed from the executive's mounting pressure on the judiciary to assist in efforts to quell the unrest. A rare 1937 instance in which a magistrate declined to indict based on dog-tracking evidence resulted in severe criticism in Westminster of the Palestine government's mishandling of the revolt. As a result, the secretary of state for the colonies demanded answers from Palestine's high commissioner.[139]

Executive pressure on the judiciary materialized in two concrete ways. First, judges who proved unwilling to aid the government or too critical of its actions faced the risk of removal. Chief Justice Michael McDonnell, who was deemed by Palestine's high commissioner to be "unhelpful," was forced to resign in 1936. His dismissal sent a clear message to his brethren concerning the limits of judicial independence and review of executive action in the colonies.[140] Second, the threat of martial law, which would allow the executive to bypass the civil judiciary altogether, was imminent (and ultimately materialized).[141] Observing brutal extrajudicial collective punishment, which sometimes relied on dog tracking, judges may have considered the admissibility of canine evidence to statutorily punish individual defendants to be the lesser of two evils. These factors may have contributed to judicial readiness to admit dog-tracking evidence deemed critical by the government, even if they found it distasteful.

## Conclusion

Developed in the colonies and embraced more fully during a time of crisis, dog tracking was a forensic method that thrived in a juridical periphery at the twilight of legalism. As anticolonial sentiment in Palestine increased and public cooperation waned, dog tracking offered an indispensable tool in British efforts to restore law and order. Initially, canines helped justify the state's otherwise capricious, extrajudicial action. Yet once employed to extrajudicial ends, dog tracking soon began figuring into the ostensibly more rational considerations of the legal system as well.

Though particular to a time and place, the history of dog tracking in Palestine offers invaluable lessons concerning the development of investigative techniques and their path toward legal admissibility. The use of dog tracking in Palestine highlights how expediency may lead to the acceptance of forensic methods despite their lack of sound, fully comprehensible scientific foundations. Particularly during times of emergency, and especially when applied in liminal spaces and to "others," the judiciary may prove an inadequate gatekeeper against such questionable evidence. That dog tracking had not been scientifically theorized or fully comprehended perhaps only contributed to its efficacy, rendering it essentially irrefutable in court. Moreover, as a case study, dog tracking emphasizes how a forensic technique's public perception may be as pertinent to its employment as its factual accuracy: by colonial officials' accounts, the canines' *perceived* abilities and the effect that dogs produced on the "native mind" were as central to their employment as the precision of their tracking skills. Like a panopticon, criminals' permanent "olfactibiliy" to the dogs contributed to the state's perceived omnipresence and omniscience.

### NOTES

1. See, generally, Shabtai Teveth, *The Arlosoroff Murder* (Jerusalem: Schoken, 1982); Horace B. Samuel, "Who Killed Arlosoroff? A Record of Crime and Justice in the Mandated Territory of Palestine" (unpublished manuscript, 1934).

2. "Arlosoroff Murder Trial—Trackers' Evidence; Plaster Casts of Footprints," *Palestine Post*, May 3, 1934, 7. See also Samuel, "Who Killed Arlosoroff?," 34.

3. Criminal Assize Appeal 7/1934, *Abraham Stavsky v. Attorney General, Law Reports of Palestine* 2 (London: Waterlow & Sons, 1937): 148, 150–51: "The evidence of the trackers was in many ways unsatisfactory. The undoubted confusion of the tracks which they followed, with tracks showing spur chains that were clearly those of a mounted constable . . . are enough to make it difficult to accept this evidence, especially in view of the circuitous route followed

by the debated tracks . . . Further, even if the evidence as to tracks on the scene of the crime were unimpeachable, I am satisfied that the foot-print parade on the beach was vitiated by the fact the trackers witnessed the identification parade in the station yard."

4. Samuel, "Who Killed Arlosoroff?," 62–76.

5. "Introduction of a Police System in Palestine (Purchase of Police Dogs)," Israel State Archives (ISA) M 335/10. See also R. G. B. Spicer, "The New Detective," *Police Journal* 9 (1936): 245–51.

6. "Police Dogs for Palestine: Officers Return with Trio," *Palestine Post*, December 24, 1934, 10.

7. "NSF Awards $15 Million to Crack the Olfactory Code," National Science Foundation, September 21, 2015, http://www.nsf.gov/news/news_summ.jsp?cntn_id=136333; "Scientists Win $6.4 Million to Crack the Code of Smell Navigation," *Berkeley News*, September 24, 2015, http://news.berkeley.edu/2015/09/24/smell-navigation-grant/. "Olfaction is one of the last frontiers of neuroscience, the least understood of the five senses."

8. Legal systems would encounter similar challenges as they began accepting automated or mechanical evidence. See Andrea Roth, "Machine Testimony," *Yale Law Journal* 126 (2017): 1972. Dog tracking relied on the agency of an independent actor, one that had been trained and "programmed."

9. Andrea Roth, "Trial by Machine," *Georgetown Law Journal* 104 (2016): 1253–54, at 1245; Roth, "Machine Testimony."

10. "Disturbances" is the term used by the British. In Hebrew, the events are commonly referred to as *P'raot Tarpat*, or "the 1929 Pogroms." In Arabic, they are known as *Thawrat al-Buraq*, or "the Wall Revolt."

11. Jews—especially from the Revisionist Party—had their own fears that the British were defying the status quo to their detriment. See Martin Kolinsky, *Law, Order and Riots in Mandatory Palestine, 1928-1935* (London: St. Martin's Press, 1993), 40–41.

12. The entire force numbered 2,100, of which 300 were Jews and 1,600 Arabs. See Martin Kolinsky, "Reorganization of the Palestine Police After the Riots of 1929," *Studies in Zionism* 10 (1989): 155–73.

13. See *Palestine Blue Book for 1929* (Jerusalem: Government Printer, 1930), 343.

14. "Report on the Scope, Character and Result of the Judicial Proceedings upon the August 1929 Riots in Palestine," ISA P 758/4.

15. See Kolinsky, *Law, Order and Riots in Mandatory Palestine*, 100–101.

16. Gad Kroizer, "From Dowbiggin to Tegart: Revolutionary Change in the Colonial Police in Palestine During the 1930s," *Journal of Imperial and Commonwealth History* 32 (2004): 115–33, at 119.

17. Kroizer, "From Dowbiggin to Tegart," 120.

18. For an analysis of the "Irish Model" and its influence throughout the empire, see David M. Anderson and David Killingray, eds., *Policing the Empire: Government, Authority and Control, 1830-1940* (Manchester: Manchester University Press, 1991), esp. chap. 2.

19. Kroizer, "From Dowbiggin to Tegart," 119.

20. Kroizer, "From Dowbiggin to Tegart," 120.

21. Douglass V. Duff, *Bailing with a Teaspoon* (London: John Long, 1953), 184.

22. Duff, *Bailing with a Teaspoon*.

23. Kroizer, "From Dowbiggin to Tegart," 121. See also Kolinsky, *Law, Order and Riots in Mandatory Palestine*, 101.

24. Edward Horne, *A Job Well Done: A History of the Palestine Police Force, 1920–1948* (Lewes, East Essex: Book Guild, 2003), 471.

25. Horne, *Job Well Done*, 471.

26. Colin Imray, *Policeman in Palestine: Memories of the Early Years* (Devon: Gaskell, 1995), 99.

27. "Synopsis of Lectures in Medical Jurisprudence, 1918–1947," ISA M 5102/11; ISA M 5102/12.

28. Wauchope to Cunliffe-Lister, June 21, 1933, National Archives of the United Kingdom (NAUK) CO 733/246/12.

29. Dorothy Kahn, "The Private Lives of Mayer and Gift: Canine Sleuths at Work and Play," *Palestine Post*, May 19, 1935, 6.

30. See, generally, Hillel Cohen, *Army of Shadows: Palestinian Collaboration with Zionism, 1917–1948* (Berkeley: University of California Press, 2008).

31. "From Town and Country: Training Boxer Dogs," *Palestine Review* 4, no. 1 (April 21, 1939): 6–7.

32. Spicer, "New Detective," 245.

33. Spicer to Chief Secretary, April 11, 1933, NAUK CO 733/246/12. Spicer, "New Detective," 245: "I tried every stratagem and trick performed by a hunted stag. I doubled my tracks, chose stony ground, went round in circles and jumped out; but the bitch never missed a yard and marked my entry into a car on the main road with no hesitation."

34. Spicer to Chief Secretary, April 11, 1933.

35. Spicer to Allan, October 16, 1934, NAUK MEPO 2/4981.

36. Spicer, "New Detective," 245.

37. J. L. Meltzer, "A Local Departure in Crime Detection: Dogs Used by Palestine Police," *Palestine Post*, May 9, 1935; "How Dogs Assist the Police: Jerusalem Journalists Receive a Lesson in Dog Training for Crime Discovery," *Doar Hayom*, May 10, 1935, 4.

38. Horne, *Job Well Done*, 458.

39. Horne, *Job Well Done*, 458.

40. Horne, *Job Well Done*, 458. See also "Police Dog Tracks Robbers: 'Kim' Follows Scent for 6 Kilometers," *Palestine Post*, September 24, 1945. The article describes how at dawn Kim followed a print from the scene where a bus had been held up the previous evening, to the middle of the village of Kaza, where the scent was lost. An identification parade of the village's eleven male inhabitants was then held, where Kim identified Ahmed Hassan.

41. "Borrowing Palestine Police Dogs," *Palestine Post*, October 14, 1935.

42. Horne, *Job Well Done*, 456. For examples of such publicity, see "Police Use Dogs," *Davar*, March 1, 1935; "Police Dogs in Salame," *Filastin*, April 16, 1935.

43. According to reports in the *Palestine Post* and *Filastin*, in attendance were journalists from *Al-Difa'a*, *Al-Jamea Al-Islamiya*, *Davar*, *Doar Hayom*, *Filastin*, *Haaretz*, *Hayarden*, and the *Palestine Post*. "How Dogs Assist the Police: Jerusalem Journalists Receive a Lesson in Dog Training for Crime Discovery" [in Hebrew], *Doar Hayom*, May 10, 1935, 4; "The Police Dogs," *Haaretz*, May 5, 1934; Meltzer, "Local Departure in Crime Detection"; "Police Dogs and Crime Detection: Important Press Demonstration," *Filastin*, May 14, 1935.

44. "Crime-Dogs Solve Crimes," *Al-Difa'a*, May 9, 1935.

45. "Police Use Dogs."

46. Meltzer, "Local Departure in Crime Detection": "The animals appear at their best in agrarian crimes or malicious injury to property." See also "Destruction of Trees," *Palestine Post*, October 18, 1935: "The cutting of the trees of one's enemy is still a common type of revenge in the Tulkarem district, despite the success of the police dogs in tracking down offenders." Though perhaps not their intended primary use, the police did anticipate using them in such investigations. See Wauchope to Cunliffe-Lister, June 21, 1933, NAUK CO 733/246/12 (mentioning particularly "treecutting [sic] and animal maiming in Arab as well as Jewish villages").

47. "Suppressing a Savage Crime," *Palestine Post*, June 4, 1935. The article was written after the dogs reportedly tracked down a man who in an act of revenge stabbed a mare belonging to the Sheikh of the Sakne tribe.

48. See, e.g., Frederic M. Goadby, *Commentary on Egyptian Criminal Law and the Related Criminal Law of Palestine, Cyprus and Iraq* (Cairo: Government Press, 1924), 320: "crimes of vengeance such as murder, wounding and malicious injuries to property are far more common in Egypt, while in the case of acquisitive crimes such as theft and the like the difference [between England and Egypt] is less startling." As Sydney Smith, principal medicolegal expert to the Egyptian government, noted, "It is perhaps a sign of civilisation and progress that in more advanced communities crimes of revenge tend to be greatly outnumbered by crimes committed for gain." *Mostly Murder* (New York: D. McKay, 1959), 65.

49. "Suppressing a Savage Crime," *Palestine Post*, June 4, 1935.

50. Sydney Smith, *Forensic Medicine: A Text-Book for Students and Practitioners* (Philadelphia: P. Blakiston's Son, 1925), 471.

51. Smith, *Forensic Medicine*, 471.

52. For a discussion of the distinction between trace and predictive evidence, albeit in the context of character evidence, see H. Richard Uviller, "Evidence of Character to Prove Conduct: Illusion, Illogic, and Injustice in the Courtroom," *University of Pennsylvania Law Review* 130 (1982): 845, at 847–48; Chris William Sanchirico, "Character Evidence and the Object of Trial," *Columbia Law Review* 101 (2001): 1227, 1234.

53. See, generally, Benny Morris, *Righteous Victims: A History of the Zionist Arab Conflict, 1881–2001* (New York: Vintage, 2001), 142–54; Rashid Khalidi, *The Iron Cage: The Story of the Palestinian Struggle for Statehood* (Boston: Beacon Press, 2006), 105–10.

54. The rebels established their own government, levied their own taxes, and established their own judicial system, which in addition to trying regular civil and criminal cases severely punished—and often executed—those suspected of collaboration with Britons or Jews. Zeina Ghandour, *A Discourse on Domination in Mandate Palestine: Imperialism, Property and Insurgency* (New York: Routledge, 2010), 99–101.

55. Morris, *Righteous Victims*, 159–60.

56. Martin Thomas, *Empires of Intelligence: Security Services and Colonial Disorder after 1914* (Berkeley: University of California Press, 2008), 244.

57. *Report by His Majesty's Government in the United Kingdom of Great Britain and Northern Ireland to the Council of the League of Nations on the Administration of Palestine and Trans-Jordan for the Year 1936* (London: His Majesty's Stationery Office, 1937), 121.

58. Horne, *Job Well Done*, 459.

59. For an analysis of the British use of airplanes in the Middle East, Iraq in particular, see Priya Satia, *Spies in Arabia: The Cultural Foundations of Britain's Covert Empire in the Middle East* (Oxford: Oxford University Press, 2008). See also David Killingray, "'A Swift Agent of Government': Air Power in British Colonial Africa, 1916–1939," *Journal of African History* 25 (1984): 429–44. This combination would be later repeated in 1950s Kenya. See "Use of Wind Scenting Dogs Provided by Colonel Baldwin, a Dog Breeder, by the Kenya Police During the Emergency," 1953, NAUK CO 822/478. See also "Mau Mau Raids near Mount Kenya," *Times of London*, September 27, 1952.

60. *Report on the Administration of Palestine, 1936* (London: His Majesty's Stationery Office, 1937), 121.

61. Lieutenant-Colonel W. Marsh (Burma), "Dogs in Jungle War," National Archives of South Africa (NASA), SAP 296/21/22/38.

62. *Report of Sir Charles Tegart and David Petrie*, January 24, 1938, NAUK CO 733/383/75742/77 (hereafter *Tegart and Petrie Report*). Some have mistakenly attributed the introduction of Dobermans to the Tegart report itself. See Laleh Khalili and Jillian Schwedler,

"Introduction," in *Policing and Prisons in the Middle East: Formations of Coercion*, ed. Laleh Khalili and Jillian Schwedler (London: Hurst, 2010), 15; Laleh Khalili, "The Location of Palestine in Global Counterinsurgencies," *International Journal of Middle East Studies* 42 (2010): 413–33, at 423.

63. *Tegart and Petrie Report*.

64. "Investigations into Murder on Beit Jibrin Track: Dogs on Trail of Highwaymen," *Palestine Post*, January 12, 1938; "Several Arrests in Starkey Murder Investigation," *Palestine Post*, January 13, 1938.

65. Four additional dogs were purchased in 1937. *Report by His Majesty's Government in the United Kingdom of Great Britain and Northern Ireland to the Council of the League of Nations on the Administration of Palestine and Trans-Jordan for the Year 1937* (London: His Majesty's Stationery Office, 1938), at 111.

66. *Tegart and Petrie Report*.

67. *Tegart and Petrie Report*.

68. For a discussion of Bedouin resistance to British rule during the Arab Revolt, see Mansour Nasasra, "The Southern Palestine Bedouin Tribes and British Mandate Relations, 1917–48: Resistance to Colonialism," *Arab World Geographer* 14 (2011): 305–35; Mansour Nasasra, "Memories from Beersheba: The Bedouin Palestine Police and the Frontiers of Empire," *Bulletin for the Council of British Research in the Levant* 9 (2014): 32–38.

69. *Tegart and Petrie Report*.

70. *Report by His Majesty's Government in the United Kingdom of Great Britain and Northern Ireland to the Council of the League of Nations on the Administration of Palestine and Trans-Jordan for the Year 1938* (London: His Majesty's Stationery Office, 1939), 113.

71. *Report by His Majesty's Government for the Year 1938*, 113.

72. Chief Secretary to Downie, August 2, 1938, NAUK CO 733/358/6. Two dogs and one bitch were purchased for a cost of 475 Reichsmarks each (the equivalent of £P.40).

73. Chief Secretary to Downie, August 2, 1938.

74. Matthew Hughes, "The Banality of Brutality: British Armed Forces and the Repression of the Arab Revolt, 1936–1939," *English Historical Review* 124 (2009): 313–54. See also "Attack on Post," *Palestine Post*, April 21, 1938; "Attacks on Traffic," *Palestine Post*, April 10, 1938.

75. Hughes, "Banality of Brutality," 327.

76. Hughes, "Banality of Brutality," 346.

77. Hughes, "Banality of Brutality," 326.

78. See Official Communiqué No. 32, reprinted in *Palestine Post*, November 27, 1945; "Desert Customs," *Palestine Post*, September 4, 1946. See also Horne, *Job Well Done*, 460: "Dogs were successfully used during the Jewish terrorist campaign to trace criminal elements from the scene of attack on government property, to the settlement that hid them from justice."

79. Horne, *Job Well Done*, 460.

80. Horne, *Job Well Done*, 460.

81. "New Official Communiqué Concerning Birya," *Davar*, March 10, 1946.

82. Top Secret Letter from Commissioner of Kenyan Police to South African Commissioner of Police, March 21, 1947, NASA SAP 21/199/26.

83. Spicer to Colonial Secretary Nairobi, September 10, 1927, NASA SAP 95/21/99/26.

84. Horne, *Job Well Done*, 459 (citing Annual Administrative Report, the Palestine Police Force, 1935, signed by R. G. B. Spicer).

85. *Report by His Majesty's Government in the United Kingdom of Great Britain and Northern Ireland to the Council of the League of Nations on the Administration of Palestine and Trans-Jordan for the Year 1935* (London: His Majesty's Stationery Office, 1936).

86. Horne, *Job Well Done*, 457.

87. Horne, *Job Well Done*, 458.

88. Certain schools in Islam decreed that a container touched by a dog be washed seven times and then sprinkled with dust before it could be used. See Richard Gauvain, "Ritual Rewards: A Consideration of Three Recent Approaches to Sunni Purity Law," *Islamic Law and Society* 12 (2005): 333–93, at 354n68. Others believed that dogs endangered not only physical but also moral purity: in the Qur'an, dogs appear as a metaphor for disbelievers. The passing of a dog was believed to negate the prayer or good deeds of a pious Muslim. Similarly, some believed that angels would not enter a house in which dogs dwelled. Muslims were commanded not to trade in dogs. Some stricter authorities mandated slaughtering all dogs not employed for herding, hunting, or protection. See Khaled Abou El Fadl, "Dogs in the Islamic Tradition and Nature," in *Encyclopedia of Religion and Nature* (New York: Thoemmes Continuum, 2005).

89. El Fadl, "Dogs in the Islamic Tradition and Nature."

90. Deuteronomy 23:18 prohibited offerings funded by the "hire of a whore or the price of a dog." Other parts of the Bible describe dogs as foolish, carcass-eating, blood-thirsty creatures. See Robert A. Rothstein, "'If a Jew Has a Dog . . .': Dogs in Yiddish Proverbs"; and Sophia Menache, "From Unclean Species to Man's Best Friend: Dogs in the Biblical, Mishnaic, and Talmud Periods," in *A Jew's Best Friend? The Image of the Dog throughout Jewish History*, ed. Phillip Lieberman-Ackerman and Rakefet Zalashik (Brighton: Sussex Academic Press, 2013). The Talmud recounts the unfortunate tale of a dog's bark causing a pregnant woman to abort, depriving the Jewish people of the missing soul needed for the *Shechinah*—the Divine Presence—to rest upon them. The Talmud therefore decrees that one should not own a "mad" or "evil" dog. Yet, notably, it contains no categorical prohibition of ownership. Tractate Bava Kama, 79B. The *Shulhan 'Arukh's* sixteenth-century interpretation, considered authoritative, deems the prohibition of dog ownership as pertaining only to dogs that intimidated or placed others at risk. Shulchan Aruch, Choshen Mishpat, 109:3. In his commentary upon the Shulhan Aruch, Rabbi Moshe Isserles (*Rama*, considered binding upon Ashkenazi Orthodox Jews) adds that because Jews live among other nations, they may own vicious dogs for protection, though they must chain them.

91. Menache, "From Unclean Species to Man's Best Friend," 44.

92. Rothstein, "'If a Jew Has a Dog,'" 135.

93. Tom Segev, *One Palestine Complete: Jews and Arabs under the British Mandate* (New York: Henry Holt, 2000), 197. In his diary, Kisch wrote that the high commissioner prohibited his waiters from wearing tarbushes for the same reason. All this changed dramatically by 1938: a dog cemetery was erected adjacent to the new government house (342).

94. "British Constable 1069 Howard Mansfield Recalls His Service in Nablus, Jerusalem and Haifa," in *The Creation of the State of Israel (Perspectives on Modern World History)*, ed. Myra Immell (Detroit, MI: Gale, 2010), 159–67. Also available at "Land of Broken Promises," accessed July 29, 2018, http://www.landofbrokenpromises.co.uk/palestine/howard.html.

95. "Allegations of Illtreatment [sic] of Arabs by British Forces in Palestine," Middle East Centre Archive, St. Antony's College (MEC), Jerusalem and East Mission (J&EM) LXV/5: "The use of intoxicants and drugs which are abhorrent to the peasants . . . Blasphemous words."

96. Meltzer, "Local Departure in Crime Detection," 3.

97. "The Criminal Investigation Department's Amazing Dogs," *Al-Difa'a*, May 14, 1935; "News of Border Crime: Dogs Detect Culprits," *Filastin*, April 4, 1935; "Crime Discovered by the Police Dogs: Tree Cutting in Ramallah," *Filastin*, March 14, 1935; "Dogs Discover Tree Cutting Crime: Track Culprit to Mosque Gates," *Filastin*, April 2, 1935. See also "Police Dogs in Salame," *Filastin*, April 16, 1935; Assad A-Shakiri, "The Barks of Our Dogs and the Activity of the Foreign Dogs," *Filastin*, June 23, 1935.

98. "Police Use Dogs," *Davar*, March 1, 1935; "The Murder near Acco: The Dogs Were Not Mistaken," *Davar*, April 5, 1935; "Police Dog Discovers Thief," *Doar Hayom*, December 17, 1935.

99. "News of Border Crime"; "Crime Discovered by the Police Dogs"; "Dogs Discover Tree Cutting Crime." See also "Police Dogs in Salame"; A-Shakiri, "Barks of Our Dogs."

100. "Crime-Dogs Solve Crimes"; "They Set Fire to Fields and Forests, Chop Down Trees, Throw Stones and Bombs—Yet the Government Remains Silent," *Davar*, May 4, 1936. Some suggested in jest that the dogs be given their share in rewards for wanted persons, also suggesting that a pension fund be established for their benefit. See "Haifa Notebook," *Palestine Post*, May 17, 1938.

101. As some scholars have recently shown, dogs' status in Islam has been contested for centuries, by some accounts dating back to the Prophet himself. Al-Jahiz, a ninth-century scholar, explained that dogs were considered impure because they were border crossers: they confused the categories of culture and nature, neither a "wild animal nor a domestic one, neither a human nor a *jinn* (spirit)." Al Jahiz, *al-Hayawan*, cited in Marion Holmes Katz, *Body of Text: The Emergence of the Sunni Law of Ritual Purity* (Albany: State University of New York Press, 2002), 19. Throughout the centuries, particularly in Arabic poetry, dogs were often depicted as the embodiment of noble virtues such as intelligence, compassion, skill, bravery, self-sacrifice, and loyalty. See notes 88–89 above.

102. Alan Mikhail, *The Animal in Ottoman Egypt* (Oxford: Oxford University Press, 2014), esp. Part II, "In Between." According to Mikhail, it was only in the nineteenth century that they ceased to be considered as productive and constructive members of society and began to be targets of eradication. See note 69 above.

103. See Rudolphine Menzel, *Dog Education and Training* (Palestine: Lanotter, 1939). In the foreword, Menzel drew on stereotypes concerning the diasporic Jew's irrational fear of dogs to explain the key role of dog training in Zionist ambitions to forge the "New Jew": "[In Europe] the dog belonged to the world of the gentiles. He served as the oppressors' companion and accomplice . . . Yet [in ancient times] . . . lived generations of free [Jewish] peasants, men of agriculture and herding, of war and of hunting. . . . In those times the dog was a companion who assisted our people. Our national revival movement is . . . strengthening its bonds with natural life . . . In our treatment of dogs we must also . . . strengthen our ties to ancient traditions of a nation of shepherds and farmers in the ancient land of Israel. The dog is like no other species . . . May he be a companion and assistant in the rebuilding of our land." For a fascinating exchange concerning the role of dogs in the public sphere in the Hebrew model town of Tel Aviv, see "Against Dog Worship, Not the Dog," *Davar*, February 27, 1935.

104. For an analysis of dog's liminal nature in the European imagination, see Aaron Herald Skabelund, *Empire of Dogs: Canines, Japan and the Making of the Modern Imperial World* (Ithaca, NY: Cornell University Press, 2011), 6–7: "As creatures of metaphor, dogs oscillate between high-status animals and low-status people. They are said simultaneously to possess admirable traits (such as bravery) that make them akin to humans and despicable attributes (such as filth) that render them unalterably inferior—or in the minds of some, like 'Other' humans."

105. On the African bushmen, see Alan Hattersley, *The First South African Detectives* (Cape Town: Timmins, 1960), 168: "Bushmen were the finest trackers in the world. Moreover they could maintain existence under conditions of extreme hardship that would kill any civilised man within seventy-two hours. In the field of detection primitive skills may yet play a significant role."

106. For a discussion of this aspect of Native American trackers, see Gina Macdonald and Andrew Macdonald, *Shaman or Sherlock? The Native American Detective* (Westport, CT: Greenwood Press, 2002).

107. For a discussion of how domestication corrupted the Arab "noble savage," see Toby Dodge, *Inventing Iraq: The Failure of Nation Building and a History Denied* (New York: Columbia University Press, 2005). For a similar discussion of how bloodhounds may have "lost

their nose" through domestication, see H. S. Lloyd, "The Value of Dogs to the Police," *Police Journal* 13 (1940): 206–22, at 210.

108. Spicer to Chief Secretary of Palestine, April 11, 1933.

109. Horne, *Job Well Done*, 454: "Inspector Ibrahim Bey Stambouli, perhaps the best person to have asked, could not recall in 1933 when taking to the Inspector General that he had ever heard of the Turks using dogs to track down criminals. 'Why would they?' he asked. 'They had village trackers, which are better than any dog.'"

110. Horne, *Job Well Done*, 454.

111. When describing the work of his Sudanese tracker Ibrahim, Len Hynds, a British military policeman serving near the Suez Canal, explained: "As I was making notes as to what had been taken, Ibrahim was circling around, sniffing the ground like a dog." See Len A. Hynds, "Ibrahim—The Sudanese Tracker," accessed July 29, 2018, http://www.the speechlesspoet.co.uk/true%20stories/desert/Ibrahim%20-%20The%20Sudanese%20Tracker .html.

112. Huw Bennett, *Fighting the Mau Mau: The British Army and Counterinsurgency in Kenya* (Cambridge: Cambridge University Press, 2012), 27.

113. "Use of Wind Scenting Dogs Provided by Colonel Baldwin, a Dog Breeder, by the Kenya Police During the Emergency," 1953, NAUK CO 822/478.

114. *R. v. Kotcho*, 1918 (Eastern Districts Local Division), 91–107. Similarly, a Canadian judge wrote: "Let it be supposed that the most skillful of these [natives] was employed to track the murderer, and that he had followed courses such as those taken by the dogs." For a similar analogy in Canada, see *R. v. White* (British Columbia) 1926, 5 D.L.R. 2.

115. Brett L. Shadle, "Cruelty and Empathy, Animal and Race, in Colonial Kenya," *Journal of Social History* 45 (2012): 1097–116, at 1099.

116. Skabelund, *Empire of Dogs*, 7.

117. See *R. v. Trupedo*, 1920 South Africa Law Reports (SALR) (AD), 58.

118. Sir Arthur Conan Doyle, "The Adventure of Silver Blaze," *Strand Magazine*, December 1892.

119. Innes's opinion echoed a distinction made by some between dog and hound. See, e.g., Lloyd, "Value of Dogs to the Police," 206: "A dog when once he gets to like his handler and is properly trained, works for the pure love of pleasing his handler . . . The hound, on the other hand, seems to work entirely on his own, by inherited instinct, and has little or no affection for his handler, or any desire to please him, but hunts to please himself."

120. See *R. v. Trupedo*, 58.

121. Meltzer, "Local Departure in Crime Detection."

122. Address by Menzel to *Haganah* staff, 1939, Haganah Archive (HA) 34/289.

123. Menzel, *Dog Education and Training*, 135.

124. Papers of Rudolfine Menzel, Central Zionist Archives (CZA) 129/67.

125. Hughes, "Banality of Brutality," 326.

126. Hughes, "Banality of Brutality, 326. See also "Re: Abdul Hafiz Abdullah et. al.," MEC, J&EM, LXVI/3: "May I humbly point out that so far in the history of investigation in Palestine, whenever police dogs taken [*sic*] to a parade after having a scent, they have never, I repeat, so far, failed to pick one or more persons out of the parade. One might argue that this is due to the shining detective mentality of our C.I.D. men. Unfortunately however I know of a case of theft, where dogs after being given scent followed the scent and arrived to a distant village, entered the room an oldbling [*sic*] man and barked at him, meaning to say that this is the person who committed the robbery."

127. Hughes, "Banality of Brutality," 326.

128. Hughes, "Banality of Brutality," 318. See also "Death Sentence for Daylight Shooting," *Palestine Post*, January 11, 1939, 2.

129. "Trial of Arab Charged with Firing at Jewish Bus," *Palestine Post*, February 24, 1938, 2; "Arab Sentenced to Death in Jerusalem: Convicted for Firing at Jewish Bus," *Palestine Post*, February 24, 1938.

130. For a similar case in which dogs provided the sole identification of the defendant, see "Death Penalty for Murderer Who Threw Bomb in Haifa," *Hatzofeh*, March 22, 1939, 4.

131. "Re: Abdul Hafiz Abdullah et. al."

132. See, e.g., "Re: Abdul Hafiz Abdullah et. al."; "Death Sentences Passed under the Emergency Regulations," ISA M 276/22. (The witness testifying at trial was Kassen Eff. Abu Ghazaleh, not handler.) Compare to situation in Kenya, where courts insisted at minimum that handlers be available to testify for such evidence to be admissible: Cr. App. 44/52 *Abdullah bin Wendo v. R.* 20 EACA 166 (1953).

133. Horne, *Job Well Done*, 456.

134. "Death Sentences Passed under the Emergency Regulations," ISA M 276/22. The witness testifying at trial was Kassen Eff. Abu Ghazaleh, a member of the Palestine Police, interrogated the defendant. Concerning the dogs, he testified: "On 17/3/37 at 9:10 in the morning, the accused was put in a parade for identification by the Police dogs. The dog picked him out from amongst 7 persons after it had smelt the odour of the place where the large blood stains were found. The dog recognized him twice. After that, I charged the accused Mohammad Yasseen with the murder. He denied killing him. He said: 'I did not kill Shmuel Gottfried and I did not see him.'"

135. The dog's techniques would be less of an issue if the defendant confessed, as they often did. See Spicer, "New Detective," 249. As South Africa's 1920 High Court suggested, even Parker and Pringle's testimony arguably would have been imperfect hearsay.

136. Criminal Assize 8/36 *Said Mustafa Abbas and Jamil Abu Imris v. Attorney General*, August 11, 1936.

137. Criminal Assize 8/36. The district court judge observed: "This is a queer evidence and it is extraordinarily not bad. The method through which the identification took place, however, is that the two accused were brought to the place of the incident, and their feet may have stepped near the place of the footsteps which are found there, and the dogs may have smelled their footsteps. The Court sees that it is a sign of extreme negligence to permit a suspect person to come near the place in which there is a trace related to the crime, and have the dogs then come and smell its odour." Defendants were sentenced to ten and fifteen years' imprisonment.

138. One possibility is that in the absence of juries, Palestine's judges trusted their professional ability to attach proper weight to such evidence, questionable as it may have been. However, given how cautious Palestine's judges proved concerning other formal evidentiary requirements—where they often displayed greater rigidity than English judges presiding over juries—this explanation seems somewhat unlikely. See, e.g., Criminal Appeal 160/37 *Ali Jarad v. Attorney General*, 5 *Palestine Law Reports* (1938), 111.

139. "Death Sentences Passed under the Emergency Regulations," ISA M 276/22. The magistrate decided not to bind the accused over for trial even though a police dog picked him out at two separate identification parades based on the smell of a blood pool and a cigarette packet found at the crime scene. Josiah Wedgwood brought the matter to the attention of parliament when questioning Secretary of State for the Colonies Ormsby-Gore, asking, "what further steps are to be taken to put an end to the terrorism in Palestine which prevents convictions being obtained by the police?" The question prompted Ormsby-Gore to take special interest in the case, ordering Palestine's high commissioner to submit a detailed report on it.

140. For the expectation that the judiciary in Palestine be more "helpful," see "Palestine and the Commission," *Times of London*, July 23, 1936: "So far the High Commissioner has not declared martial law; but he may have to adopt this drastic measure, especially if the attitude

of some of the Palestine Judicature is not more helpful than it has been hitherto." For the exchange between the Secretary of State for the Colonies and Chief Justice Michael Mc-Donnell concerning the judiciary's obligation to be "helpful," see Letter from Ormsby-Gore to McDonnell, July 9, 1936, NAUK CO 733/313/1. For press coverage of McDonnell's forced resignation, see "The Chief Justice," *Palestine Post*, October 22, 1936; and newspaper clippings from October 22, 1936, onward, NAUK CO 733/313/1.

141. Government of Palestine, *Ordinances, Regulations, Rules, Orders and Notices, Annual Volume for the Year 1937*, vol. 2 (Jerusalem: Government Printing Press, 1937), 268.

# CHAPTER EIGHT

# "DNA Evidence Cannot Lie"

## Forensic Science, Truth Regimes, and Civic Epistemology in Thai History

QUENTIN (TRAIS) PEARSON

On September 15, 2014, two British backpackers were found brutally murdered amid the bleached monoliths that grace the beaches of Koh Tao, an island off the coast of southern Thailand. The female victim, Hannah Witheridge, had been raped and bludgeoned to death. The male victim, David Miller, had been beaten, stabbed, and left to drown in the surf. The subsequent investigation, which was managed under the sole authority of the Royal Thai Police, was plagued from the start by what foreign press reports described as grave missteps in handling the crime scene, testing forensic evidence, and identifying potential suspects.[1] Guided by the facile assumption that no Thai was capable of committing such a violent act—a notion voiced by senior police officials involved in the case[2]—the Royal Thai Police initially focused their investigation on British tourists, friends of David Miller. Once they had been cleared, the police eventually trained their sights on the usual suspects: the island's community of Burmese migrant workers. They identified two twenty-one-year-old Burmese men, Zaw Lin and Wai Phyo, and charged them with the killings.

In December 2015, after a trial that received at least as much critical press attention as the investigation itself, the accused were found guilty and sentenced to death. Widespread protests broke out in Myanmar, accompanied by vocal calls at home and abroad for the Royal Thai Police to conduct a complete review of the case. Major General Piyaphan Pingmuang, the deputy spokesman for the Royal Thai Police, rebuffed the calls, declaring, "The DNA evidence cannot lie."[3] Redirecting attention away from the pervasive questions that continued to plague the police investigation of the murders, he warned of groups who intended to politicize the outcome of the case and encouraged the public to resist their efforts.

This chapter looks to the past to contextualize the Koh Tao murder case and the status of forensic science as public knowledge in contemporary Thailand. Specifically, it considers forensic science within the broader configurations of Thai civic epistemology, a term employed by Sheila Jasanoff to denote "shared understandings about what credible claims should look like and how they ought to be articulated, represented, and defended in public domains."[4] To do so, I first consider the broader political context of the historical moment when the Siamese state first turned to forensic expertise, in the late nineteenth century. Second, I take up the work of scholars attentive to the connections between the social and epistemic order in Thai society to highlight the iterative nature of attempts to assert control over forms of expert knowledge.[5] These insights provide an essential background for making sense of the failings of the Koh Tao investigation and its aftermath. Before addressing the peculiarities of the Thai case, however, it is necessary to consider other historical examples that might serve as comparative models for the consideration of forensic science as a form of public and political knowledge.

## Forensic Science and Its Politics: Three Models

Well over half a century ago, Lewis Mumford asked productively whether some kinds of technology might be inherently more democratic than others. He juxtaposed "the small scale method of production, resting mainly on human skill and animal energy but always, even when employing machines, remaining under the active direction of the craftsman" with the "herculean feats of mechanical organization" that characterized "totalitarian technics."[6] The question might equally be posed with respect to forensic science, especially institutions and forms of expertise concerned with the investigation of sudden or unnatural death. A review of recent historical scholarship on the introduction and deployment of forensic investigations into cases of unnatural death suggests two basic but radically divergent models corresponding to the prevailing political conditions of the day. These models might tentatively be called democratic and colonial, and each seems to implicate forensic science in its own peculiar political projects.

The medicolegal investigation of cases of unnatural death, an institution known as the inquest, is deeply imbricated in the history of constitutional politics in British history.[7] Inquests initially began as a safeguard for the king's financial interests; the investigation of cases of unnatural death was intended to guarantee that the crown would receive its financial due in the form of forfeiture of property, levies, or other forms of "revenue deriving from the administration of justice."[8] Over time, however, the institution eventually became a part of civic life and a guard against the excesses of autocratic rule. Inquests were held in public spaces, usually a pub, and allowed for a sort of local sovereignty over the legal and economic repercussions of cases of unnatural death.[9] By the middle years of the nineteenth century, the inquest had come to be viewed by many as a crucial facet of constitutional politics, and a central feature in radical democratic political rhetoric.[10] This moment represented something of an apex in the democratic career of forensic medicine, however, as local magistrates began to see the inquest as a manifestation of the growing power of medical professionals. They were able to tap into the "culturally profound and socially diffuse concern at medicine's perceived transgressions against the physical integrity of death" in order to halt the efforts of medical professions to expand both the purview of the inquest and the role of the medical professional in the inquest.[11]

The relationship between forensic medicine and politics was even more salient in the colonial world, where medical science was an important channel for articulating imperial ideologies. Throughout the colonial world, arguments about European racial and moral superiority were substantiated through medical discourses about indigenous bodies. David Arnold, for example, has identified a significant shift in medical discourse in British India over the course of the nineteenth century, whereby environmental explanations of disease and ill health were supplanted by "explanations which gave greater prominence to the peculiar characteristics of Indian society, morality, and culture."[12] In this way, medicine became a seemingly objective means of giving voice to politicized and racialized "truths" about native bodies. In a medicolegal context, such discursive truths were operationalized in the service of empire in two crucial ways: (1) as one facet of the more or less all-encompassing efforts of the colonial state to render its subjects visible as part of what has been aptly called "the conquest of illegibility"[13] and (2) as tools of indemnification for white settlers who might commit acts of violence against native bodies.[14] Scholarship on the colonial world thus suggests a much more diverse and ambivalent picture of forensic science and its politics than in some metropolitan contexts.

This rough taxonomy, of course, is not intended to capture the specificity of individual historical cases; the aim is rather the most modest one of suggesting something of the fluid but significant connections between different forms of forensic knowledge and politics. Just as there were states whose sovereignty waxed and waned during their interactions with imperial powers throughout the era of European imperialism, so, too, were there intermediate cases between the metropolitan democratic model and the multifaceted case of colonial forensics. The history of forensic science and medicine has seemingly articulated with different political realities in the metropolitan and colonial worlds, but what of the history of forensic science in semi- or peri-colonial states?

Independent states seeking to forestall imperial intervention provide a third model of forensics and politics. Such states turned to forensic science as part of a broader effort to demonstrate effective and enlightened governance of the local population to preempt the intervention of imperial powers. Khaled Fahmy demonstrates precisely this model of science as strategic action in his work on forensic science in nineteenth-century Egypt, where Mehmed Ali adopted forensic sciences as part of a bid to appeal to European powers in his efforts to assert Egyptian independence from Ottoman rule.[15] Siam was similarly an independent state working to stave off imperial interventions by the British and French when it began to implement forensic investigations into cases of unnatural death in the late nineteenth century. And while in some respects the "forensic turn" in Thai history does appear to function as the same sort of pragmatic move, the reality is somewhat cloudier. To begin, one must reckon with theories of the origins of forensic science advanced by Thai scholars that seem to track the democratic model of development. Beyond that, however, lies the more contentious and complex question of Siam's status in the colonial world and the ways in which early efforts to institute forensic medicine intersected with its semi-colonial past.

In a thesis on the development of forensic medicine in Thailand, Thai scholar Niwat Thungthong points to a law promulgated by the Siamese Ministry of the Capital in 1891 (*ro. so.* 110 [2434]).[16] The law stipulates that an inquest should be performed to determine the cause of death in all cases where prisoners died in the custody of police or in correctional facilities under unnatural or suspicious circumstances. It further specified that if no cause of death could be readily discerned, then a surgical autopsy should be performed. Evidence from the Archives of the Ministry of Justice confirms that such deaths did occur, and that they incited public outrage and forced state officials to pay heed. One such case occurred in June 1894, when the Ministry of Justice was informed of the deaths of two prisoners under suspicious circumstances while in custody in the city of Songkhla in southern Thailand.[17] The deaths in question included one death by caning and one by decapitation. In the former case, the mother of the deceased had registered a complaint with local authorities about the circumstances of the death of her son. Her pleas went unanswered, so she appealed to the local agents of the central Siamese state. The newly formed Ministry of Justice took up the case, and with the cooperation of the head civil servant of Songkhla, it brought about the interrogation of witnesses, including other corrections officers at the jail, and eventually the arrest and trial of the accused.

In the abstract, this case suggests a developmental logic akin to that of the English inquest, where concerns about the abuse of power by the state and its representatives and assertions of local sovereignty over death drove the historical spread of forms of forensic concern for the dead and related institutions. In the mother's plea, which reached across the kingdom to the ears of the minister of justice, we see not only an appeal for personal justice but also something more: a cry for independent oversight of overzealous local authorities. From this perspective, the case does seem to support Niwat's hypothesis about the development of forensic medicine in Thai history. But this explanation amounts to what can only be deemed an *internalist* account: it employs an abstracted vision of medicolegal knowledge as an objective form of expertise that is immune to the historical and cultural context of its deployment. This sort of internalist account of the historical development of forensic science requires bracketing other causal factors, including the deep grammar of knowledge and authority in Thai society on the one hand (a topic discussed below), and the broader political context of Siam's semi-colonial status on the other. It fails to consider, for example, the fact that Songkhla was a former Malay sultanate, and although it had been brought under Siamese suzerainty in the late seventeenth century, it was a region fraught with imperial competition with the British, and one that would only officially become a Siamese territory by treaty agreement in 1909.[18] These variables offer an alternative account of the origins of forensic science in Thai history.

## Arch Anxieties of the Semi-Colonial State

Christopher Hamlin has recently made the case that our contemporary fixation on forensic science as a matter of investigating "isolated crimes of individual against individual" has obscured the ways in which forensic cultures develop in response to the perception of systemic issues. Such perceptions, according to Hamlin, can coalesce into "arch anxieties" that drive states to invest resources in forensic technologies.[19] Much like in Europe, the history of modern forensic science in Siam dates to the end of the nineteenth century, and "it reflects the perception of a broadly based failure to act against pervasive crimes."[20] In the Siamese historical case, however, I have argued elsewhere that these two were—perhaps uniquely—conflated: isolated incidences of individual violence were seen as an indicator of a systemic asymmetry between Siamese and foreigner.[21] The semi-colonial status of the Siamese state might thus be understood to have dictated the nature of its engagement with forensic science in crucial ways.

By the waning years of the nineteenth century, the members of the Siamese royal elite who controlled the state were rightly fixated on the constrained sovereignty of the kingdom. Since the mid-nineteenth century, a series of unequal trade treaties had dictated important aspects of the kingdom's fiscal, legal, and territorial realities.[22] These impositions incited a turn to new forms of expertise—most notably mapping,[23] law,[24] and medicine[25]—that were intended to shore up the Siamese state's claims to sovereign rule and to persuade the foreign powers to abrogate the unequal treaties. Yet these very concerns about sovereignty were also translated into the peculiar idiom of bodily crimes against Siamese subjects.[26] Although these were likely among the most marginal of crimes in a demographic sense, they were nevertheless highly visible and demonstrative indicators of the disadvantaged status of Siamese subjects with respect to foreign residents, who enjoyed extraterritorial legal privileges and the consular institutions that guarded their rights. The Siamese elite turned to forensic medicine as a way to combat this injustice: they invested resources in forensic expertise in response to the arch anxiety of diminished sovereignty. Forensic medicine thus should take its place alongside what have become the canonical forms of expertise such as mapping, law, and medicine that allowed for the assertion of Siamese sovereignty in the age of imperialism.

Beyond the subjective anxieties of the Siamese elite, however, the forensic turn in Thai history was crucially shaped by Siam's semi-colonial status in other, more direct ways. The specter of foreign violence against Siamese subjects was likewise linked to a broader conflict between Siamese and foreign legal systems within the plural legal arena created by the introduction of extraterritorial law. The institutions that administered justice to foreign residents in Siam, known as consular courts, were local outposts of foreign legal systems. They represented comparatively liberal forms of legal culture, especially those like the British consular court, which employed a system of trial by jury in the adjudication of civil and criminal cases. Trial by jury—a system that effectively empowered "lay jurors [to make] up their own minds about the weight of expert and other testimony after hearing the cases brought

in by opposing counsel"—was an institution completely foreign to the traditional Siamese system of justice.[27] In the inquisitorial Siamese legal institutions, by contrast, adjudication as well as questions of evidence and expertise
were left to the sole discretion of judges.[28] Thus the clash of forensic cultures
was redolent of a broader clash of epistemological cultures: one in which the
many had been empowered to judge for themselves, and another where truth
was the preserve of the elite few.[29]

The forensic turn in Thai history was indelibly shaped by these institutional and epistemological conflicts, which differed in important respects
from both the democratic and colonial models outlined above. Distinct from
colonial state concerns focused on a population that it deemed anonymous,
hostile, and untrustworthy, and more attuned to the imperial challenge as an
internal threat, Siam's forensic turn was in some respects unique. But the historical forensic turn—and likewise its legacies—was shaped by more than
just these contemporaneous legal, political, and social realities. In the re-
mainder of this chapter, I take up the epistemological background of these
developments and make the case for the existence of durable patterns of epistemic authority and their relation to public forms of knowledge.

## Public Knowledge: From Truth Regimes
## to Civic Epistemology

"Each society has its régime of truth, its 'general politics' of truth: that is,
the types of discourse which it accepts and makes function as true; the
mechanisms and instances which enable one to distinguish true and false
statements, the means by which each is sanctioned; the techniques and
procedures accorded value in the acquisition of truth; the status of those
who are charged with saying what counts as true."[30]

Truth, as Michel Foucault posited, is a matter of consensus. But the
process of achieving consensus is also subject to mechanisms of discern-
ment, sanctioning, and disputed techniques of acquisition. Moreover, claims
to authority over the truth are inherently connected to the operations of
social status. These insights have been put to good use by scholars of Thai
intellectual history who have attempted to identify and analyze its distinctive formations of power/knowledge through hypostatized notions such as
"state poetics,"[31] "truth regimes,"[32] and "systems of knowledge."[33] The following discussion attempts a synthesis of this scholarship on the relations
between truth and power in Thai society, and the durability and persistence
of the patterned historical relations that take hold among them. It moves

beyond synthesis, however, to consider science as another node in the relations between knowledge and society, between social and epistemic order.

Craig Reynolds and Andrew Turton were among the first to take up the call to attend to the relations between truth and power in Thai society, struggles over "who can say what, how, where, when, and to whom."[34] In his insightful analysis of the history of a poetic account of a military expedition in the late nineteenth century, Reynolds explores how the poem was deemed to have violated the "cultural code" of its literary form, which dictated that poetry should be free of politics.[35] He uses the poem and its contentious reception to develop the notion of "state poetics," the discourse that allowed the royal elite to proscribe and enforce conventions governing forms of cultural production. State poetics provided a discursive space for censorship, often through the mechanism of denying the beauty, truth, or logic of subversive visions of social life, for example, as ugly, false, or irrational.[36] Thus, for Reynolds, there was much more at stake in the aesthetic conventions of poetry than mere questions of beauty—nothing less than the standards of morality and truth that were both propagated by and in turn helped to keep the royal elite in power.

Foucault's understanding of the intricate and pervasive relations of knowledge and power is also the starting point for an ambitious collaborative project undertaken by Craig Reynolds and Tony Day. Writing against the tendency to view the colonial period as one of irrevocable disruption in indigenous intellectual life, Day and Reynolds identified patterned "relationships between knowledge and power [that] seem to recur at different times and places" throughout Southeast Asia.[37] According to their findings, the Thai elite—as well as elites in other indigenous states in the region—performed their political power in part through the systematic appropriation of new forms of totalizing knowledge or "cosmologies." In spite of the shifting forms of universalizing knowledge paradigms over time, Day and Reynolds argue that the mode of appropriation remained stable: "Potentially opposing truths, combatants in knowledge wars between the foreign and the indigenous, have been made to seem harmonious corollaries of an even greater Truth, to which the state lays claim."[38] In sum, Day and Reynolds view state power in Southeast Asia as a product of elite efforts to assert ownership over new and potentially disruptive forms of knowledge. Ultimately, the power of the elite rests upon their claims to superior and ultimate forms of truth.

In his provocative study of defamation law in Thailand, David Streckfuss similarly highlights the correlation between social standing and epistemological privilege. For Streckfuss, the organization of Thai society is predicated on a differential vision of access to truth, one that has its roots in Thai understandings of Theravada Buddhism. Truth, in Thai society, "is only accessible to spiritually powerful people."[39] In effect, this creates "a hierarchy of truth-holders . . . one that reflects rather well the class structure of Thailand."[40] There is a rather insidious circularity to this reasoning, of course: only the virtuous can claim to have access to truth, and you may know the virtuous in part by their high social standing, itself purported to be the natural outcome of their virtuous conduct in past lives.

As anomalous as these conditions adhering between the social and the epistemic might seem, they are by no means unique. In his "social history of truth," Steven Shapin argued that in the early days of modern scientific inquiry, when empirical methodologies existed in nascent and as yet largely unproven forms, scientists relied on claims to social standing as a warrant for epistemological claims.[41] "Preexisting gentlemanly practices," according to Shapin, "provided working solutions" to the pressing issues of trust and credibility associated with the emergence of a new epistemic order.[42] Thus it is hardly surprising to see truth claims in Thai society as residing in what Shapin calls a "moral field"[43]—that is, as predicated on the very sorts of aesthetic proscriptions that made possible Reynolds's "state poetics," for instance, or claims to ownership of the kinds of universalizing forms of knowledge discussed by Day and Reynolds, or of access to higher forms of truth based on religious standing.

Insightful as these studies of knowledge and power are, they suffer from what Sheila Jasanoff has diagnosed as a common flaw of studies of science and the modern Democratic state. The tendency, according to Jasanoff, is to depict science and the state as having a dyadic relationship when in fact it is triadic: "Citizens after all are the primary audience for whom the state enacts its scientific and technological demonstrations. As a play could not exist without spectators, so the grand narrative of progress through science and technology demands assenting publics to maintain its hold on the collective imagination, not to mention the collective purse-strings."[44] This is a departure from the top–down vision of power/knowledge, and one that admits broader forms of contestation, opening up as well the question of the *audience* of claims to the possession of authoritative forms of knowledge. And, in fact, one can see Day and Reynolds moving in this direction in their attempt to distill a "methodology for understanding the way in which knowledge is produced, appropriated, represented, and contested in particular historical moments" as a product "of power relations at work in the respective social formations that produced it."[45]

Developments in realms such as law and science, which pertain to issues of public welfare, do not occur in a vacuum. Rather, they "play out against backdrops conditioned by human expectations about what constitutes adequate knowledge, what counts as justice, and how the two are linked."[46] Jasanoff further points to the ways in which "these expectations . . . are consolidated and continually reperformed by powerful institutions" to forge "shared understandings about what credible claims should look like and how they ought to be articulated, represented, and defended in public domains."[47] She calls these paradigms of public knowledge "civic epistemologies."[48] Even authoritative forms of knowledge have an audience whose expectations help to shape what counts as both public truth and public good. One can see the pervasive entanglements of knowledge and social standing in traditional Thai society as forming a crucial part of that backdrop. Although social elites have long claimed privileged access to truth in Thai society, those claims have nevertheless helped to condition what counts as public knowledge. They have, in short, helped to dictate the nature of Thai civic epistemology in ways that have important consequences for forms of public knowledge like forensic science and for scientific knowledge in general.

## Civic Epistemology: The Case for Continuity

Following Jasanoff's insights, it seems possible to conceive of the specific historical configuration of forensic science, state, and society that was forged in the late nineteenth century as but one iteration of broader patterned relations. But the question remains: to what extent are these configurations durable, capable of diachronic persistence? Might they be said to constitute the sorts of "residues and repetitions" that Day and Reynolds perceive in the history of state power/knowledge?[49] Or is each assemblage of science, state, and society unique and historically specific?

David Streckfuss's work on the defamation regime in Thai law helps to answer this question. His work suggests how and why the traditional patterns of relations between knowledge and power in Thai society might have survived in the transition from traditional to modern forms of state power. Thailand's colonial exceptionalism—the fact that it was the only country in Southeast Asia to escape European imperialism—implies for Streckfuss that the traditional Thai Theravada Buddhist truth regime was allowed to endure and even thrive as it was enshrined in formal legal codes and institutions. As the Thai nation-state successfully centralized its power over the kingdom and confronted the threats posed by European imperialism over the course of the late nineteenth and early twentieth centuries, this distinctive "system of knowledge" was further entrenched "in part through the creation of a modern law code and 'rationalization' of the judicial system."[50] Streckfuss thus provides a powerful case study of the ways in which traditional elite claims to epistemological authority not only survived processes of modernization but also received new stature in modern state institutions.

A recent contribution from the field of medical anthropology further corroborates the notion of the durability and persistence of traditional Thai patterns of civic epistemology. Daena Funahashi's study of the local Thai capture of the World Health Organization's Health in All Policies Initiative attests to the survival of the traditional configurations of knowledge and power in Thai society in the contemporary moment.[51] Following in the footsteps of the WHO, Thai health experts elevated health as a public good and as a (sufficient) criterion for good governance. They then asserted their bioscientific authority to effectively rule in the name of good health—albeit as the assignees of a junta regime "that forcibly ousted an elected prime minister and violently suppressed a popular protest."[52] Funahashi explores how these moves in fact relied on Thai Buddhist notions of righteous leadership, which in turn were predicated on the very sorts of claims to epistemic privilege based on wisdom born of morality that have long characterized the epistemic order in Thai society. By claiming "moral immunity" to what elite and authoritarian commentators viewed as the base economic motivations and misguided political interests of the masses, Thai health experts construct an epistemic subject position that is above the mundane fray, and claim privileged access to a globally defined notion of public good, namely, health. Their claims evince a clear parallel to earlier eras of Thai history when truth was firmly linked to social status, as well as claims to religious and moral truth, and when public discourse was carefully guarded by "state poetics." The royal elites of yesteryear, who claimed privileged access to the higher truth of the day, have given way to today's (medical) technocrats, who claim the mandate to rule based on their "moral immunity" to greed and political demagoguery.

But what about the specific case of forensic knowledge? In his innovative analysis of modern policing in Thailand, Samson Lim provides evidence of the iterative nature of state efforts to control the process of forensic knowledge production. In *Siam's New Detectives*, Lim identifies two historical moments when the production of knowledge about crime became a paramount issue for the state: in the early twentieth century and in the immediate post–World War II era.[53] Lim highlights the parallels between the efforts of the state to assert epistemic control in these two historical moments of widespread unrest by promoting "the adoption of the most current knowledge-production techniques" and by stressing "scientific methods" in police work.[54] Lim's work therefore seems to suggest that the historical efforts of the Royal Thai Police to implement forensic expertise might be said to mimic in significant ways the broader patterns and logic of Thai civic epistemology. The Koh Tao murders provide a fitting contemporary case study for testing this relationship.

When the bodies of Hannah Witheridge and David Miller were discovered on the morning of September 15, 2014, the initial investigation fell to local police officers with "rudimentary training and apparently no idea how to seal off a crime scene."[55] Police efforts to identify suspects were clearly informed by a sense of xenophobia: operating on the assumption that no Thai was capable of the attacks, they initially focused instead on other foreign tourists. After the earliest suspects—fellow Brits and traveling companions of David Miller—were cleared, the police were forced to confront the possibility of Thai involvement. They turned their attention to Worat (Nomsod) Tuwichian and Montriwat (Mon) Tuwichian, the son and brother, respectively, of a prominent local politician, as potential suspects.[56] Soon after the announcement of their status as "persons of interest" in the case, however, the chief of the local region of the Royal Thai Police, Lt. Gen. Panya Mamen, was "promoted" out of the jurisdiction, and a replacement was hastily brought in from the nearby resort city of Phuket.[57] Following the abrupt change in oversight of the case, the police refocused their search on the community of Burmese migrant workers on the island, eventually identifying the two twenty-one-year-old Burmese men who were charged with the killings.

The case against the two Burmese migrant workers hinged largely on their confessions, which were corroborated by a farcical reenactment of the murders conducted under police direction, and DNA evidence, both of which came with substantial caveats. The confessions, which were allegedly coerced, were immediately retracted once the defendants consulted legal counsel. Questions soon arose about the processing of the DNA evidence as well, which included trace evidence recovered from the body of the female victim and from two cigarette butts found discarded near the murder scene, the testing of which was handled under police authority. As described by police during the criminal trial, the testing process was a remarkably—if not impossibly—efficient affair. Police testified that DNA evidence was recovered from the body of the female victim at 8:00 a.m., and a viable DNA profile of two suspects was announced by 10:00 p.m. In that brief interval, the police claimed to have identified three distinct DNA profiles (including that of the victim). The records of the pathologist, however, seem to throw the police time line into question, as the forensic examination of the bodies did not begin until 11:00 a.m.[58] Discrepancies with the DNA evidence did not end there. According to original reports, the police performed the DNA extraction on September 17, but the report submitted to the court as evidence was dated October 5, two days after the police had claimed a positive identification with DNA samples taken from the two Burmese defendants.

Even more problematic for expert observers was the accompanying documentation that the police entered into evidence. The documentation of the DNA analysis was limited to a one-page summary, partially handwritten, with visible redactions, and a four-page report corroborating its findings.[59] Forensic experts, including an Australian, Dr. Jane Taupin, and Dr. Pornthip Rojanasunand, a leading Thai forensic scientist, were skeptical. They called for the release of the complete lab reports documenting the methodology of both the extraction and the matching process. Others pointed to the obvious lack of oversight and independent confirmation of the findings by experts outside of the police. Unlike in other jurisdictions, where scientific evidence requires independent vetting, the Thai Police did not allow any independent oversight in the processing and testing of forensic evidence in the case. The top forensic agency in Thailand, the Central Institute of Forensic Science—an agency headed by Dr. Pornthip that operates independently of and often antagonistically with the police—was excluded from the investigation and analysis.

At the trial in July 2015, the defense concentrated their efforts on perceived issues surrounding the collection, testing, and preservation of the DNA evidence. In her testimony, Dr. Pornthip highlighted gross failures in police efforts to secure the crime scene.[60] Photographs of the grisly scene taken soon after the discovery of the bodies, for example, showed Montriwat (Mon) Tuwichian, one of the Thai suspects, mingling with police behind the crime scene barrier, raising the prospect that evidence may have been tampered with.[61] In light of these overarching concerns, the defense demanded the retesting of all DNA evidence collected at the crime scene. The police responded that the earlier testing of trace evidence found on the female victim and on cigarette butts discarded near the murder scene could not be replicated because it had all been "used up."[62] They eventually handed over DNA evidence collected from the supposed murder weapon, a garden hoe. After independent testing, however, Dr. Pornthip concluded that the object did not yield DNA that matched either of the two Burmese suspects. Moreover, despite international attention and the willingness of foreign forensic experts like Dr. Taupin to testify at trial, the defense chose not to call them as witnesses. Media reports speculated that the defense counsel feared that any appeal to foreign experts might be viewed as an attempt to undermine the police and the Thai justice system at large, thereby provoking the ire of the judges who would ultimately decide the case.[63]

The Koh Tao murder investigation affords important lessons about the nature and status of forensic science in contemporary Thailand and its relation to the historical relations of knowledge and power in Thai society. First, the exclusion of foreign experts from the trial signals that truth is evidently a matter to be decided within the constraints of a domestic epistemic sphere, where the views of outside observers are taken as a threat to an established hierarchy of truth holders. Second, scientific processes like DNA analysis are "blackboxed" to such an incredible extent that they are deemed immune to question, let alone duplication.[64] When Police Major General Piyaphan Pingmuang asserts that "DNA evidence cannot lie," he is effectively claiming not only the perfect transparency of forensic evidence, but also, implicitly, the absolute irreproachability of the police. A claim to both epistemic and moral privilege, it should thus be understood as a move anchored in a traditional vision of Thai civic epistemology, where police and forensic analysts—like royal elites of old, or health policy workers—claim access to higher forms of truth that are not accessible to the masses. Moreover, claims to epistemic privilege that once relied on assertions of superior moral status are now bolstered by scientific forms of analysis that are beyond the ken of laypeople. This, once again, is a testament to the iterative nature of the Thai civic epistemology, namely, its capacity to incorporate and assimilate new epistemic technologies (in the form of DNA analysis).

In highlighting the historical persistence of such authoritarian and statist visions of epistemic order, however, I do not mean to suggest that they necessarily go unchallenged. Lim notes that recurring efforts to reform the production of forensic knowledge in twentieth-century Thailand, which "were all seen as a way to keep up with international standards and aid in the delivery of justice through the law," had profound if unintended consequences for public knowledge in Thailand. In the post–World War II era, as political violence was on the rise in Thailand (for many domestic and international reasons that are beyond the scope of this chapter), Lim charts the rise of what might be deemed a new civic epistemology: the conspiracy theory. Conspiracy theories were employed by both agents of the state—as a means "to justify the political violence they so regularly employed"—and by "Thais of all stripes" who utilized them to envision "the 'real' truth behind the visible reality of everyday life."[65] The opacity of widespread political violence thus gave rise to a Janus-faced discourse: the conspiracy theory was the idiom of both official narratives and subversive truths. Conspiracy theories, according to Lim, provided "a convenient and potent way for people, from factory workers and street vendors to doctors and professors, to make sense of the violence and inequity of modern life," while likewise serving as "a technique of modern statecraft in response to (and taking advantage of) international capitalism and technological modernity."[66] In effect, a powerful police force that claimed epistemic authority over crime and violence helped produce a public—and a vernacular press—committed to a surreptitious yet methodical skepticism.

Although authoritative epistemologies have proven durable in Thai society, we may take heart that though they command assent, they can in practice give rise to civic epistemologies anchored in dissent. Funahashi recognizes this tension when she relates the predicament of her health policy worker-informants who must confront a Thai public that is skeptical of their claims to a "righteous place within the hierarchy of *panya* [wisdom]" on the one hand, and their status as beneficiaries of a military regime that unseated a democratically elected government and used violence to suppress public protests on the other.[67] It is also fundamental to Reynolds's reading of the "seditious poem" from the nineteenth century, which employs an ostensibly apolitical medium of cultural expression to level a critique of elite officials and to challenge oppressive "state poetics." In the concluding section, I again consider the status of dissenting epistemologies and evidence of alternative regimes of public knowledge in formation.

## Forensic Science and Prospects for a New Civic Epistemology

Some twenty-five years ago, Andrew Turton looked hopefully at the prospect that a burgeoning Thai civil society might provide forms of "civil rhetoric" to counteract the repressive "state poetics" that had for so long stifled expressions of dissent.[68] A year later, when hundreds of thousands of Thais took the Bangkok streets in May 1992 to protest a military coup, they lent to his words a sense of prescience, and for a time validated their optimistic tone. Since then, however, Turton's heralding of the triumph of civil rhetoric over state poetics has proven to be unfounded, as evidenced in particular by the resurgence of prosecutions under Article 112 of the Thai Criminal Code, which pertains to the crime of *lèse majesté*, or offenses against the dignity of the monarchy.[69] Ongoing efforts to reassert state poetics through the vehicle of Article 112 have prompted calls for reform by Thai scholars—who protest at great personal and professional risk—and the public at large.[70] But largely lost in the furor over increasingly assertive enforcement of what Streckfuss has called the defamation regime in Thailand is a spate of deaths under suspicious circumstances of those accused of the crime. These deaths further illustrate the problematic position of forensics in Thai civic epistemology.

In October 2015, a police officer and a well-known fortune-teller were arrested and charged with *lèse majesté* for having allegedly used the Crown Prince's name in a scheme to profit from charity events. Police Major Prakrom Warunprapha was said to have hanged himself in his prison cell two days after his arraignment before a military tribunal. At the time of his death, rumors were already circulating that one of the co-defendants, Suriyan Sucharitpolwong, a fortune-teller who was more popularly known by his professional name, "Mo [Dr.] Yong," had also died in police custody.[71] Police quickly quashed the rumors, noting that although Suriyan did suffer from chronic diseases, he was undergoing treatment while in custody.[72] The fortune-teller's good fortune would not last: in a matter of weeks, he too would be dead, having succumbed to what police later diagnosed as a case of acute blood poisoning.[73] Others involved in cases of *lèse majesté* have died under similarly suspicious circumstances, including a senior police official who died by suicide in late 2014.[74] In each case, the bodies were hastily cremated without the benefit of religious rites or public scrutiny of the official autopsy results. The circuit connecting forensic investigation and civic truth was severed.

Yet strikingly in each case, the official "truth" of the cause of death also elicited the kinds of dissident forms of knowledge identified by Lim as conspiracy theories. The truths about such deaths voiced furtively at street-side *som tam* (papaya salad) stalls evince a civic epistemology that is increasingly at odds with that of the state.[75] Thus while such deaths speak volumes about the status of forensic medicine as a form of public knowledge, one that remains mired in traditional forms of Thai civic epistemology, they also suggest the continuing possibility of the emergence of subversive knowledge regimes. They provide hope that perhaps the transparent, independent investigations of such deaths will provide an occasion for the reassertion of public interest in forensic knowledge, and reinvigorate the third node in the triad of science, state, and society.

Finally, an unexpected coda to the Koh Tao murder case suggests another possible vector of change in Thai civic epistemology. In response to the continued furor over the Koh Tao case, including persistent allegations of police incompetence or worse—the suggestion of an intentional scapegoating intended to save the reputation of the region as a tourist paradise—the international consortium of hackers known as Anonymous took an interest in the case. In early January 2016, they hacked the web pages of the Royal Thai Police and posted a video challenging police findings in the case and demanding a full review and retrial of the Burmese migrant workers convicted in the case.[76] The group's actions raise the prospect that the cultural constraints on public forms of knowledge in Thailand might be broken by the interest of a global audience with its own civic epistemology. Perhaps such acts of piracy will help to upset the iterative re-performance of authoritative forms of civic epistemology and institute new repertoires for the performance and assertion of public truths under the watchful eye of a global public.

## NOTES

1. See, e.g., Jonathan Head, "Thailand Beach Murders: A Flawed and Muddled Investigation," *BBC News*, December 24, 2015, http://www.bbc.com/news/uk-35170419.

2. Jonathan Head, "Thailand's Paradise Island Murder Mystery," *BBC News*, September 19, 2014, http://www.bbc.com/news/uk-29262496.

3. "Cops Reject Call for Koh Tao Case Review," *Bangkok Post*, December 28, 2015, http://www.bangkokpost.com/news/general/808360/cops-reject-call-for-koh-tao-case-review.

4. Sheila Jasanoff, *Designs on Nature: Science and Democracy in Europe and the United States* (Princeton, NJ: Princeton University Press, 2005), 249.

5. The nation-state Thailand was known as Siam until 1939.

6. Lewis Mumford, "Authoritarian and Democratic Technics," *Technology and Culture* 5, no. 1 (1964): 1–8, 2–3.

7. The following discussion relies on Ian A. Burney, *Bodies of Evidence: Medicine and the Politics of the English Inquest, 1830–1926* (Baltimore: Johns Hopkins University Press, 2000).

8. Burney, *Bodies of Evidence*, 23.

9. Burney, *Bodies of Evidence*, 23–28. This element of local sovereignty could also extend to the moral repercussions, for example, in cases of apparent suicide. See Elisabeth Cawthon, "Thomas Wakley and the Medical Coronership: Occupational Death and the Judicial Process," *Medical History* 30, no. 2 (1986): 191–202, 193n11; and Thomas R. Forbes, "By What Disease or Casualty: The Changing Face of Death in London," *Journal of the History of Medicine and Allied Sciences* 31 (1976): 395–420, 416–17; and Burney, *Bodies of Evidence*, 70–73.

10. Burney, *Bodies of Evidence*, 40–51.

11. Burney, *Bodies of Evidence*, 54.

12. David Arnold, *Colonizing the Body: State Medicine and Epidemic Disease in Nineteenth-Century India* (Berkeley: University of California Press, 1993), 42. A similar shift occurred in the American empire in the early twentieth century; see Warwick Anderson, *Colonial Pathologies: American Tropical Medicine, Race, and Hygiene in the Philippines* (Durham, NC: Duke University Press, 2006), 74–103.

13. James Scott, John Tehranian, and Jeremy Mathias, "The Production of Legal Identities Proper to States: The Case of the Permanent Family Surname," *Comparative Studies in Society and History* 44, no. 1 (2002): 4–44, 7; quoted in Clare Anderson, *Legible Bodies: Race, Criminality and Colonialism in South Asia* (Oxford: Berg, 2004), 7.

14. On the problematic role of medical jurisprudence in determining causality and assigning blame for death in colonial jurisdictions, see Jordanna Bailkin, "The Boot and the Spleen: When Was Murder Possible in British India?," *Comparative Studies in Society and History* 48 (2006): 462–93; Elisabeth Kolsky, *Colonial Justice in British India* (New York: Cambridge University Press, 2010).

15. Khaled Fahmy, "The Anatomy of Justice: Forensic Medicine and Criminal Law in Nineteenth Century Egypt," *Islamic Law & Society* 6, no. 2 (1999): 224–271.

16. Niwat Thungthong, "Kan-phatana rabop kan-nitiwet kieo kap kan-chana sut phlik sop" [The development of the system of legal medicine with special reference to the autopsy] (LL.M. thesis, Chulalongkorn University, 1997 [2540]), 24–25.

17. National Archives of Thailand, Documents from the Fifth Reign, Ministry of Justice (*Krasuang Yutitham*), Division 13.4 (*Khadi khwam: tham rai rang kai, khatakam* [Cases: physical assault, murder]), item #5, *Khadi ruang luang phichaisena tat sisa nak thot* [Case of Luang Phichaisena decapitating a prisoner]. This case is not discussed by Niwat, whose analysis focuses on codified law to establish the chronology and motives behind the early history of forensic medicine in Thailand.

18. Thamsook Numnonda, "The Anglo-Siamese Negotiations, 1900–1909" (PhD thesis, University of London, 1966).

19. Christopher Hamlin, "Forensic Cultures in Historical Perspective: Technologies of Witness, Testimony, Judgment (and Justice?)," *Studies in History and Philosophy of Biological and Biomedical Sciences* 44 (2013): 4–15, 12.

20. Hamlin, "Forensic Cultures in Historical Perspective," 10.

21. Quentin (Trais) Pearson, "Morbid Subjects: Forensic Medicine and Sovereignty in Siam," *Modern Asian Studies* 52, no. 2 (2018): 394–420.

22. Pasuk Phongpaichit and Chris Baker, *Thailand: Economy and Politics*, 2nd ed. (Oxford: Oxford University Press, 2002), 95–107.

23. Thongchai Winichakul, *Siam Mapped: History of the Geo-Body of a Nation* (Honolulu: University of Hawai'i Press, 1994).

24. Tamara Loos, *Subject Siam: Family, Law, and Colonial Modernity in Thailand* (Ithaca, NY: Cornell University Press, 2006).

25. Davisakd Puaksom, "Of Germs, Public Hygiene and the Healthy Body: The Making of the Medicalizing State in Thailand," *Journal of Asian Studies* 66, no. 2 (2007): 311–44; and Davisakd Puaksom, *Chua rok rang kai lae rat wetchakam: prawatisat kan-phaet samai mai nai sangkhom thai* [Disease, the body, and the medicalizing state: the history of modern medicine in Thai society] (Bangkok: Chulalongkorn University Press, 2007).

26. Pearson, "Morbid Subjects."

27. Hamlin, "Forensic Cultures in Historical Perspective," 9.

28. David M. Engel, *Law and Kingship in Thailand during the Reign of King Chulalongkorn* (Ann Arbor: Center for South and Southeast Asian Studies, University of Michigan, 1975), 60.

29. This conflict was taking place at a moment when emerging forms of legal and medical expertise were asserting themselves through processes of professionalization, adding yet more uncertainty to the debates surrounding unnatural death in the Siamese capital. See Trais Pearson, *Sovereign Necropolis: The Politics of Death in Semi-Colonial Siam* (Ithaca, NY: Cornell University Press, in press).

30. Michel Foucault cited in Tony Day and Craig Reynolds, "Cosmologies, Truth Regimes, and the State in Southeast Asia," *Modern Asian Studies* 34, no. 1 (2000): 1–55, 18.

31. See Andrew Turton, "State Poetics and Civil Rhetoric: An Introduction to 'Thai Constructions of Knowledge,'" and Craig J. Reynolds, "Sedition in Thai History: A Nineteenth-Century Poem and Its Critics," in *Thai Constructions of Knowledge*, ed. Manas Chitakasem and Andrew Turton (London: School of Oriental and African Studies, University of London, 1991), 1–14, 15–36.

32. Day and Reynolds, "Cosmologies, Truth Regimes, and the State in Southeast Asia."

33. David Streckfuss, *Truth on Trial in Thailand: Defamation, Treason, and Lèse-majesté* (London: Routledge, 2011).

34. Turton, "State Poetics and Civil Rhetoric," 10.

35. Reynolds, "Sedition in Thai History," 26–28.

36. Reynolds, "Sedition in Thai History," 30–31.

37. Day and Reynolds, "Cosmologies, Truth Regimes," 3.

38. Day and Reynolds, "Cosmologies, Truth Regimes," 18–19.

39. Streckfuss, *Truth on Trial*, 36.

40. Streckfuss, *Truth on Trial*, 36.

41. Steven Shapin, *A Social History of Truth: Civility and Science in Seventeenth-Century England* (Chicago: University of Chicago Press, 1994).

42. Shapin, *Social History of Truth*, xxi.

43. Shapin, *Social History of Truth*, xxvi.

44. Sheila Jasanoff, *Designs on Nature: Science and Democracy in Europe and the United States* (Princeton, NJ: Princeton University Press, 2005), 248.

45. Day and Reynolds, "Cosmologies, Truth Regimes," 48–9.

46. Sheila Jasanoff, "Bhopal's Trials of Knowledge and Ignorance," *Isis* 98, no. 2 (2007): 344–50, 348.

47. Jasanoff, "Bhopal's Trials."

48. For a further elaboration of this concept, see Jasanoff, *Designs on Nature*, 249.

49. Day and Reynolds, "Cosmologies, Truth Regimes," 49.

50. Streckfuss, *Truth on Trial*, 36.

51. Daena Aki Funahashi, "Rule by Good People: Health Governance and the Violence of Moral Authority in Thailand," *Cultural Anthropology* 31, no. 1 (2016): 107–30.

52. Funahashi, "Rule by Good People," 124.

53. Samson Lim, *Siam's New Detectives: Visualizing Crime and Conspiracy in Modern Thailand* (Honolulu: University of Hawai'i Press), 119.

54. Lim, *Siam's New Detectives* 117.

55. Jonathan Head, "Thailand Beach Murders: A Flawed and Muddled Investigation," *BBC News*, December 24, 2015, http://www.bbc.com/news/uk-35170419.

56. Terry Fredrickson, "Koh Tao Murders: 'Person of Interest,'" *Bangkok Post* September 30, 2014, http://www.bangkokpost.com/print/434177/.

57. "Phuket Police Chief Says 'No Regrets,' Names Replacement," *Phuket Gazette* September 30, 2014, http://www.phuketgazette.net/phuket-news/Phuket-police-chief-says-regrets-names-replacement/36049#ad-image-0.

58. Head, "Thailand Beach Murders."

59. Lindsay Murdoch, "Australian Scientist Jane Taupin Questions Koh Tao Death Penalty Evidence," *Sydney Morning Herald*, January 7, 2016, http://www.smh.com.au/world /australian-scientist-jane-taupin-questions-koh-tao-death-penalty-evidence-20160106 -gmo5af.html.

60. Jon Fernquest, "Koh Tao Murders: No DNA Match Says Dr. Pornthip," *Bangkok Post*, September 11, 2015, http://www.bangkokpost.com/learning/learning-from-news/689964/koh -tao-murders-no-dna-match-says-dr-pornthip.

61. Andrew Drumond, "Koh Tao Murders Revisited—What Happened to the Evidence They Left Out?," May 31, 2016, http://www.andrew-drummond.com/2016/03/so-what-happened -to-footprints-of.html.

62. "Thailand Backpacker Murders: DNA Evidence 'Lost'—Police," *BBC News*, July 9, 2015, http://www.bbc.com/news/uk-33457038. There are no trials by jury in the Thai justice system.

63. Head, "Thailand Beach Murders."

64. Bruno Latour, *Pandora's Hope: Essays on the Reality of Science Studies* (Cambridge, MA: Harvard University Press, 1999), 304.

65. Lim, *Siam's New Detectives*, 114. Cf. Andrew Alan Johnson, who argues that conspiracy theories do not present an alternative truth, but merely hint at its possibility but ultimate inaccessibility; see his "Moral Knowledge and Its Enemies: Conspiracy and Kingship in Thailand,' *Anthropological Quarterly* 86, no. 4 (2013): 1059–86.

66. Lim, *Siam's New Detectives*, 115.

67. Funahashi, "Rule by Good People," 120, 124.

68. Turton, "State Poetics and Civil Rhetoric."

69. Streckfuss, *Truth on Trial*.

70. Duncan McCargo and Peeradej Tanruangporn, "Branding Dissent: Nitirat, Thailand's Enlightened Jurists," *Journal of Contemporary Asia* 45, no. 3 (2015): 419–42.

71. Teeranai Charuvastra, "Further Autopsy of Lese Majeste Inmate Unnecessary, Justice Minister Says," *Khoasod*, October 26, 2015, http://www.khaosodenglish.com/detail.php ?newsid=1445858351&typecate=06&section=.

72. Charuvastra, "Further Autopsy of Lese Majeste Inmate Unnecessary."

73. "Second Thailand Lese Majeste Detainee Dies in Military Custody," *BBC News*, November 9, 2015, http://www.bbc.com/news/world-asia-34764619.

74. This is to say nothing, of course, of the countless other deaths in police custody by those suspected or accused of other crimes, from drug violations to political insurgency.

75. Sanitsuda Ekachai, "Mor Yong's Death Breeds Mistrust," *Bangkok Post*, November 11, 2015.

76. "Koh Tao Murder—Thai Police, Their Scapegoats and Tourist Trade—Exposed by Anonymous," accessed May 1, 2016, https://www.facebook.com/video.php?v=567635633385675.

# PART THREE

# TRAINING AND TRANSMITTING

# CHAPTER NINE

# Cleaning Out the Mortuary and the Medicolegal Text

## Ambroise Tardieu's Modernizing Enterprise

BRUNO BERTHERAT

In the preface to one of his major works, *Étude médico-légale sur l'infanticide*, published in 1868, Parisian forensic pathologist Ambroise Tardieu (1818–79) presents his methodology in the following terms: "Former teachings and the classic texts are inadequate in providing the practical elements of expert testimony and the specific concepts the doctor . . . requires in order not to fall short in his work."[1] His approach was based on his rejection of the knowledge produced, in the past (even recent), through forensic science, as the sole source of medicolegal knowledge. Thus he indicated that, with few exceptions, he would not engage in historical and critical discussions on this literature, adding that he would form his opinions based solely on his own experience.

This rejection—and, even more, the tone used to express it—may seem radical, even for a treatise on legal medicine, at a time when most authors were used to discussing the work of their predecessors. This traditional attitude was widespread in medical encyclopedias of that era, such as the *Dictionnaire encyclopédique des sciences médicales* (1864–89), edited by Amédée Dechambre, a major reference work in France in the latter part of the nineteenth century. Doctors seemed eager to subscribe to a teleological history in which they could associate their own names with established names, at a time when the cult of great men was valued, especially within the field of knowledge.[2]

Analyzing Tardieu's rejection is interesting for two reasons. The first reason is that Tardieu was the most famous mid-nineteenth-century forensic pathologist in France and one of the great men in the history of French medicine.[3] Consequently, his works and lectures had an impact on forensic medicine, representing a specific moment in the history of the French forensic medicine. The second reason is that this denial forces historians to question their own methods. In the history of medicine, the issue of heritage—because it raises the question of continuity and ruptures, as well as that of the linearity of evolution—is important, just like the comparison with other geographical areas.[4]

This contribution should be viewed as no more than a starting point in an attempt to uncover the reasons for Tardieu's rejection of previous literature and its limitations. I first examine the reasons for his rejection of a medico-legal legacy before identifying the bodies of knowledge and influences that nevertheless surface in Tardieu's approach.

## Rejecting a Legacy

Three main reasons can be identified for Tardieu's reluctance to rely on previous literature. The first can be traced to Tardieu's professional positioning in the scientific community and in society, the second is related to his adherence to a specifically Parisian forensic model, and the third has to do with his belief in the anatomic-clinical method.

At the time his work was published, Tardieu held a powerful position in French forensic medicine.[5] He was a professor of legal medicine at the Medical School of Paris from 1861 until his death in 1879, dean from 1864 until 1866, member of the Academy of Medicine since 1859 (of which he became president in 1867), and president of the General Association of French Medical Practitioners (AGMF)—the first great union of physicians in France—from 1867 until 1875. By acquiring those prestigious positions, Tardieu reached the pinnacle of the medical profession: he was a "mandarin."[6] But Tardieu did not belong to the Society of Forensic Medicine of Paris (which would later become the Society of Forensic Medicine of France), founded in 1868, the year of publication of his book on infanticide. His work as a forensic expert, particularly in Paris, was considerable: he authored over five thousand forensic reports during his career. In the preface to his book on infanticide, Tardieu claims to have almost twenty-five years of experience and to have studied "over eight hundred cases."[7] He published a series of studies or monographs on the main subjects of legal medicine from the early 1850s onward. His study on infanticide was therefore not his first. Tardieu was also a recognized public health specialist: he wrote a reference dictionary on the subject[8] and pioneering works on occupational hygiene.[9] He was a member of the Paris City Council until the end of the Second Empire and contributed to the Paris sanitation project headed by Haussmann.[10] He became chairman of the advisory committee on public hygiene in 1867. The eminent position he occupied in the field of legal medicine and public hygiene is reflected in his active collaboration with the *Annales d'hygiène publique et de médecine légale*, the famous French medical journal founded in 1829, in which he published his main works. His close ties with the political powers—namely, Napoléon III's regime—may have facilitated his rise,[11] but they were also, at times, the cause of tension with his students.[12] Thus, despite a few discordant notes, the 1860s were the peak of Tardieu's career.[13]

This powerful position probably had repercussions in his practice. Here was a man who was sure of his own academic knowledge. In his book, Tardieu recalls an important finding he had presented at the Academy of Medicine in 1855 and published the same year in the *Annales d'hygiène publique et de médecine légale*.[14] He demonstrated that spots found on lungs, called "subpleural ecchymoses," were indications of suffocation[15] (particularly in newborns), and proved that the victim had been murdered.[16] The importance of Tardieu's work in this domain was recognized early by both his French colleagues and those around the world, as detailed below. Subpleural ecchymoses are known, even today, as "Tardieu's spots."[17] Tardieu became a celebrity, recognized in France and abroad, owing to his position and his works. In the preface to his book, he wrote that he had received numerous letters from his professional colleagues asking for his opinion on cases of infanticide.[18] He is quoted in his foreign colleagues' books, such as the *Traité de médecine légale pratique*, by Johann Ludwig Casper, professor of forensic medicine at the University of Berlin (d. 1864), which was translated into French in 1862.[19] Some of Tardieu's books were (or were going to be) translated and adapted into other languages.[20] In 1864, as a form of recognition by his foreign colleagues, he was also appointed an honorary member of the German Medical Society of Paris, which was composed of Paris-based German-speaking doctors.[21] His fame extended beyond academic circles into the public domain in major criminal cases when his interventions were closely followed by newspapers. Tardieu attracted press attention because of his brilliance.[22] According to one judge, who had dealings with him in his professional capacity, he was a "marvelous orator": "He asserted himself so confidently that he seemed almost infallible."[23]

One has the same impression reading his literature on infanticide: Tardieu did not seem to doubt himself. This self-confidence can be seen in the results of his autopsies of newborn infants, particularly those conducted at the Paris Morgue, the center for judicial autopsies in Paris (his colleagues also carried out autopsies there).[24] The statistics that Tardieu established, the "perfect accuracy" of which he was certain, show that the number of infanticides in relation to the number of autopsies performed and bodies brought to the Paris Morgue increased between 1837 and 1866. Regarding legal autopsies, which he performed on precisely 804 infants between 1844 and 1868, Tardieu confirms a high number of infanticides (69% of all infant deaths were infanticides) and a high rate of deaths from suffocation (50.63% of all infanticides), far ahead of other causes of death.[25] Tardieu claimed that by applying his theory of subpleural ecchymoses, he could confirm that the leading cause of death in cases of infanticides was suffocation. And he writes: "By identifying and revealing, twelve years ago, the characteristics of death from suffocation, I contributed a crucial element to the history of infanticide forensics."[26] Thus it is tempting to say that Tardieu's main reference source was his own work.

The second reason for Tardieu's rejection of the knowledge produced in the past by forensic science is that Tardieu represented a French and Parisian forensic model that emerged in the Revolutionary and Napoleonic era. First, it was a judicial model, founded on an inquisitorial process (carried out prior to the trial) that had been widely used in continental Europe since the end of the Middle Ages,[27] and on an accusatory process (the trial itself) inherited from England and the Enlightenment. This model conferred an important role to the expert[28] as the justice system expected him to provide scientific evidence. Although an expert was called to the stand as a simple witness, his voice carried more weight than that of other witnesses, especially in Paris, and even more so when the expert in question was famous. The specificity of this judicial process, which differed from the English legal culture but was close to that of other continental states, allowed for the emergence of "a new judicial objectivity that prepared the 19th-century advent of positivist criminology and anthropology."[29] It was also an academic model: forensic studies had been taught at the university since the Revolution. There were only three medical schools in France at the time, in Paris, Montpellier, and Strasbourg.[30] New medical schools, such as those in Lyon and Lille, would subsequently—from the 1870s onward—be founded.[31] Following the annexation of Alsace-Lorraine by Germany, the Medical School of Strasbourg was relocated to Nancy. For most of the century, however, the chair of legal medicine of the Paris Medical School, founded in 1794, was the most famous in France; this placed Tardieu in a more eminent position than his colleagues.

Lastly, it was a practical model. In Paris, judicial autopsies were not performed in hospitals or in universities but in a police building called the morgue, where Tardieu often worked as the main expert (at the time, Alphonse Devergie, another forensic pathologist, was the chief medical inspector of the institution until his death in 1879). The Paris Morgue had two functions: the identification of unknown bodies and judicial autopsies. From the beginning of the century onward, it gradually emerged as a site for scientific work. In 1868, the morgue was a new and modern building. It had been inaugurated in 1864 and was located in the center of Paris, on the eastern tip of the Île de la Cité, behind the apse of the Notre-Dame Cathedral.[32] It was situated at a short distance from the medical school (Rue de l'École de Médecine, on the left bank) where Tardieu taught forensic medicine, and from the Palace of Justice (on the western side of the Île de la Cité), where he testified as an expert during trials. The morgue had a dedicated area for performing autopsies as well as an office for magistrates. It had another advantage for forensic doctors: it housed numerous corpses of unknown and unclaimed persons that could be used for research in various fields,[33] particularly the study of the decomposition of corpses (conducted by Orfila and Devergie in the 1820s).[34] Thus the practice of forensic medicine in France was different from that of other countries.[35] The Paris Morgue served as a model, reproduced in some French cities such as Lyon (though with much fewer resources)[36] and in some other countries at the end of the century. For example, Berlin built its own morgue in 1886, Bucharest in 1892.[37]

Was this Parisian model open to other forensic cultures? Was Tardieu open-minded? We can provide an initial answer to these questions by observing Tardieu's stance toward the German forensic culture. In his preface to *Étude médico-légale sur l'infanticide*, Tardieu wrote: "The number of PhD dissertations that have been defended in German universities for half a century now, would try the patience of the most avid collector."[38] Tardieu's words are ambivalent: on the one hand, he recognizes the productivity and vibrancy of the German forensic culture; on the other, he implies that he is not interested in it. Other questions arise: Had Tardieu read those theses? Could he read or speak German? We return to this point in the second half of this chapter.

The third reason is related to methodology. In certain respects, Tardieu appeared to be biased—and even unjust—but this also reveals a profound truth: that of the triumph, in nineteenth-century Europe, of the anatomic-clinical method, whose origins date back to the seventeenth century.[39] Bodies, both living and dead, were the basis of this method, which was used in hospitals with patients and with the bodies of the deceased. The doctors observed the symptoms of a disease present on and inside the body. The disease became visible when the corpse was dissected and autopsied. The corpse was therefore an essential object of study for the physician.[40] It was in large cities, which housed many hospitals, that this scientific approach was developed. The links between the hospital and the university were fundamental, and they were reinforced by selective competitive examinations and internship training. Internship programs gave students access to hospitals (mobility was a requirement) and helped them increase their medical knowledge (through contact with patients and access to training in diagnosis). This explains why Paris then became one of the main centers of European medicine.[41]

Parisian doctor Xavier Bichat (1771–1802), professor at the medical school[42] and considered the founder of pathological anatomy, summarized this approach with these famous words: "Open up a few corpses: you will dissipate at once the darkness that observation alone could not dissipate."[43] He was echoed at the end of the same century by Dechambre's definition of an autopsy: "To perform an autopsy *is to place directly under the eyes of the physician the organs situated more or less deeply*, so as to allow him to *see first-hand* the lesions or alterations they may present, and to deduce from this examination solutions to a plethora of problems related to either pathology or forensic medicine."[44]

Clinical observation finds its continuation in the internal exploration of the corpse. Hence the importance attributed by Tardieu, as well as his colleagues elsewhere,[45] to "research and . . . personal observation," which he applies to the study of sub-pleural ecchymoses to justify the importance of his discovery. According to him, "[the essential features of death by suffocation] do not consist of external signs, which are of great value when they do exist, but which can be completely absent. It is in the respiratory and circulatory organs that those signs are to be found."[46] In this approach, the anatomist's gaze played a central role, especially in the medicolegal field. The physician's gaze presents similarity with and complements that of the police officer, particularly at the morgue: the object of their gaze is the corpse, which the former autopsies and the latter identifies (or exposes for identification by witnesses). The anatomic-clinical and panoptic models somehow echo each other with their own sets of knowledge regarding bodies that are brought under control.[47]

It was through the external and internal observation of the body that one came to truth, to certainty, which was what a court of justice expected of experts. But did the expertise of a forensic pathologist enable him to provide a judge with certainty, which was what the expert was appointed for? The key issue here is to define the frontier between doubt and certainty, a frontier that varied depending on the experts and whose delimitation was related to the very essence of the expert's profession. There was no clear-cut definition of certainty in the French medicolegal field. Experts such as Devergie and Orfila disagreed on the subject and were engaged in a fierce quarrel between 1830 and 1840: Devergie criticized Orfila's excessive caution and opposed him to Fodéré.[48] Tardieu did not expand on this point. He even opposed his experience as an expert to "the discussions some casuists of forensic medicine have indulged in."[49] This raises the question of the training of experts. Why was it not until 1877 that a practical training course in forensic medicine was officially created in Paris, in this case at the morgue? Here, "practical course" implies that the students could observe the autopsies done by the professor. Such practical courses already existed in Germany. What influence did the Germanic model have on this creation? These last remarks prompt us to reevaluate the question of legacy and influences. Tardieu's work was set in his time and was therefore influenced by it.

## Heritage, Influences, and the Dissemination of Knowledge

Whether or not he admitted it, Tardieu was not the only one conducting research on infanticide: he was part of a scientific network. Three aspects reveal that Tardieu was indeed connected with his scientific environment, an environment with an undeniably international dimension: first, his book was published amid controversy; second, he had to take into account the long-term history of forensics; and, finally, his expertise was the result of a legacy of training in forensic practices.

An examination of the controversies, because they were accompanied by tension in the fields of knowledge and power, helps reveal the networks at play in the medical world and their impact. Tardieu's book provides a good example thereof. He alludes in his book to recent controversies pertaining to "Tardieu's spots." Tardieu received criticism, which he rejected, but it seems to have had a significant impact on Tardieu because he mentions it in his book: "I have however to report a number of misconceptions and errors of fact which are the most dangerous since they are to be found in some recent works and reputed classics both in France and abroad."[50] The "casuists" he is contemptuously referring to, without naming them, seem to designate some of his skeptical colleagues (and he may have been thinking of specific authors, settling scores with them in the process). Tardieu then responded extensively and precisely to those criticisms, especially in the sections of the book devoted to infanticide by suffocation. Thus the Parisian expert's book was published in a context of international controversies.[51]

We examine those controversies using three French reference sources: Dechambre's dictionary,[52] the *Bulletin de la Société de médecine légale de Paris*,[53] and Paul Brouardel's published lectures.[54] Brouardel (1837–1906) was Tardieu's successor to the chair of legal medicine in Paris, dean (from 1887 until 1901), and chief medical inspector of the morgue (from 1879 until his death). Therefore his was a voice of authority. Although these sources acknowledge Tardieu's pioneering role in the discovery of the significance of subpleural ecchymoses as signs of suffocation (which Dechambre called "Tardieu spots" early on), they highlight the many criticisms leveled at his theory by forensic experts, in France and abroad. In this regard, Dechambre's dictionary even makes mention of a "movement of opinion" against Tardieu's theory.[55] Famous names are mentioned, among them Casper; Carl Liman, his successor in Berlin; Eduard von Hofmann, professor of forensic medicine at the University of Vienna; and Alfred Swayne Taylor, professor of forensic medicine at Guy's Hospital in London. In their own works, those critics quote other authors to support their arguments. It would be useful here to build a genealogy of the criticisms. Carl Liman seems to have been one of the first to challenge Tardieu's theories in a German publication, edited by his master Casper in 1861. The article was translated into French in 1867, and the translated version is cited in Dechambre's dictionary.[56] What exactly was Tardieu criticized for? For presenting subpleural ecchymoses as certain and infallible signs of suffocation. And Liman was highly critical of a "doctrine" he considered "erroneous and dangerous," arguing that "no anatomic lesion in internal organs is sufficient in itself to make a post mortem diagnosis of the causes of asphyxia. It is on the basis of this principle that I disagree with M. Tardieu. . . . Those lesions are in no way specific to any type of violent asphyxia."[57]

As for Tardieu, he certainly did not entirely refuse to discuss the question, nor did he refuse to consider what other researchers did. Throughout his book, Tardieu frequently quotes his French and foreign colleagues, particularly the Germans—Casper being the most cited[58]—and he mentions his amicable relationship with Casper's successor ("my knowledgeable friend Professor Liman"). Tardieu knew the international forensic literature, particularly that of his German colleagues. We cannot be sure, however, that he could read German. In any case, almost all the quotes he took from Casper were drawn from French translations of his publications.[59] Thus Germanic forensic medicine appears to have been an important reference for Tardieu. The somewhat contemptuous remarks he made in his preface concerning the German forensic school do not expressly refer to the authors or reference works he later mentions in the book. The pages devoted to the subject of subpleural ecchymoses illustrate how the art of citation can be put at the service of pro domo advocacy. Tardieu aimed his attacks at the Italian and German forensic physicians, especially at the two Berlin professors, Casper and Liman, because they gave authoritative support to the criticisms made against him.[60] But the art of citation has its limitations. Tardieu could apparently claim no support from any of the big names in French and European forensic medicine: he mentions none.[61] Thus, placed in his historical context, Tardieu appears to be one actor among many others, of various ranks, in the vast European academic world, fueled by information flows via books and major journals—whether in their original or translated versions[62]—but also via meetings between all these actors.

This flow of information and knowledge must be situated within a specific temporal framework combining the short and long term. This dimension is present in Tardieu's 1855 memoir, in which he presents a classic history of the question (which he did not do in his book published thirteen years later, but his position had changed by then) describing "the exact state of scientific knowledge on the subject."[63] He recognized at the time that he was not the first to have discovered those lesions, citing French and foreign authors of the nineteenth century, granting them his approval but each time emphasizing discrepancies with his theory, which he presents a few lines later. Subsequently, and regardless of what he said, Tardieu did take into account the criticisms of his colleagues and nuanced certain elements of his theory in his book and its posthumous republication in 1880.[64] He recognized, for example, that subpleural ecchymoses could be found in many cases of death other than by suffocation. He also acknowledged that in some cases of suffocation, few or even no subpleural ecchymoses were found. But those are, according to him, "rare exceptions" that could not bring into question the validity of his theory. Tardieu wrote a few lines earlier: "While the value of this sign may not be absolute, it is nevertheless very positive and very high and . . . when it can be rigorously interpreted, forensic pathologists can entirely rely upon it."[65] He did not really have an error rate for his methods but simply relied on his own judgment. Thus, throughout the development and defense of his discovery, Tardieu contrasted his "personal findings" with those of the academic environment of his time and interacted with the latter. His pugnacity can probably be explained by a desire to not lose face in the academic world and perhaps also to maintain the reputation of infallibility he acquired in the courts.

But despite his best efforts, Tardieu could not prevent the establishment of a new creed that the *Bulletin de la Société de médecine légale*, Dechambre's dictionary, and Brouardel's work, in France, contributed to propagating and that is still relevant today: subpleural ecchymoses are a frequent sign of suffocation, a sign that must be corroborated by other clues. Those criticisms must have had consequences on the relations between Tardieu and his French colleagues and may explain why Tardieu was never a member of the Legal Medicine Society.[66] Those criticisms mostly had consequences on the conclusions of the experts' reports, particularly at the morgue. Following Tardieu's death, the departure of his disciple and closest collaborator, Georges Bergeron, and the arrival of Brouardel and of a new generation of forensic physicians, the observed percentage of infanticides slumped, and this sudden fall was related to that of infanticides by suffocation. This resulted in an increase in the number of deaths by uncertain causes. On the basis of the absolute figures drawn from the morgue's records, I calculated the number of infanticides reported for the following years: 1876 (56 infanticides), 1886 (54 infanticides), 1896 (49 infanticides), and 1906 (5 infanticides). In 1876, Bergeron performed almost all the autopsies. The high number of infanticides observed for 1876 compared to the numbers for the following years seem to suggest that Bergeron adhered to his master's theory. This decline in the number of infanticides observed from 1886 onward occurred at the time when Brouardel and his collaborators (Socquet, Vibert, Toinot) took over the autopsy department. This trend is confirmed by the figures provided by Brouardel, who also produced statistics on the different forms of infanticide and compared them to Tardieu's figures.[67] According to the statistics provided by Brouardel, 43 percent of all infant deaths (531 infants) were infanticides, and according to Vibert's statistics, infanticides represented 26 percent of all infant deaths (434 newborns). The difference between these figures and those provided by Tardieu lies essentially in the different estimates of the rate of death from suffocation (approximately 11.50%); suffocation was no longer the leading mode of infanticide.

Although the abovementioned authors continued to extensively cite Tardieu even after his death, and he was presented as a great master of forensic pathology,[68] the infallibility of his theory was definitively rejected. But the outcome of the controversy seems to have had a deeper significance, as indicated in Dechambre's dictionary: "Despite all his knowledge, perspicacity and tremendous experience in forensic medicine, didn't Tardieu commit a philosophical error by thinking he had found a sign of absolute certainty in the field of biology?—Is there such a thing, in the clinical domain, as a truly pathognomonic sign?—Of course not. The same is true of forensic medicine, where one can only find probabilities, probabilities sometimes so numerous that they lead, especially in practice, to near-certainty. But to aim at absolute certainty almost inevitably leads into error."[69]

This criticism is significant. Tardieu's succession by Brouardel was not a mere change of person. The new head of forensics initiated a critical turn in Parisian forensic medicine, which was reflected in his teaching and in his forensic reports and was characterized by caution in his conclusions. Regarding infanticide, Brouardel's testimony deserves to be quoted at length: "I am convinced that Mr. Vibert and I have allowed many women to be vindicated, women who had actually killed their child; but we could not prove scientifically that they had committed infanticide. We did not have the right to use our personal opinion as anatomic evidence."[70] A comparison between Tardieu's and Brouardel's autopsy reports in their respective books provides interesting insight. Both experts mention subpleural ecchymoses, but their interpretations often lead to different conclusions. At the turn of the century, Brouardel's work became the main reference in the field of legal medicine. Doctor Thoinot, his student, summarized his importance in the statement: "Legal medicine was going astray: he set it back, and kept it, on the right path. He taught us to doubt."[71] Even the judge (quoted above) who used to praise Tardieu expressed a preference for Brouardel.[72] Tardieu was then seen as an omnipotent and omniscient— but nonetheless fallible—expert. Controversies developed again at the end of the century around miscarriages of justice and the role of forensic experts in those miscarriages. And although Tardieu was spared,[73] his disciple Bergeron became the symbol of the bad expert.[74]

But Brouardel's approach also gave rise to new debate in the early twentieth century following the case of the "Ogress of the Goutte d'Or" (1905–8), often cited as the gravest medicolegal error committed at the time.[75] It involves a large part of the Paris forensic team. A woman, Jeanne Weber, is suspected of strangling several children. Thoinot, who is the expert in charge of the autopsies, believes that he cannot prove that strangulation was the cause of death. Jeanne Weber is acquitted in 1906, partly thanks to his report, which is validated by Brouardel. But she is again accused of infanticide a year later. And again, Thoinot, joined by Socquet, disproves the strangulation theory defended by provincial doctors. Jeanne Weber obtains a court dismissal in early 1908 but kills another child shortly thereafter. This time, the theory of death by strangulation prevails despite Thoinot's opinion. Thus the critical turning point mentioned above must be viewed from the perspective of the long history of medical expertise and of the question of diagnostic certainty. As a magistrate pointed out, certainty in science is relative: "A scientist can only know what is known during his time; his affirmation cannot reach absolute truth but is limited to the state of science at the time of his claim."[76] This temporal dimension can also be applied to the question of training in forensic medicine.

Tardieu constantly referred to his experience and practice to show that he was autonomous in his reasoning. But his expertise was also the result of a legacy, of a learning process, that fits in a long tradition. This learning process occurred through studies, at the faculty, in the dissection amphitheaters, in the hospitals, in his work environment, and with multiple people: students, magistrates, dieners, teachers, colleagues, and so on. The acts performed by the forensic pathologist also have their own temporality, which he has to learn. All books on legal medicine emphasized the need to conduct autopsies quickly (because corpses decay and clues can disappear). But the speed required to perform an autopsy was not comparable with that required of a surgeon operating on a live patient,[77] and manuals on forensic medicine warned against excessive haste. The question of training took a particular turn in Tardieu's time with the creation of a practical course of forensic medicine at the morgue in 1877. The first requests dated back to the beginning of the century: forensic pathologists in Paris wanted forensic medicine to be taught in both theoretical and practical classes, as was the case for other specialties. In the 1830s, informal practical courses were temporarily launched by Devergie, Brouardel's predecessor at the morgue.

In this context, why did it take forty more years for those practical courses to be officially reinstated on a permanent basis? In the absence of relevant sources, one may consider that the inertia of university authorities and of the judicial system, as well as a lack of resources, are plausible causes. Was the inertia of university authorities also that of Tardieu? The latter had, in 1866, written a report for the General Association of French Medical Practitioners asking for the creation of a special diploma in forensic medicine.[78] But Tardieu's lack of interest in using the morgue to train successors may have arisen from his feeling that his expertise was embodied in himself, and his own experience, which could not be translated into a curriculum. Another factor of inertia was related to the choice of facilities. The morgue had a special status, and the creation of a practical training course in the establishment was far from a given. Although the morgue had the advantage of housing a large number of corpses, particularly those intended for forensic examination, it was not suited for the teaching of students; furthermore, the morgue was not affiliated with the faculty, but with the police headquarters. And yet it was there that the first practical lectures were given, that a post of senior lecturer was created and assigned to Brouardel—appointed chair of forensic medicine two years later—and that an amphitheater was built a few years later. The first practical lecture (the official term at the time was "conference") was held on January 9, 1878, in the presence of Devergie and Tardieu. Thus Brouardel taught theoretical and practical forensic medicine (by becoming chair). The title of his highly didactic fourteen-volume publication *Cours de médecine légale*, released between 1895 and 1909 and produced with the help of collaborators, is significant. Regarding the practical classes, Brouardel summarized his philosophy as early as 1879 in an interesting statement: "As my teacher, Professor Lasègue once said: at the hospital, the student is the protector of the patient. This saying could be applied to forensic examinations. The presence of someone—even with little competence—who controls you, and to whom you have to demonstrate the value you attribute to lesions, forces you to constantly refine and revise, with the progress of science, the determination of the signs we rely on."[79] Thus Brouardel's teaching philosophy was in line with his approach to forensic expertise: caution and teamwork. The opposite of Tardieu's conceptions.

How does one explain such proactiveness in creating practical training at the morgue? The shift appears to have been initiated by Devergie.[80] But it was the defeat by Germany in 1871 and the regime change in France that probably marked a turning point. The 1870s were characterized by a deep crisis in France, one that also affected the French scientific community, even though the latter had been aware since the 1860s of the competition from other emerging scientific centers, particularly in Germany.[81] Germany, the victorious enemy, was also a model that the new republican regime wanted to match or even surpass, including in the field of legal medicine. And practical training in forensic medicine had existed for many years in German and Austrian universities.[82] In France, this model was probably disseminated through mediators, whether men or institutions. This was the case of Strasbourg's Medical School, at which, under the auspices of Gabriel Tourdes (1810–90), then chair of forensic medicine, a practical course was introduced and hosted until the 1870–71 war. Nancy, which inherited the Medical School of Strasbourg, took over the program.[83] This openness to foreign models is reflected in "a journey of scientific exploration," undertaken by Paul Brouardel between July and September 1878, with a view to modernizing the forensic examination equipment. Germany formed the main part of his trip.[84] On this occasion, Brouardel observed that other countries had a similar approach. He met colleagues who were taking or planning the same type of trip, whether in their private or official capacity.[85] Among other things, Brouardel stressed the need to create laboratories that were subsequently partly set up at the morgue.[86] Equally significant was the work undertaken to translate foreign reference treatises. After Casper's treatise in 1862, Hofman's and Taylor's were to be translated in 1881. The Parisian forensic medicine world was more than ever connected to its international environment. As for Brouardel, who had links with both the Pasteurians and the republican elites, he was able to establish his power in the field of forensic medicine and public hygiene and to build his network, becoming a key figure in French medicine at during the Belle Époque.[87] He had taken Tardieu's place.

The above is intended to initiate a reflection that should be extended and deepened. But relations were intense between European forensic experts, especially between the two leading countries in the field of medicine at the time, France and Germany. The dissemination of knowledge is not a one-way affair, and it points to mutual influences at the theoretical and practical levels of forensic medicine.[88] While the value of Tardieu's discovery was lessened by criticism from European colleagues, and particularly from the German school, and although French forensic medicine sought, after 1871, to emulate the German model, Germany and other countries did draw inspiration from the morgue's model, and students and doctors went to Paris to be trained in forensic medicine.[89] This puts in perspective the scientific hegemony of Germany and the appeal it generated among foreign students and physicians.[90]

More interest should be taken in individuals. French historians have long neglected the biographical approach, chastened by the criticisms expressed by sociologist Pierre Bourdieu concerning the "biographical illusion."[91] There is no biography of Tardieu, nor of most other French forensic experts of that era.[92] The general public and historians appear today to be more interested in criminologists[93] than in forensic doctors, perhaps because the former are associated with the world of investigation, which represents a form of modern-day adventure, whereas the latter are confined to the small space of an autopsy room.[94] Moreover, the biographical approach cannot be restricted to "great men": the collaborators, mediators,[95] and experts who worked in secondary institutions as well as the students, morgues, and hospital staff are all essential to understanding how the world of expertise operated.

Thus there is a need to open up time and space by broadening the temporal scope of analysis and multiplying perspectives. We have examined Tardieu's progression and tried to understand his stance and reactions as well as his interactions with the French and European academic world. We should now further extend the chronology to include older and more recent periods, and widen the geographic scope of study so as to build a comparative history of forensic cultures. To understand the particularities of forensic practice in a specific location such as the Paris Morgue, comparison with other forensic cultures—for example, with Austria, England, Germany, Italy—is essential. But the comparison should not be limited to the leading countries in this field. Moreover, the circulation of ideas and people should be traced with more accuracy.[96]

### ACKNOWLEDGMENTS

The author expresses thanks to Olivier Faure, Serenella Nonnis, Lynne Oyama, Isabelle Renaudet, Sue Reid, Jérôme van Wijland, and David Worall. Special thanks go to Jérôme Bertherat, Vanessa Guignery, Delphine Silberbauer, and Jennifer Smith-Daye.

### NOTES

1. Ambroise Tardieu, *Étude médico-légale sur l'infanticide* (Paris: J.-B. Baillière et fils, 1868), vi. The preface is dated January 1868. The book was written the previous year.

2. See Mona Ozouf, "Le Panthéon," in *Les Lieux de mémoire*, vol. 1, *La République*, ed. Pierre Nora (Paris: Gallimard, 1984), 139–66; François Dosse, *Le Pari biographique: Écrire une vie* (Paris: La Découverte, 2011 [2005]), 181–200.

3. E.g., he has a particularly complimentary biographical entry in the *Dictionnaire encyclopédique des sciences médicales*—see Louis Hahn, "Tardieu (Ambroise-Auguste)," in *Dictionnaire encyclopédique des sciences médicales*, ed. A. Dechambre et al. (Paris: Masson, Asselin et Houzeau, 1864–89, 3rd ser., 15:746–48)—and even in Larousse (see note 12).

4. On the history of French and Parisian legal medicine, see Pierre Darmon, *Médecins et assassins à la Belle Époque: La médicalisation du crime* (Paris: Le Seuil, 1989); Frédéric Chauvaud, *Les Experts du crime: La médecine légale en France au XIXᵉ siècle* (Paris: Aubier, 2000); Bruno Bertherat, "La Morgue de Paris au XIXᵉ siècle (1804–1907): Les origines de l'institut médico-légal ou les métamorphoses de la machine" (PhD thesis, Université de Paris 1,

2002) (the third part is dedicated to forensic medicine in Paris); Frédéric Chauvaud and Laurence Dumoulin, *Experts et expertise judiciaire: France, XIX^e et XX^e siècles* (Rennes: PUR, 2003); Bruno Bertherat, "L'élection à la chaire de médecine légale: Acteurs, réseaux et enjeux dans le monde universitaire," *Revue historique* 644 (December 2007): 823–56. For a long-term international perspective, see Catherine Crawford and Michael Clark, eds., *Legal Medicine in History* (Cambridge: Cambridge University Press, 1994); Vincent Barras and Michel Porret, eds., "Homo criminalis: Pratiques et doctrines médicales (XVI^e-XX^e siècles)," *Équinoxe: Revue de sciences humaines* 22 (1999); Michel Porret, ed., "La médecine légale entre doctrines et pratiques," *Revue d'histoire des sciences humaines* 22 (2010): 3–144; Katherine D. Watson, *Forensic Medicine in Western Society: A History* (London: Routledge, 2011); Christopher Hamlin, "Technologies of Witness, Testimony, Judgment (and Justice?)," *Studies in History and Philosophy of Biological and Biomedical Sciences* 44 (2013): 4–15.

5. There is no file on Tardieu in the staff records of the Medical School of Paris in the Archives Nationales (AN), but I obtained his file from the Légion d'Honneur (AN, LH 2568 9). See also AN, AJ16 6259, séance du 16 janvier 1879, 159–65. In addition to the biographical entries in dictionaries mentioned earlier, see, more recently, Françoise Huguet, *Les Professeurs de la Faculté de médecine de Paris: Dictionnaire biographique, 1794–1939* (Paris: Institut national de la recherche pédagogique, CNRS, 1991), 459–62; and Jean-François Lemaire, "Tardieu (Ambroise)," in *Dictionnaire du Second Empire*, ed. Jean Tulard (Paris: Fayard, 1995), 1237. See also the preface to Georges Vigarello's "La violence sexuelle et l'œil du savant," in Tardieu's book about indecent assaults: *Les Attentats aux mœurs, 1857* (Grenoble: Jérôme Millon, 1995), 5–28.

6. George Weisz, *The Medical Mandarins: The French Academy of Medicine in the Nineteenth Century and Early Twentieth Century* (New York: Oxford University Press, 1995).

7. Tardieu, *Étude médico-légale sur l'infanticide*, vi.

8. Ambroise Tardieu, *Dictionnaire d'hygiène publique et de salubrité, ou Répertoire de toutes les questions relatives à la santé publique complété par le texte des lois qui s'y rattachent*, 3 vols. (Paris: J.-B. Baillière, 1852–54 [2nd ed., 1862]).

9. Gérard Jorland, "L'hygiène professionnelle en France au XIX^e siècle," *Le Mouvement social* 213 (December 2005): 71–90.

10. Jacques Léonard, *La Médecine entre les savoirs et les pouvoirs: Histoire intellectuelle et politique de la médecine française au XIX^e siècle* (Paris: Aubier, 1981), 226.

11. Tardieu was one of the consulting physicians to the emperor. He was also a regular visitor, during Napoléon III's regime and the Third Republic, at the famous literary salon hosted by Princess Mathilde, cousin of Napoléon III and a convinced Bonapartist. Tardieu was not the only physician who supported Napoléon III's regime. Some supported the social aspects of Bonapartism (public hygiene, construction of infrastructure). Generally speaking, the regime of Napoléon III sought the support of scholars by honoring them (Bernard, Pasteur) and in particular supported the creation of the General Association of Mutual Insurance and Aid for the Physicians of France, or AGMF. See Léonard, *Médecine entre les savoirs et les pouvoirs*, 224–35.

12. In 1866, Tardieu failed to support students who were sentenced for causing political unrest at the university. This led to his resignation as dean of the medical school. In 1870, during the trial of Pierre Bonaparte, the emperor's cousin, accused of the murder of journalist Victor Noir, Tardieu once again caused the wrath of students following his testimony, as an expert, at the High Court of Justice. Discontent among students was such that the lectures were temporarily interrupted. See, in particular, Pierre Larousse, "Tardieu (Auguste-Ambroise)," in *Grand dictionnaire universel du XIX^e siècle* (Paris: Larousse, 1866–77), 14:1471–72, and Léonard, *Médecine entre les savoirs et les pouvoirs*, 224–25.

13. He was awarded the Legion of Honor under three different regimes: he was made Knight of the Legion of Honor 1846 under the July Monarchy, received the title of Officer of

the Legion of Honor under the Second Empire in 1860, and was commander under the Third Republic in 1876. (Admittedly, the nature of the regime had then not yet been fixed because the power was divided between conservatives and republicans). To my knowledge, however, no public monument was erected in honor of Tardieu after his death.

14. *Bulletin de l'Académie nationale de médecine* 20 (1855): 897–99 (his report is summarized by Piorry, Londe, and Adelon; Adelon was chair of forensic medicine at the time). See also Ambroise Tardieu, "Mémoire sur la mort par suffocation," *Annales d'hygiène publique et de médecine légale*, 2nd ser. 4 (1855): 371–441 (see 378–79 for the description of subpleural ecchymoses); idem, *Étude médico-légale sur l'infanticide*, 101–33 (See 101–19 on subpleural ecchymoses).

15. "Death by suffocation includes . . . all cases in which a mechanical obstacle, other than strangulation, hanging or drowning, is brought violently to the entrance of air into the respiratory organs." Tardieu, "Mémoire sur la mort par suffocation," 377.

16. Tardieu resumed and complemented this 1855 study in *Étude médico-légale sur la pendaison, la strangulation et la suffocation* (Paris: J.-B. Baillière, 1870).

17. See Bertherat, "La Morgue de Paris," 538–41. Tardieu is still known today for other findings, particularly battered child syndrome (Tardieu's syndrome) and his theories on homosexuality (Watson, *Forensic Medicine in Western Society*, 118, 127).

18. Tardieu, *Étude médico-légale sur l'infanticide*, v–vi.

19. J. L. Casper, *Traité pratique de médecine légale rédigé d'après des observations personnelles* (Paris: Germer Baillière, 1862). It is the translation of the third German edition, *Practisches Handbuch der Gerichtliche Medicin nach eigenen Erfahrungen bearbeitet* (Berlin: August Hirschwald, 1860), in which Casper cites several other forensic doctors from different countries.

20. See, for instance, Ambrosio Tardieu, *Estudio médico-forense de los atentados contra la honestidad* (traducida de la tercera edicion por D. Nemesio Lopez Bustamante y D. Juan de Querejazu y Hartzenbusch) (Madrid: Imprendamédica de Manuel Alavarez, 1863); idem, *Die Vergiftungen in Gerichtsärtzlicher und klinischer Beziehung (Der gerichtlich-chemische Theil bearbeitet von Z. Roussin)* (Erlangen: Ferdinand Enke, 1868); Alexander Wynter Blyth, *A Dictionary of Hygiène and Public Health, Comprising Sanitary Chemistry, Engineering, and Legislation, the Dietetic Values of Foods, and the Detection of Adulterations: On the Plan of the "Dictionnaire d'hygiène publique" of Professor Ambroise Tardieu* (London: C. Griffin, 1876).

21. Also called "Der Verein der deutschen Aertze in Paris" or "Societas medicorum Germanicorum Parisiensis," it was founded in 1844. Jean-Marie Mouthon, "Les médecins de langue allemande à Paris au XIX^e siècle, 1803–1871" (PhD diss., École pratique des hautes études, 2010). Jules Béclard and Jean-Martin Charcot were, like Tardieu, professors at the Paris School of Medicine. It is not known, however, whether other German organizations took him seriously with honorary memberships.

22. See, e.g., the Praslin (1847), La Pommerais (1864), and Troppmann (1870) cases. These were not cases of infanticide, which generally do not draw the attention of the press. The great criminal trials had been reported upon in dedicated sections of newspapers since the beginning of the century. The 1860s were marked by the emergence of wide-circulation newspapers (*Le Petit Journal* was founded in 1863) that devoted increasing column space to miscellaneous events (*les faits divers*) and their actors. See Dominique Kalifa, *La Culture de masse en France*, vol. 1, *1860–1930* (Paris: La Découverte, 2010 [2001]); Chauvaud, *Experts du crime*, 89; Bertherat, "La Morgue de Paris," 569–70 and 634–35; Frédéric Chauvaud, "Le théâtre de la preuve: Les médecins légistes dans les prétoires (1880–1940)," in Michel Porret, ed., "La médecine légale entre doctrines et pratiques," *Revue d'histoire des sciences humaines* 22 (2010): 79–97.

23. Bérard Des Glajeux, *Souvenirs d'un président d'assises: Les passions criminelles. Leurs causes et leurs remèdes* (Paris: E. Plon, Nourrit et Cie, 1893), 2:149–50.

24. Scientific autopsies were also conducted at the morgue, but they were mostly performed in hospitals.

25. See Tardieu, *Étude médico-légale sur l'infanticide*, 10–11 (for changes in the number of infanticides reported at the morgue between 1837 and 1866), and 99 (for the causes of infanticide in general, according to his own reports; statistics established for the period 1844–68). The causes of death are also indicated in the morgue's annual registries stored at the Archives of the Police Headquarters (corpses of newborn infants found in Paris and its surroundings; see Bertherat, "La Morgue de Paris"). The figures are slightly different from those established by Tardieu and are probably less reliable owing to errors in expert reports. But they confirm the increase in the number of autopsies and in the number of infanticides. Thus in 1866 (the year I have studied), almost all corpses of newborns were autopsied; among them, 62% were found to have been victims of infanticide.

26. Tardieu, *Étude médico-légale sur l'infanticide*, 101–2.

27. This inquisitorial process had its origins in a judicial system based on Roman law.

28. See, in particular, Michel Porret, "Crimes et châtiments: L'œil du médecin légiste," *Dix-huitième siècle* 30 (1998): 37–50; Michèle-Laure Rassat, *Institutions judiciaires* (Paris: Presses universitaires de France, 1996), 9–13; Jean-Marie Carbasse, *Histoire du droit pénal et de la justice criminelle* (Paris: PUF, 2000), 395–425. Autopsies were regulated by the rules and laws enacted during the Napoleonic period: the Civil Code of 1804 (Article 81), the Code of Criminal Procedure of 1808 (Articles 43 and 44), the Criminal Code of 1810 (Article 475), and the decree of June 18, 1811, which sets the experts' fees and sessions.

29. Porret, "Crimes et châtiments," 48.

30. Olivier Faure, *Histoire sociale de la médecine (XVIII^e–XX^e siècles)* (Paris: Anthropos-Économica, 1994). For more information on medical teaching at the Medical School of Paris, see A. Corlieu, *Centenaire de la Faculté de médecine de Paris (1794–1894)* (Paris: Imprimerie nationale, 1896), 82–90 and 365–69; A. Prévost, *La Faculté de médecine de Paris: Ses chaires, ses annexes et son personnel enseignant de 1794 à 1900* (Paris: A. Maloine, 1900), 29–31 and 37–39.

31. These new medical schools had previously been secondary and then preparatory medical schools.

32. The Memorial to the Martyrs of Deportation is now found at this location; the morgue, now called the Forensic Institute, was relocated to Quai de la Rapée in 1923. See Allan Mitchell, "The Paris Morgue as a Social Institution in the Nineteenth Century," *Francia* 4 (1976): 581–96 and 992–93; Vanessa R. Schwartz, *Spectacular Realities: Early Mass Culture in Fin-de-Siècle Paris* (Berkeley: University of California Press, 1999), 45–88.

33. In 1866, for example, 734 corpses or parts of corpses were delivered to the morgue. Of these, 287 corpses or parts of corpses could not be identified, and 194 identified bodies were not claimed. In the course of the century, the proportion of unidentified and unclaimed corpses declined, but the number of corpses delivered to the morgue increased significantly. In addition, scientific studies (besides autopsies) were conducted on identified and claimed corpses.

34. From this point of view, the morgue could be considered as a "body farm" before its time.

35. This specificity had already been highlighted a long time before by Erwin H. Ackerknecht, "Legal Medicine Becomes a Modern Science [19th century]," *Ciba Symposia* 11 (1950–51): 1299–1305. For a comparison with England, see Ian A. Burney, *Bodies of Evidence: Medicine and the Politics of the English Inquest, 1830–1926* (Baltimore: Johns Hopkins University Press, 2000).

36. The Paris Morgue is an eponymous institution. Its name was eventually given to similar institutions established in France.

37. F. Strassmann, "Das Institut für Gerichtliche Medizin an der Universität Berlin," in *Methods and Problems of Medical Education*, 9th ed., *Institutes of Legal Medicine* (New York: Rockefeller Foundation, 1928), 56–62; M. Minovici, *Discurs tinut cu ocasia deschiderei Morgei*

*in diua de 20 decembre 1892* (Bucharest: Imprimeria statului, 1892); Louis Dausset and Georges Lemarchand, *Rapport sur la reconstruction de la Morgue et la création d'un Institut médico-légal* (Paris: Imprimerie municipale, 1908), 61–64. The founder of the Bucharest morgue, Mina Minovici (1858–1933), is considered the restorer of Romanian forensics. He completed his studies in Paris, where he presented his PhD dissertation on forensic medicine. See Dan L. Dumitrascu, Marc A. Shampo, and Robert A. Kyle, "The Institute of Forensic Medicine established by Mina Minovici," *Mayo Clinic Proceedings* 8 (1995): 776, http://www.minovicifoundation.com/mina-minovici/.

38. Tardieu, *Étude médico-légale sur l'infanticide*, vi.

39. See Michel Foucault, *Naissance de la clinique: Une archéologie du regard médical* (Paris: PUF, 1963); Léonard, *Médecine entre les savoirs*, 133–38; Guenter B. Risse, "La synthèse entre l'anatomie et la clinique," in *Histoire de la pensée médicale en Occident*, vol. 2, *De la Renaissance aux Lumières*, ed. Bernardino Fantini and Mirko D. Grmek (Paris: Le Seuil, 1997), 177–97; Faure, *Histoire sociale*, 43–50, 67–71; Rafael Mandressi, *Le Regard de l'anatomiste: Dissections et invention du corps en Occident* (Paris: Le Seuil, 2003).

40. Anne Carol, *Les Médecins et la mort, XIXᵉ-XXᵉ siècle* (Paris: Aubier, 2004), esp. 235–69.

41. See Erwin H. Ackerknecht, *La Médecine hospitalière à Paris, 1794–1848* (Paris: Payot, 1986 [1967]); Faure, *Histoire sociale*, 80–83. The importance of hospitals was such that a city like Lyon, in which there was no medical school until the end of the century, outranked Montpellier and Strasbourg in terms of training of the medical elites.

42. During that period, it successively bore the names of "School of Healthcare" and then "School of Medicine." The title of faculty for the French denomination ("Faculté de médecine") was reinstated after the revolutionary period.

43. Xavier Bichat, *Recherches physiologiques sur la vie et la mort avec les notes de M. Magendie* (Paris: J. Vrin, 1981 [reprint of the 1855 edition]), 1.

44. Marc Sée, "Autopsie (anatomie)," in *Dictionnaire encyclopédique des sciences médicales*, 1st ser., 7:411. Italics original.

45. This is evidenced by the title of Casper's book, cited in note 19 above, in the original and in French versions.

46. Tardieu, *Étude médico-légale sur l'infanticide*, 103.

47. Michel Foucault, *Naissance de la clinique* and *Surveiller et punir: Naissance de la prison* (Paris: Gallimard, 1975).

48. "Fodéré's Forensic Medicine was based on an approach whereby a solution was sought for every difficulty. This desire to solve everything was rightly opposed by M. Orfila; but this scientist may himself have fallen in the other extreme, by seeking to assess the real value of every isolated fact, without then giving all these facts an overall value. M. Orfila endeavored to guard the doctors against the errors which they were at risk of committing, by constantly showing them the table of possible mistakes; this resulted too often in a fatal and destructive uncertainty of science itself." Alphonse Devergie, *Médecine légale, théorique et pratique* (Paris: Germer-Baillière, 1836), 1:iii–iv.

49. Tardieu, *Étude médico-légale sur l'infanticide*, vii.

50. Tardieu, *Étude médico-légale sur l'infanticide*, vii.

51. For a comparison, see Ian Burney, *Poison, Detection, and the Victorian Imagination* (Manchester: Manchester University Press, 2006).

52. Georges Morache, "Suffocation," in *Dictionnaire encyclopédique des sciences médicales*, 3rd ser., 13:221–41. Georges Morache was a professor of forensic medicine at the Medical School of Bordeaux.

53. The journal's table of contents includes an entry for subpleural ecchymoses (comprising the detail of referenced articles), which testifies to the importance of the debate. See *Bulletin de la Société de médecine légale de Paris*, 2nd ser., 4 (1907): 58–60.

54. P. Brouardel, *Cours de médecine légale de la Faculté de médecine de Paris*, vol. 3, *La Pendaison, la Strangulation, la Suffocation, la Submersion* (Paris: J.-B. Baillière, 1897), 17–27, and *Cours de médecine légale de la Faculté de médecine de Paris*, vol. 5, *L'Infanticide* (Paris: J.-B. Baillière, 1897), 77–82.

55. Morache, "Suffocation," 224.

56. Dr. Liman, "Ueber die forensische Bedeutung der sogenannten punktförmigen Ecchymosen unter der Pleura und dem serosen Ueberzuge anderer Organe," *Vierteljahrsschrift für gerichtliche und öffentliche Medicin*, 73–102; M. Liman, "Quelques remarques sur la mort par suffocation, par pendaison et par strangulation," *Annales d'hygiène publique et de médecine légale*, 2nd ser., 28 (1867): 388–402.

57. Liman, "Quelques remarques sur la mort par suffocation," 389–90.

58. Casper's name is cited thirty times. The other authors are cited a few times (the following list is not exhaustive): A. Taylor (2), Robert Froriep of Berlin (24), Elsaesser (27, 32), senator of Berlin (64–65), Liman (108, 114). The German speaking authors are by far the most cited. Tardieu mentions authors of other nationalities—Italian or Dutch, for example (127)—but does not name them.

59. He cites articles published in German journals, however, as well Taylor's book, which was written in English.

60. Tardieu had written a first, conciliatory reply to Liman at the end of the French version of the latter's article (1867), in which he referred to the forthcoming publication of his book (Liman, "Quelques remarques sur la mort par suffocation," 402). See Tardieu, *Étude médico-légale sur l'infanticide*, 101–19. Their names are cited both in the text and in the footnotes (the sources mentioned by Tardieu are the French versions of Casper's treatise and of Liman's article). He also once cites Hecker and Hoogeweg's text, quoted by Casper.

61. In addition to one collaborator mentioned in an autopsy report, Tardieu cites two secondary French forensic doctors: Dr. Toulmouche, from Rennes (but Tardieu did not think Toulmouche supported his theory strongly enough), and Dr. Dégranges, from Bordeaux (who fully adhered to his theory).

62. Tardieu might have played an active role in this field. Thus the French translator of Casper's treatise thanked him for his support in his preface: Gustave-Germer Baillière, "Préface du traducteur," in Casper, *Traité pratique de médecine légale*, xiii. A qualified physician, Gustave-Germer Baillière was a member of a famous family of publishers of medical books and journals founded by Jean-Baptiste Baillière, who also translated French publications into foreign languages. See Danielle Gourevitch and Jean-François Vincent, eds., *J.-B. Baillière et fils, éditeurs de médecine* (Paris: Bibliothèque interuniversitaire de médecine et d'odontologie, De Boccard, 2006). This publisher is cited several times throughout this study.

63. Tardieu, "Mémoire sur la mort par suffocation," 373–77.

64. Ambroise Tardieu, *Étude médico-légale sur l'infanticide*, 2nd ed. (Paris: J.-B. Baillière, 1880), 101–33. The author of the preface (none other than the publisher) supported the master's theory (vii).

65. Tardieu, *Étude médico-légale sur l'infanticide*, esp. 106–9 (quote on p. 107).

66. The society's criticisms against Tardieu's theory on subpleural ecchymoses are undoubtedly not the only possible cause. According to Professor Thoinot, the relations between Tardieu and Devergie, who was a member and president of this society, were "not the most cordial." "Paul Brouardel, 1837–1906: Éloge prononcé à l'Académie de médecine dans sa Séance annuelle du 13 décembre 1910," *Mémoires de l'Académie de médecine* 42 (1911): 6. These tensions, whose origins are unknown (it is nevertheless known that Devergie and Tardieu opposed each other regarding a criminal case in 1855; Chauvaud and Dumoulin, *Experts and Expertise*, 94–95), were added to older ones.

67. Brouardel, *Cours de médecine légale de la Faculté de médecine de Paris*, 5:77–78.

68. His book on infanticide was even translated into foreign languages, for example, *Estudio medico-legal sobre el infanticidio* (Barcelona: Daniel Cortezo, 1883). I have not conducted any exhaustive research concerning the translations of this book.

69. Morache, "Suffocation," 234.

70. Brouardel, *Cours de médecine légale de la Faculté de médecine de Paris*, 5:78.

71. "Discours de M. le Professeur Thoinot," in *Inauguration du monument élevé à la mém moire de P. Brouardel, 20 juillet 1909* (Corbeil: Imprimerie Crété, 1909), 22–27.

72. Bérard Des Glajeux, *Souvenirs d'un président d'assises*, 2:150–51.

73. Yet a detective novel published in 1877 depicts a forensic doctor who bears a strong resemblance to Tardieu and points out his shortcomings: "Mr. Ravinel was then a man in the prime of his age; He occupied a high position in the medical profession, his reputation as a devotee to science was justly earned. The only reproach that could be levelled against him was his excessive confidence in his own knowledge, an unlimited confidence in his deductions, a constant need to profess and put himself forward, and also what sometimes diverted him from his purpose: An exaggerated imagination." René de Pont-Jest, *Le N° 13 de la rue Marlot* (Paris: E. Dentu, 1877), 268. As for Brouardel, he writes: "In the cases in which Tardieu concluded that suffocation was the cause of death, because he had found sub-pleural ecchymoses in the newborn, I would not say that the mother was condemned unjustly. But at the Palais, a sign whose value is not recognized by all authors as incontestable, a sign which has not been recognized by all forensic pathologists, must not be treated as an absolute sign." *Cours de médecine légale de la Faculté de médecine de Paris*, 5:78.

74. Frédéric Chauvaud, "Un 'sujet de deuil': La fabrique des erreurs judiciaires," in *L'Erreur judiciaire de Jeanne d'Arc à Roland Agret*, ed. Benoît Garnot (Paris: Imago, 2004), 153–70; Bruno Bertherat, "L'élection à la chaire de médecine légale," and "Les mots du médecin légiste, de la salle d'autopsie aux Assises: L'affaire Billoir (1876–1877)," *Revue d'histoire des sciences humaines* 22 (2010): 117–44.

75. A reference source that stands as a veritable indictment against Thoinot is E. Doyen and Fernand Hauser, *L'Affaire Jeanne Weber: L'ogresse et les experts* (Paris: Librairie universelle, 1908). See Darmon, *Médecins et assassins*, 254–71, 273–74; Chauvaud, *Experts du crime*, 65–68.

76. Adolphe Guillot, "Les erreurs judiciaires et leurs causes," *Séances et travaux de l'Académie des sciences morales et politiques* 148 (1897): 112–20.

77. Marie-Jeanne Lavilatte-Couteau, "Le privilège de la puissance: L'anesthésie au service de la chirurgie française (1846–1896)" (PhD diss., Université de Paris I, 1999); Marie-Christine Pouchelle, *L'Hôpital ou le théâtre des opérations: Essais d'anthropologie hospitalière—2* (Paris: Seli Arslan, 2008).

78. Jacques Léonard, *Les Médecins de l'Ouest au XIXᵉ siècle*, Atelier de reproduction des thèses Lille III (Paris: Diffusion Honoré Champion, 1978), 3:1355.

79. P. Brouardel, *Organisation du service des autopsies de la Morgue: Rapports adressés à Monsieur le Garde des Sceaux* (Paris: Imprimerie de E. Martinière, 1879), 7.

80. Prévost, *La Faculté de médecine de Paris*, 39; A. Devergie, "Enseignement de la médecine légale," *Gazette hebdomadaire de médecine et de chirurgie*, January 7, 1876, 14.

81. Léonard, *Médecine entre les savoirs et les pouvoirs*, 138–40.

82. Bertherat, "La Morgue", 658–59. My sources are mainly French, e.g., Paul Loye, "L'enseignement de la médecine légale en Allemagne et en Autriche-Hongrie," *Annales d'hygiène publique et de médecine légale*, 3rd ser., 21 (1889): 55–64.

83. Gabriel Tourdes and Edmond Metzquer, *Traité de médecine légale, théorique et pratique* (Paris: Asselin et Houzeau, 1896), 11–12; Ministère de l'Instruction Publique et des Beaux-Arts, *Enquêtes et documents relatifs à l'enseignement supérieur*, vol. LXXV, *Facultés de médecine. Enseignement de la médecine légale* (Paris: Imprimerie nationale, 1900), 37–39.

84. Brouardel indicates that he went not only to Berlin, Bonn, Dresden, Heidelberg, Leipzig, Munich, and Strasbourg—recently reintegrated into Germany—but also to Prague and Vienna (Austro-Hungarian Empire), Geneva and Zurich (Switzerland), and even Lyon (where, in the same year, Alexandre Lacassagne was appointed professor of forensic medicine in the new medical school) and Nancy. He was accompanied by two former German-speaking students. Missions of this kind had taken place in the 1860s in German universities but had not been devoted to forensic medicine.

85. He cites Professor Brunetti, from Rome, Doctor Hardwick, from London, and Doctor Gosse, from Geneva.

86. Brouardel, *Organisation du service des autopsies de la Morgue*.

87. About the medical context of Belle Époque, see, e.g., Léonard, *Médecine entre les savoirs et les pouvoirs*, 241–327, and Faure, *Histoire sociale*, 177–98.

88. Bernard Spilsbury (1877–1947), the leading expert in London during the interwar period, provided a rare counterexample in this regard, that of the choice of solitary practice. See Keith Simpson, *Forty Years of Murder: An Autobiography* (London: Harrap, 1978); Ian Burney and Neil Pemberton, "Bruised Witness: Bernard Spilsbury and the Performance of Early Twentieth-Century Forensic Pathology," *Medical History* 55 (2011): 41–60; and idem, *Murder and the Making of English CSI* (Baltimore: Johns Hopkins University Press, 2016), chaps. 3 and 4. He shared characteristics with Tardieu.

89. Also worth mention is the emergence of the Lyon center chaired by Alexandre Lacassagne (1843–1924).

90. Christophe Charle, *Paris Fin de siècle: Culture et politique* (Paris: Le Seuil, 1998), 21–48; Christian Bonah, *Instruire, guérir, servir: Formation, recherche et pratique médicales en France et en Allemagne pendant la deuxième moitié du XIX^e siècle* (Strasbourg: Presses Universitaires de Strasbourg, 2000), 119–24. Thus Sigmund Freud attended Brouardel's practical classes in the mid-1880s and sang his praises; see his "Préface à l'édition allemande," in John Gregory Bourke, *Les Rites scatologiques* (Paris: PUF, 1981), 31. The closing of the German Medical Society of Paris in 1871 did not prevent the pursuit of exchanges, nor did it stop German speaking physicians from coming to the French capital after the war.

91. Pierre Bourdieu, "L'illusion biographique," *Actes de la recherche en sciences sociales*, 62–63 (June 1986): 69–72; Dosse, *Pari biographique*, 227–34.

92. There are two recent exceptions. The first is Lacassagne; see Philippe Artières, Gérard Corneloup, and Philippe Rassaert, *Le Médecin et le criminel: Alexandre Lacassagne (1843–1924)* (Lyon: Bibliothèque Municipale de Lyon, 2004), and Muriel Salle, "L'avers d'une Belle Époque: Genre et altérité dans les pratiques et les discours d'Alexandre Lacassagne (1843–1924), médecin lyonnais" (PhD diss., Université Lumière Lyon 2, 2009). The second is Orfila, of whom José Ramón Bertomeu-Sánchez is preparing a biography.

93. Bertillon, Galton, or Reiss, for example.

94. See Burney and Pemberton, *Murder and the Making of English CSI*.

95. For example, translators or editors.

96. Thus a comprehensive study could be undertaken of all the translations of Tardieu's work so as to evaluate how widely they were disseminated. Another relevant question is that of French pathologists' levels of fluency in foreign languages and more generally the role of language in scientific exchanges (Latin occupies a special place in this regard).

# CHAPTER TEN

# The Strange Science

## Tracking and Detection in the Late Nineteenth-Century Punjab

GAGAN PREET SINGH

The relationship between colonialism and folklore can be explored through a study of the forensics of "tracking." In this chapter, I engage with Carlo Ginzburg's famous 1978 article, in which he showed how, from the sixteenth century onward, knowledge forms were differentiated into oppositional categories of "high" and "low," "science" and "lore," and "formal" and "informal." Ginzburg's analysis outlined a generic model for the history of knowledge, in which he considered the roots of connoisseurship, detective reasoning, psychoanalysis, philology, medicine, and history—all of which derived from what he categorized as "medical semiotics." These disciplines, he argued, originated in the hunter's lore, by which a hunter studied traces to track an unknown quarry. These disciplines dealt with particulars, as there could be no universal model that would be applicable to all situations. Further, this knowledge could not exist separate from the physical presence of its author, as only his embodied experience could serve to guarantee its authoritative status.

In contrast, Ginzburg argued that from the sixteenth century onward, an alternative, "Galilean" paradigm of knowledge grew to prominence, whose associated disciplines—later called "sciences"—came to be categorized as "high" forms of knowledge. This paradigm was rooted in universal laws, dealt with abstract forms, and over time relegated the "human" sciences to the rank of "low" knowledge. But Ginzburg also suggested that toward the end of the nineteenth century, at least some of the "low" forms of knowledge were becoming "scientific" in orientation. In this process, he argued, from the eighteenth century, the low forms of knowledge were colonized and presented as "scientific." To corroborate his formulation, he provided examples of anthropometry and fingerprinting. In these scientific practices, he seems to suggest, one may locate the hunter's method.

These binaries serve as a framework for situating the forensics of the footprint, popularly known as "tracking." I argue that in the everyday techniques of the hunter and in the methods of the detective, in the pastoralist's world and in the laboratory analysis of plaster of paris molds, and indeed in the colonies and at "home," the study of the footprint straddles the categories of forensic "lore" and "science." This presents us with the following question: Where should we locate the "origins" of the modern forensics of the footprint. Was it statist or subalternist, hegemonic or non-hegemonic, oral or codified, "local" or "global," quotidian or specialist? This chapter suggests that the history of the forensics of the footprint is more complicated than Ginzburg's generic model proposed. It does so by providing an outline of the complex process by which the forensics of the footprint was normalized and routinized within scientific policing in colonial Punjab. Like the well-known case study of fingerprinting, this excavation of the forensics of the footprint in the context of colonial India suggests that the appropriation of knowledge into modern forensics offers a richer and more significant story than commonly thought.

In the history of science and technology in colonial India, it was the imperial science that dominated. In a seminal work in this historiography, Daniel R. Headrick argued that the spread of Western technology and imperialism were interlinked, and that the technology brought "progress" to colonies.[1] More recent studies provide us with a more nuanced understanding.[2] Against the benevolent impact of "quinine" in curing malaria in the subcontinent, for example, historians have shown how colonial projects of modernization, such as the introduction of canals, caused environmental and health hazards that resulted in the rise of malarial tracts in areas that were previously free from this disease.[3] Recent studies have also foregrounded subalternist forms of knowledge that had not figured in the earlier discussions on the impact of Western technology. Historians of south Asian medicine Projit Mukherji and David Hardiman, for example, have emphasized the nontextual, non-statist, and non-hegemonic forms of medicinal practices that were dismissed as "quackery" within the elite traditions but whose survival testifies to their resilience and importance. Moreover, historians have identified subaltern forms of knowledge that either ran parallel to colonial technology or were colonized by the state.[4] Smritikumar Sarkar argues that technology had a complicated impact on nineteenth-century colonial Indian society, as innovations such as railways, steamers, and rice mills brought both prosperity and misery for the colonial subjects. Moreover, there was no linear victory march:

in most cases, it took decades for Western technology to replace existing forms of technology.[5] Clive Dewey's work on introduction of steamers to the Indus River shows that they failed to replace the local *Dhow* boats, as local boatmen possessed knowledge that mechanical steamboat operators could not grasp.[6] They understood the riverbed, its flow, and its course, and their mastery of the *Dhow* enabled them to use this knowledge to navigate safely, whereas steamboats operated on the basis of rigid principles that could not adapt to local conditions.

This chapter contributes to this scholarship by focusing on a nontextual form of forensic knowledge that coexisted alongside statist forms of knowledge, namely, the practice of tracking, which thrived in nineteenth-century colonial India. While the recent studies of technology in colonial India have emphasized the resilience of local forms of technology, this study suggests how a local form of knowledge took over imperial knowledge and became "normalized" within the imperial governance, laws, and science. Perhaps surprisingly, however, this effort at normalization in significant respects failed, as the trackers' mastery of their form of valuable knowledge enabled them to successfully resist it.

## Tracking and Society

*Khoj*, or tracking, has been documented in several parts of colonial India, especially in the dry, arid northwestern provinces, which may be roughly identified as the basin of the Indus River. Khoj literally means "search," and it has a similar meaning to the English words "tracing" and "tracking." In Punjab, a practitioner of tracking was called *Khoji*, and in Gujarat and Sind, he was called *Puggi*.[7] Khoj could mean the act of tracking, but it could also refer to the physical footmarks that formed the tracker's evidentiary material. In the nineteenth century, another term used for a tracker was *sooragi*, where *soorag* meant "clue."[8] When detective fiction was introduced in colonial north India, some translators and the authors of detective fiction found the word *sooragi* to be the nearest equivalent to the English word "detective," and detective science came to be called *soorag-rehsan*.[9] There may be still other names in Indian languages for tracking, many of which might have remained unrecorded. But the use of *sooragi* for both "tracker" and "detective" suggests that in his vernacularization, the figure of the detective was not imagined much differently from the tracker.[10]

Colonial sources suggest that tracking was prevalent at least in the four modern-day states of the Union of India, which include Gujarat, Haryana,

Punjab, Rajasthan, and most likely some parts of central India. In the present-day map of Pakistan, it existed in the states of Punjab and Sind. There may be other regions in which tracking existed and thrived, but again, owing to absence of any documented references, this is difficult to determine. This chapter focuses on tracking as a practice in the region of Punjab, though I have also referred to other regions where it provides further useful detail. The region of Punjab had a loose geographical identity until the British conquest of the region in 1849, when it became a province of British India; therefore Punjab as referenced in this chapter implies the undivided British administrative province of Punjab, which existed till the 1947 partition of India and Pakistan.

I begin with the first provincial survey of tracking in colonial Punjab. In 1861, George Hutchinson, the provincial chief of police, asked his district police superintendents for a report on the extent to which tracking was known in their respective districts. Several responded, and though the detailed survey report is no longer available, we have the gist of it, as published in the annual police administration report of 1862. A brief discussion of this report, the nature of this survey, and the underlying motive behind it may help us to understand the evolving relationship between tracking and the imperial policing in mid-nineteenth-century Punjab.

In the northwestern districts, most of the district superintendents acknowledged the practice of tracking. The superintendent of the district of Jhelum reported that tracking was "in full force as a system." The districts of Gujrat, Gujranwala, Jhung, Montgomery, Lahore, Multan, Muzaffargarh, and Shahpur—which were inhabited by nomadic pastoralists—reported that the system was well established there. The superintendent of Gujrat reported that "the very foundation of the system" was that the people resorted to it themselves, understood it, and acknowledged its justice. The superintendent of Lahore also observed that the system "is one the people thoroughly understand, and are attached to."[11] Hutchinson's report also provides information on tracking in the southeastern districts of Hissar and Sirsa. The superintendent of Sirsa wrote that the people of the district "place peculiar faith in the tracking system, and every village of consequence has two or more professional trackers." In Hissar, the system was also in full force.[12] Hutchinson's interest in documenting the system of tracking was shaped by his drive to utilize tracking in the policing system. Knowledge production was thus linked with imperial interests and flourished within the new imperial policing. Police officers like Hutchinson introduced it even in the areas in which tracking was previously considered unknown. Tracking in colonial Punjab is a testament to the fact that imperial knowledge does not always lead to the disappearance of folk knowledge; it rather reintroduced it in a new form.

## Practice and Rules: Tracker's Evidence

Chapter 7 in this volume, "From Bedouin Trackers to Doberman Pinschers: The Rise of Dog Tracking as Forensic Evidence in Palestine," by Biniyamin Blum, highlights how the colonial authorities in 1930s Palestine turned from reliance on Bedouin trackers to Doberman Pinschers as their favored means of hunting criminals. Blum argues that the extraordinary political turmoil in Palestine at the time in large part explains this turn to animal evidence. For one thing, while evidence from Bedouin trackers could be scrutinized, and thus disputed, the dogs' evidence was beyond scrutiny, thereby leaving it unfettered as an instrument of "legitimate" colonial suppression. Blum also suggests that the introduction of animal trackers "bore symbolic, moral, psychological and religious significance that went far beyond the method's probative sway."[13] Blum's study shows that imperial forensic practices were to a significant degree shaped by political anxieties.

Did political anxieties shape the relationship between the British and Khojis of Punjab? Unlike the monolith British distrust of Bedouin trackers in 1930s Palestine, the British response to Khojis in Punjab may be characterized as polyvalent. Like the Bedouin trackers, Khojis on the one hand were subject to distrust throughout colonial Punjab. Judicial officers believed that the evidence of Khojis was unreliable, and the police could falsely implicate innocent people by invoking it. In contrast with Palestine, however, tracking also remained a part of imperial policing in Punjab until the final days of the British rule. In this section, I explore how tracking was jettisoned as well as accommodated within Punjab's imperial policing constellation.

In colonial India, these anxieties over the reliability of "native" judicial evidence was shaped by the British colonial construction of the Indian subjects as mendacious and unreliable. Historian Vinay Lal has summed up the core assumption of this discourse: "In England, the witnesses understood it as their sworn duty to help in the illumination of truth; in India, by way of contrast, each native took it upon himself to obfuscate the truth and—more modestly—embellish the story."[14] Similarly, David Arnold has recently argued that "there was a widespread belief among colonial officials that Indians' verbal and written testimony could not be trusted." From the late nineteenth century, colonial officials in India resorted to technical devices like fingerprinting "to bypass false witnesses or overturn unreliable testimony." Arnold argues that the increasing emphasis on new forensic technologies like postmortems was expected to resolve this problem.[15] This orientalist construction of the native "other" seems to have further reinforced suspicion against Khojis, whose verbal evidence could be dismissed as inadequate and misleading.

The state of Western forensic knowledge further reinforced bias against trackers' evidence. By the 1860s, there were few Western theoretical studies to substantiate the forensics of the footprint. The existence of such scientific studies may have made judicial officers more positively disposed toward tracking. But it was only toward the late nineteenth century that the forensics of the footprint gained importance in forensic literature and the use of "the footprint" was normalized. Moreover, the other forensic aids with which the evidence of a tracker could be corroborated, like the use of plaster of paris to take molds of footprints, were not yet in common use. It was only toward the end of nineteenth century that the use of plaster of paris as a means of turning ephemeral, scene-bound signs into preservable and transportable evidence became well established.[16] The case of tracking thus suggests a situation in which practice ran ahead of principle: tracking had long been used in solving crimes in colonial India, and yet there was no scientific treatise that explained its method. Consequently, judicial officers did not know how to cross-examine a tracker's evidence, and police had not established procedures to record their evidence. Unable to deal with Khojis' evidence, judicial officers were ill equipped, and disinclined, to accept it.

Even when we recognize this general aversion of the colonial judiciary to accept tracker evidence, the reality may well have differed from the written record. For example, the acceptance of a Khoji's evidence depended on the orientation of a judge. While a sympathetic judge readily accepted the evidence of Khojis, an unsympathetic judge could easily dismiss the evidence of Khojis as baseless. A good example of a judicial officer's sympathetic attitude toward Khojis comes from the Karnal Cattle Lifting Case of 1913–14.[17] In this case, specially appointed judge F. L. Brayne relied heavily on the testimonies of Khojis. Though a detailed study of this case is the subject of another paper, it is pertinent to mention here that during the case proceedings a Khoji, Ram Nath, contradicted himself before Brayne. While commenting on these inconsistencies, Brayne wrote, "his memory is charged with so many stories of cattle he has tracked and panchayat he has attended that we cannot expect it to be always perfect."[18] During the proceedings, Brayne tried to defend the evidence of Ram Nath and other trackers. The judgment of Brayne was later upheld by the Chief Court of Punjab at Lahore. The fact that this case, which was highly dependent on the evidence of tracking, comes from a much later period shows that throughout the colonial rule in Punjab, the attitude of the judicial officers toward evidence of Khojis remained polyvalent.

The judicial and police officers remained critical of the evidence of tracking, and yet in Punjab more sympathetic officers were able to combine tracking with modern forensic practices. Because of their efforts, by the late nineteenth century, as the following pages reveal, tracking had been normalized and routinized within the forensic culture of Punjab.

## Normalizing Tracking: The Case of Major General George Hutchinson

While the judicial officers, who showed concern for the legality of the procedure with which the results were achieved, were usually reluctant to accept Khojis' evidence, police officers, who valued end results over the legality of the procedure, emphasized the usefulness of trackers in the detection of crime. Yet the police officers made several efforts to normalize the deployability of Khojis in the detection of crime. The most important efforts of a police officer in this regard is that of Major General George Hutchinson, the inspector general of the Police of Punjab, who made several attempts to normalize the deployability of trackers in the Punjab force.[19] This section explores how Hutchinson attempted to absorb tracking within the everyday policing practices of the province.

Hutchinson's most important contribution to provincial policing was his preparation of a training manual called *Punjab Police Catechism* in 1861, translated into Hindustani as *Hidyatnama-i-polis*.[20] In preparing his catechism, Hutchinson drew upon the contemporary British manual *Police Catechism*, by William C. Harris, and as a result, there is a remarkable similarity in structure between the two texts. Hutchinson's *Catechism* achieved considerable popularity among police constables, and a later generation of policemen attested that every literate constable possessed a copy.[21] Hutchinson devoted a short chapter to instructions for the investigation of an occurrence of "unnatural death," in which he outlined the expectations of a constable when he reached the crime scene: "Allow no one to come near the body or to touch it, or any thing near it; do his utmost to keep things exactly as they were when he arrived, cover up all footmarks, but be careful not to touch the body or any thing near or attached to it."[22] Here Hutchinson drew attention to the importance of footmarks, exhorting constables to "Allow no crowding round the spot as thereby all foot-steps are obliterated," and to "Cover up carefully all marks of foot-steps until the trackers arrive."[23]

Hutchinson's recognition of the importance of footprints in forensic investigation indicates the importance of tracking in his scheme of investigation. He expected constables to be keen observers of footprints, and to preserve them as potentially valuable forensic evidence in court: "Record carefully on paper any peculiarity of the footmarks discovered; their correspondence with the shoes supposed to have made them, recollecting that a mere similarity in the sole of a shoe with a footmark is of little value, and that some striking peculiarity, noticeable in both, should be sought for."[24] Hutchinson's instructions suggest that the use of footprints had been already normalized within his force's model of investigation and indicates that, in a case of murder, Khojis were as likely to visit a crime scene as a medical surgeon. The use of footprints in the Punjab had thus become an important routine well before it was recognized in Britain. This becomes clear from a comparison of Hutchinson's catechism to Harris's: in his chapter on the case of unnatural death, Harris emphasized the role of surgeon and coroner's inquest but made no reference to footprints or footmarks.[25] Given that Hutchinson used Harris as a model for his manual, his discussion of footprint protocols is a clear instance of innovative addition based on local practice. It was not until several decades later that we see analogous instructions included in manuals aimed at British police officers.[26]

The British in Punjab had been using and relying on trackers since their arrival in Punjab, and in this Hutchinson repeatedly emphasized the importance of Khojis. For example, he recalled an 1847 episode near the southeastern Punjabi town of Karnal in which property was stolen from his camp. Hutchinson was able to recover the property with the help of Khojis. Despite its history of use in colonial policing, in his annual police reports during the 1860s, Hutchinson argued that the system of tracking would become extinct unless the government encouraged it. In his 1866 report, reflecting on the reports that tracking had "fallen into disuse" in the districts of Gujaranwala, Montgomery, and Multan, he wrote, "I consider it most desirable that, this very valuable system should not fall into disuse in those districts where it has for ages existed, and where its evidence is readily admitted by the people themselves, who are all more or less able to follow up a 'track.' . . . There can be no doubt of the value of the system as a detective measure, and I think every effort should be made to keep it up in full force where the people are accustomed to it."[27]

"The use of trackers is on the decline," he again wrote in 1869. He believed there was little hope of reviving it, for "as cultivation increases, more roads made, and the country becomes more populous, the difficulties in tracking increase enormously." He also felt that the supply of trained trackers was declining because, as he wrote, it required half a lifetime to become an adept at it, "and men will not take up what is now but a precarious livelihood."[28]

The system of tracking also suffered from the adverse opinion of judicial officers on the reliability of Khojis. Hutchinson used his annual police reports to condemn this attitude. He quoted several local police officers' reports that showed that the judicial officers were unsympathetic to the evidence of tracking. As one police officer from the Amritsar District reported in 1867, "The Khoj system is falling into disuse . . . partly because little weight is attached to a track by the Magistrate."[29] Another police officer from the Gujaranwala District wrote in 1866, "owing to the more strict proof now required both by the Magistrates and the Appellate Courts," it was difficult now to hear of "a really good tracking case."[30] In 1872, Lt. Col. Orchard, district superintendent of Amritsar, wrote, "On occasions when trackers have been brought to bear in aiding the police inquiries, the Magistrates have not felt disposed to place any confidence in the trackers' dispositions."[31]

Such reports must have shaped the understanding of Hutchinson, who mentioned several reasons for the judicial officers' adverse attitude. In the 1860s, as Hutchinson noted, the coming into force of the Criminal Procedure Code and Laws of Evidence had shaped this prejudicial attitude of the judicial officers.[32] The judicial officers, as per the laws, demanded that "the fact of the theft, the amount of property, and culpability of the villagers" should be "distinctly proved." Hutchinson argued that under "a more careful sifting of evidence and of the application of a nicer discrimination," it was likely that "the rough-and-ready test of tracking will carry less weight."[33]

Beneath the laws of evidence lay the judicial officers' suspicion of the native, especially the poor and illiterate. Hutchinson captured courts' reluctance "to credit the recognition of a foot-print by a poor uneducated creature," adding that there was "a feeling" among Europeans that "the man may have been bought over, may have testified to save his reputation in his craft," and therefore he was unreliable. In other words, the court reposed more trust in the testimonies of educated, rich, and respectable classes, especially if they were white. Hutchinson also lamented that a tracker's evidence was of "a nature to suffer more than other evidence from the modern habit of requiring corroboration." Moreover, because there was "no guarantee beyond the tracker's experience," and when questioned on the basis of "probabilities," "mistakes" were possible, Hutchinson concluded that the problem lay with "us," rather than the tracker: the court expected the tracker's evidence to be in agreement with "European standards." "Wherever it is indigenous, there it flourishes amongst the people," he noticed, adding, "with them it has lost none of its value, but with us undoubtedly it is not so much resorted to."[34] To counter this, Hutchinson urged the adoption of local notion of evidence: "Less acquaintance with the people, less observation of the trust reposed by them in these trackers, and stronger sympathy with our ideas of what is likely to be true than with the people's judgment as gathered from their daily conduct—all these combine to place tracking in a position more subordinate than before."[35]

Against this grim estimation, Hutchinson proposed a system by which tracking would remain relevant in detection. He argued that the time would come "when tracking will only be cultivated as a possible aid to the detective," where a tracker's knowledge would only be an "aid" in the detection of crime, and the tracker's evidence would no longer be furnished as "the best evidence in a case." Hutchinson advised his colleagues to "be prepared for the change," so that they shall learn how to utilize "to the uttermost the assistance of the tracker" as well as "to base . . . cases on evidence which . . . [was] more likely to convince."[36] By doing so, the system of tracking would not succumb to the challenges; rather, it would become a cornerstone of detection in the province.

Hutchinson backed his claims by concrete examples. In a special section of his annual police reports, Hutchinson provided examples of investigations in which Khojis had played a crucial role. Notwithstanding the general adverse attitude of the judicial officers, Hutchinson, by providing these examples, illustrated how the tracker may be used as a detective agency. Hutchinson was, in a way, suggesting that Khojis could be highly useful as mechanical tools in investigation. Rather than looking for Khojis to provide judicial evidence, he deployed them to chase criminals, to recover the stolen property, and to gather the material evidence. An analysis of some of Hutchinson's case summaries reveals this dimension of his program.

On September 30, 1867, a shepherd named Sahara discovered the corpse of a woman buried in the Thull Desert near the village of Meerwallee in the Muzaffargarh District. The case summary mentioned that the villagers along with Khojis found footprints at the scene: "[They] discovered near the body the tracks of four people who seemed to have come into the Thull; there were traces of their having rested themselves close to the spot where the shoes were first discovered. The tracks of only three people were discernible as having left the spot; these foot-steps were of three different descriptions, which made them suppose that the largest one was that of a man, and the other two those of a woman and of a child; besides this they appeared to have possessed a donkey, heavily laden." After following the footprints for twelve miles, the tracking team came across a fakir, who was traveling with his wife, son, and a donkey. On determining a resemblance in footprints, they arrested the fakir and his party. When the fakir protested, the tracking team escorted him along the trail they had followed until they reached the crime scene, and from there to a further set of tracks from the scene to the village of Lungur Saraie. There the villagers immediately recognized the fakir as Deendar and enquired about the whereabouts of Fakeernee Fatma. It emerged that Fatma, a wanderer, had planned to visit a local holy place but was scared to cross the desert alone. When Deendar offered to accompany her, she willingly agreed. As his son confessed, however, Deendar had strangled Fatma en route. The villagers also recognized goods in possession of Deendar that belonged to the slain wanderer.

This case summary does not provide any judicial proceedings from the trial of this case. Had those proceedings been available, they would have undoubtedly informed us of the importance of the Khoji's evidence in identifying the culprit. But in the plot summary provided by Hutchinson, it was the statements of villagers of Lungur Serai, discovery of Fatma's goods in the possession of Deendar, and confession of his son that provided the conclusive evidence at trial, which concluded with an execution sentence. The evidence of tracking, in Hutchinson's telling, is thus not the conclusive evidence, and this arguably reflects an awareness of the prevailing anxieties over Khoji evidence: his presentation of the case suggests the benefits that a police officer might gain by employing trackers while at the same time indicating that this information alone has been insufficient to establish guilt.

There are numerous other cases that appear in Hutchinson's police reports.[37] By providing these case summaries, he suggested the usefulness of tracking as a detective aid. In these cases, the evidence of trackers is again only of marginal importance. The case summaries are written to show at once the importance of tracking and the authorities' ability to resolve the case within the framework of modern rational explanations. In this sense, they attempted to reconcile the intuitive knowledge with rational scientific knowledge. None of the cases are based only on the tracker's evidence, which suggests that these case summaries were carefully selected to satisfy the imperial anxieties over the infallibility of evidence.

Hutchinson sought to normalize and routinize tracking in colonial policing: by recognizing tracking as an important detective aid in the police training manual, he ensured that tracking would become a standard police procedure in the investigation of homicide and other crimes like burglaries. By appropriating tracking as a detective aid, he reconciled tracking with the modern forensic culture. And by providing examples of successful cases in which tracking had played a vital role, he silenced his critics who would have liked tracking to be dismissed. His understanding dominated the forensic culture of the police until the last days of British rule in Punjab. Hutchinson's legacy, in short, was to have successfully combined two alien systems of forensic practices, which were also rooted in different paradigms of evidence, thereby laying the foundation of a coherent forensic culture in Punjab.

## Tracking in the New Policing

The British response to Khojis in colonial Punjab raises several questions. While trackers in Australia, the Middle East, and South Africa were considered savage and unreliable, and were gradually purged from the police forces, the response of British police officers in Punjab to an extent celebrated Khoji skills.[38] The trackers enjoyed respectable status within the hierarchies of the police. The fact that other provincial forces—for instance, those in neighboring North Western Provinces and the eastern provinces of Bihar and Bengal—could not enroll trackers was a matter of pride for the Punjab Police. In several cases, we learn that the police from Punjab lent their trackers to other provinces, and that some of the receiving police officers were so impressed by trackers that they learned tracking from their visitors.

This celebration of trackers in colonial Punjab, in contrast to its dismissal in elsewhere in the British Empire, comes as a surprise. Blum's chapter, for example, shows that Bedouin trackers were distrusted and considered unreliable for detection largely on racial grounds. The situation in Punjab was reverse: racial difference did not impede the trustworthiness of trackers. Moreover, the later development of nationalism does not seem to have affected this patron-client relationship. D. Gainsford's narrative, written after independence—which may be taken as representative of the views of his fellow elite British officers who worked in Punjab—demonstrates a continued fondness for Khojis until the last days of the empire. These cordial relations between the British and Khojis were shaped both by the practical achievements of tracking and by the respect accorded the practice in wider Punjabi society. For example, we are told that some villages took keen pride in having a renowned tracker living in their midst.[39]

This local prestige presented the British with a problem, however. As well-known Khojis enjoyed considerable power and leverage, British police officers often found it impossible to lure them into police careers. The record of the recruitment of Khojis to the Punjab Police Department clearly reveals that it was the Khojis who shaped the conditions of their employment, and that the British had to accommodate their idiosyncrasies. From the very inception of the colonial rule in Punjab (1849), the British had employed Khojis in tracing the footprints of animals and men. In 1861, Khojis were absorbed into the new police force. Major General Hutchinson, the first chief of the Punjab Police, especially encouraged the employment of trackers in and for the police. But the recruitment of Khojis was a provincial matter: the all-India scheme of police organization, which was based on the British models of policing, made no special room for the recruitment of "trackers" in the police. The Police Act of 1861, for example, did not prescribe any rule for the employment of trackers within provincial police forces. Thus the recruitment of trackers in the Punjab Police remained a difficult matter. Constrained by the provisions of the 1861 act, the government in Punjab could not create a separate office for the tracker in its police department; it was left to the police department how it would absorb Khojis within the scheme of the police organization, which was prescribed by the act. Because the natives were recruited only at the lowest ranks in the police, and Khojis were also invariably illiterate, they were absorbed as constables, at the lowest positions in the police department.

Appointing Khojis at this subordinate level presented a problem, as they loathed the hierarchy, discipline, working conditions, police duties, and poor remuneration of the police employment. They preferred private tracking errands to the police employment, by which they could earn more than the government salaries. In districts where cattle theft was common, the best Khojis could, as an official report tells us, "earn more money by private tracking" than the police salary—as much as double the salary from the government. As a result, Khojis could negotiate the conditions of their service in the police.[40]

There are numerous instances in which the police officers mentioned their difficulties in recruiting Khojis. In 1872, George Hutchinson noted: "The universal experience of the [past] . . . eleven years" suggested that "good trackers will not take service in the police even on Rs. 10 a month." Hutchinson concluded on a despairing note: "My belief is that no amount of wages that this department could afford to pay would satisfy trackers so that they would give up private practice, and be always ready for any work required from them." These comments, repeated in other reports, reflected the reality that an average policeman's salary was hardly ever equal to what the tracker could earn in private practice. Relaxation of their duties had produced no better results: even when they were "kept solely for tracking and relieved from all [other] duty," the report continued, the trackers persisted in their lack of interest.[41]

As a consequence, the police department had to be satisfied with the supply of less efficient trackers, which in turn resulted in the kinds of complaints that, if we did not appreciate the British desire for expert trackers, might be taken as evidence of race-based criticism. "I have observed that these half experienced Kojrees [Khojis]," wrote F. H. Oliver, a district superintendent of the Gurgaon District Police, "never attempt to carry through any difficulties but on the contrary give up at the very first obstruction." Oliver also suggested that the government should preserve the tradition of Khojis by providing them a handsome income: "It must be remembered that the trade must be learnt from infancy and unless the income is secure, it is really not worth a man's while to devote himself to it or bring up their children to the pursuit. If more of these men were employed, say one at each [police station], with the full understanding that the appointment would be hereditary, provided they were honest and well behaved, the services of a valuable body of detectives would be secured to Government."[42]

The British had to relax their service rules to entice Khojis to enlist. Ordinarily, recruitment to the police was restricted to men under the age of twenty-five; however, the government exempted trackers from this rule. Trackers could be recruited in the police "up to the age of 35 years, and beyond that at the discretion" of the chief of the police, on the condition that trackers were efficient, which means there was no age limit.[43] Furthermore, they were excused from standard training and duties. They were not, for example, to be "compelled to attend [drill] school or learn the subjects taught to ordinary police officers," nor were they required to wear police uniforms.[44]

Even with these exemptions, Khojis remained largely uninterested, and this led to an adjustment of police rules allowing their hiring on a temporary basis, and at a rate that sought to compete with "local" market conditions. In private practice, though there were no fixed wages, clients typically paid the tracker according to the value of the property involved in a case. With this in mind, police rules provided a rate of 8 annas for a day or for the fraction of a day—a rate higher than the fixed salaries of the police-employed Khojis. This again demonstrates the extraordinary efforts made to accommodate the services of Khojis.[45]

By the early decades of the twentieth century, the Punjab Police employed, on average, 150 Khojis in the police force.[46] Government reports reiterated that it was hard to find competent Khojis to be a police employee. The demand for Khojis in early twentieth-century Punjab shows that their profession continued to thrive throughout the duration of colonial rule. It appears that the British police officers in Punjab had normalized the presence of Khojis in the police force. Khojis had been given special concessions, and thus they were accommodated in the police department. It appears that several British police officers—like Oliver mentioned above—had established a patron-client relationship in which they patronized their favorite Khojis. The British officers' appreciation for Khojis did not suffer from the anxieties of anticolonialism, nationalism, or racial prejudice. In short, as the next section shows, a different description of trackers' relationship with colonialism appeared in case of colonial Punjab.

## Forensic Modernization and Tracking: Testimonies of British Officers

In this section, I consider how forensic modernization in the late nineteenth century influenced tracking. A provincial police training school was set up in 1891 in the historic fort at Phillour to act as a center for imparting education in the latest detective techniques. In the coming decade, specialized departments for the development of more modern and "scientific" forms of forensic technique like anthropometry and fingerprinting were introduced in Phillour. In addition, and in accordance with the London Metropolitan Police model, a special museum was set up in the school to preserve material evidence from well-known cases for training purposes. The purpose of this discussion is first to outline these developments, focusing on the case of fingerprinting, and second to analyze how these new "scientific" forensic practices accommodated "traditional" tracking.

In India, anthropometry was first introduced in Bengal. In 1892, the police in Punjab dispatched a special team to study the anthropometry department set up in Bengal. The team provided a positive account of the Bengal anthropometry experiment and recommended that such a bureau also be established in Punjab. The police invested resources in preparing an anthropometric record of prisoners and criminals of Punjab. In the reports, it was mentioned that anthropometry had been highly successful in some cases that would not have been solved without the help of anthropometry. But the later shift to fingerprinting in Britain and elsewhere led to the discontinuation of this program. In fact, the whole experiment survived only for a few years.

Over the course of the decade, Phillour's bureau became a well-known center for fingerprinting in the region. Samples were sent from other provinces and princely states for analysis, and each year it received visitors from across the country seeking instruction in fingerprinting. The courts readily welcomed fingerprinting evidence, as it could be measured, reproduced, verified, and preserved. The police department boasted that they could secure convictions based on the scientific forms of evidence.

This regime of forensic modernization also included the use of plaster of paris. In the past, the evidence of footprints was considered wanting because of its ephemerality. Plaster of paris for the first time turned footprints into concrete, and preservable, evidence. One might expect that the introduction of the practice of taking of molds of footprints would have replaced trackers. In reality, this was not the case. In fact, like Hutchinson's absorption of trackers in the police catechism in the 1870s, in the 1890s, the police reconciled the expertise of trackers with new forensic techniques. As late as 1930s, the police department recognized the importance of trackers. The *Punjab Police Rules* (1934) mentioned in Rule 25.26 (1) that "Footprints are of the first importance in the investigation of crime . . . For this reason . . . when any crime occurs all footprints and other marks existing on the scene of the crime should be carefully preserved and a watch set to see as few persons as possible are permitted to visit the scene of the crime." The next rules explained how, "in the presence of magistrate of the highest available status," the footprints of the suspects were to be identified by "the tracker or other witnesses." As per the procedure,

> ground shall be prepared for the tests. On this ground the suspect or suspects, and not less than five other persons shall be required to walk. The magistrate, or in his absence the police officer conducting the test, shall record the names of all these persons and the order in which they enter the test ground. While these preparations are proceeding the tracker or other witness who is to be asked to identify the tracks shall be prevented from approaching the place

or seeing any of the persons concerned in the test. When all preparations are complete the witness shall be called up and required to examine both the original tracks and those on the test ground, and thereafter to make his statement.

The last paragraph of this section mentions that "The evidence of a tracker or other expert described in the foregoing rule can be sustained by the preparation of moulds of other footprints of the criminal or criminals found at the scene of the crime." It also mentioned that the method of taking molds with plaster of paris was taught at the Police Training School at Phillour, but it required practice "before an officer can become proficient."[47] This one example shows how the tracker's expertise had been reconciled with other scientific forensic practices. The rule did not make it mandatory for police officers to take molds. Instead, it only recommended that the evidence of a tracker or other expert "*can be* sustained by the preparation of moulds."[48] The involvement of the magistrate from the beginning in the making of the evidence further suggests the normalization of the evidence of trackers in the courtroom.

Arguably, however, the actual practice of policing is not accurately captured by these rule books, as their formality may belie actual forensic activity. Consequently, we may never know the full extent to which tracking was used in Punjab. But another set of sources—the reflections of British officers from as late as the 1930s—provides us with alternative insight into tracking practice. These writings are valuable for several reasons. First, they are based on firsthand observations and actual cases, thereby providing a glimpse of actual functioning of police. Second, they suggest how the forensic investigation of crimes like burglary and cattle theft involved tracking. Third, they reveal the relationship of the British police officers with Khojis that persisted into the final decades of colonial rule.

In 1928, J. A. Scott, a police officer, wrote an article in the London-based *Police Journal* titled "Burglary in an Indian Province," in which he emphasized his role in suppressing burglary in the Lyallpur District of Punjab. The purpose of the article was not to discuss the role of Khojis in his campaign, but to reflect upon how his "systematic and scientific treatment" of burglary had led to the decline in this crime. Like Morelli's method of recognition of painters from the inimitable flourishes on their canvases, as described by Ginzburg, Scott recognized that each burglar had a unique "signature" that might reveal his identity from his idiosyncratic style of committing burglary.

In this self-laudatory "scientific" approach, Scott also asserted the value of footprints: "The stiff clay soil of parts of the Punjab retains excellent impressions of tracks which, with the settled conditions of atmosphere obtaining here, will last for days. Orders were therefore issued for the dimensions of footprints to be taken and entered in a form to be attached to the first case diary."

Scott then turned to some cases, and in doing so unwittingly suggested the importance of tracking. For example, he mentioned how in one 1925 case, Khojis had identified a thief from his footprints. At the time, the Lyallpur District had received several reports of cases of burglary, all of which appeared from the modus operandi as the work of a single gang. The gang, on moonless nights, burgled government property, leaving at the scene of each crime a footprint that was "specially noticeable owing to its being long and thin."[49] Prompted by these discoveries, Scott states, "All the local trackers for miles around were called out, and a unanimous verdict was given that it belonged to a certain village ne'er-do-well called Shahu." On further enquiry, the police found that Shahu had absconded from his house, and when the police questioned his companions, they confessed that they had committed the burglaries with him. The police arrested all the members of the gang and recovered the stolen property. In this case, while footprints played a central role, there is no mention of print "moulds."[50]

Scott mentioned other cases in which tracking played a crucial role. In one, thefts were reported from three villages, which appeared to be the work of a single gang of four men. Because "flour and sugar" had been stolen, the police believed that the members of the gang belonged to a menial caste. Scott appointed four police parties to search for strangers with footprints similar to those found at the burglaries. Within a range of five miles from the scene— after which tracks often disappeared—the search parties were directed to "examine all men of the castes who perform menial duties for footprints similar to those of the gang."[51] While three search parties returned unsuccessful, the fourth found a suspicious track that resembled the track found at the crime scene. The suspect, whose footprint resembled the track, believed that the police had discovered all the information and he readily confessed to the crime and mentioned the names of the other gang members. In this case, the police arrested the other thieves along with stolen property.[52] Though Scott did not mention Khojis by name, it is likely that Khojis had been at the forefront in the investigation.

## A Critical Gaze: The British Criticism of Tracking

Though from the mid-nineteenth century onward colonial officials both criticized and celebrated tracking, there was rarely a systematic attempt to understand its scientific basis. From the late nineteenth century, we come across occasional critical attempts to explain tracking. The authors of these comments argue that tracking was useful, but that it was a mechanical art that could be learned with some practice. They dismissed the idea that a Khoji's knowledge was rooted in intuition and that only a Khoji had the special ability to understand the footmarks.

Perhaps the most systematic attempt to provide a scientific explanation for tracking comes from G. W. Gayer, another police officer. Gayer made a scientific study of tracking in colonial India and published it in 1909 as *Foot Prints: An Aid to the Detection of Crime for the Police and Magistrate*. Gayer was inspired by the efforts of the Austrian criminologist Hans Gross, who had devoted a large section of his monumental work *System der Kriminalistick* to tracking. Gayer was especially influenced by Gross's attempt to use modern science in understanding tracking. In the "Prefatory Note," Gayer argued: "The idea of making use of human tracks in criminal work has interested the author in a vague way since 1893, when he had a short spell of duty in Rajputana, and saw Khojis at work."[53] Gayer was simultaneously trying to record and integrate Khoji knowledge within a well-developed scientific system of studying of human footmarks. For example, to understand the anatomy of the human foot, he used the knowledge of Khojis. As per this knowledge, human feet had twelve anatomical features: Anguta, Barri assi, Chab, Chichi, Chothi, Chotti assi, Dusri, Eri, Pab, Talli, Tesri, and Zanjeri. "Every one of these twelve features," Gayer argued, "has as much character as the features of a face, and, once a pupil learns to distinguish them at a glance, he is able to recognize a person from his foot impression as easily as he can from his face."[54] In the introduction, he wrote that this important knowledge had awaited scientific study: "It is true that in certain parts of India professional trackers are used to track down criminals; but these trackers are entirely illiterate, and consequently, are unable to improve on the methods handed down from the past; moreover, the people who employ them, being altogether ignorant of the craft, are unable to utilise to the full powers of the instrument they employ."

Gayer offered an orientalized understanding of the institution of tracking, which involved a master "guru" and disciple "sishya." According to him, a "*Khoji* explained the system of training" to him in the following manner:

> The *Khoji's* teacher or *guru* gets some person in the village to have a clear impression of his naked foot on suitable soil, and carefully protects it by placing an inverted box over it; the pupil is then sent for and the characteristics of the foot—or its "chhin"—are explained to him in detail, and he is set to study the foot-mark until he thinks he will be able to distinguish it from others; when he feels confident, he is made to cover it again and told to try and find another impression made by the same foot. When he thinks he has succeeded, he has to confirm his conclusions by referring to the original impression. The conditions are gradually made more difficult until he is able, quickly and certainly, to commit the features to memory, to obliterate the original impression, and some days later, to commence his search for a corresponding impression; he then has to track the maker to his home.[55]

In the name of tradition, and by placing a Khoji as his source of information, Gayer clearly emphasized that anyone can learn tracking with practice, and he thus rebuffed the dominant understanding that tracking was an uncanny art that cannot be learned. Moreover, he insisted that there is a method by which tracking can be learned, the crux of which consisted of everyday practice. Gayer argued, "It not infrequently takes years of practice before the *guru* is satisfied with his pupil, for the latter must finally be able, not only to track down his quarry, but also to say, after studying the track, whether it is that of a man or woman, whether the person is old or young, fat or lean, burdened or unburdened; whether he or she was running, walking fast, walking ordinarily or loitering, and what he or she was probably doing while moving along."[56]

This description of a Khoji's training reminds us more of the scientific system presented in abstracted terms by Hans Gross than of an account seeking to convey the actual practice of tracking. The institution of a master Khoji who trained a pupil for several years in tracking seems an innovation of Gayer, as none of the other sources mention it. In his report, on the basis of the institution of Guru-sishya, Gayer developed an elaborate set of instructions with which a person could be trained in tracking. In a self-congratulatory note, he argued that he had tried this method on two batches of sub-inspectors in the Central Provinces and had achieved remarkable rate of success:

Two small classes of Sub-Inspectors have been taught the system. Each of these officers passed a two month's course of instruction. An ordinary way-farer was stopped, his name and address taken, and he was asked to hide his tracks, by going across ground that did not take impressions, before going on with his occupation. The Sub-Inspector, who of course had not been permitted to see the man, or to have any hint as to who he might be, was then shown the impressions on the road. He took trackings of them, and, after an interval, sometimes running into the third day, produced the person who had made the impression before the Examiner.[57]

Gayer had reduced the Khoji's knowledge to a mechanical skill that could be learned within two months. Perhaps the most important fault in the Khoji's knowledge was the lack of recording the footprint, which could be preserved and reproduced. Gayer invented a method by which the records of footprints could be obtained. As he argued, "the author has devised a transparent slate made of plate glass etched off into sections exactly like 1/8th sectional paper, the inch lines being enamelled red to show them up clearly and the fractional lines uncoloured. . . . These slates—if set in thin frames with low legs at the four corners, so arranged as to raise the under face of the glass about half an inch above the surface of the ground—allow a sufficiently clear view of an impression on the ground to be obtained to permit an accurate drawing being made."[58] Toward the end of his report, Gayer also included various "plates" in which he illustrated the differences between various types of feet. Apart from the different types of human feet, he also illustrated the footprints of animals—especially cow, bull, goat, stag, cat, hyena, wild dog, and leopard—perhaps reflecting how hunters' knowledge could inform modern science. He differentiated not only between the male and female prints of an animal but also between the old and young.

I end this section with the second and final assessment of tracking by a British officer. But because this assessment is a based on both the South African and Indian experience, it provides a broader response to tracking by British colonial officers, and thus suggests possible parallels across the empire that may be worth future exploration. The author's identity is also suggestive: Col. R. S. S. Baden-Powell, founder of the Boy Scouts Association. In the work *Aids to Scouting for N.C.O.s & Men*, Baden-Powell presented tracking as an essential part of training a scout: "Scouting without tracking is like bread and butter without the bread." Tracking, he explained, was essential for survival in the wilderness as well as during war—two of the central testing grounds for the adventurous British male. Linking the art of tracking with his politics, Baden-Powell observed, "All our uncivilized enemies make continual use of tracking when campaigning against us." By contrast, the British made "very little use of the art," even "when the ground is lying like a book before us—full of information."[59] Like Gayer, Baden-Powell believed that tracking was a mechanical skill that could be learned by practice:

> In some parts of India every village has one or two trained trackers in it. When a robbery occurs anywhere, the trackers in that village track the thief till he gets into the country belonging to another village; the trackers of the new village then take up the spoor, and if they fail to follow it out of their country their village has to pay up for the stolen property. Thus the trackers are very highly trained, and can follow up a spoor over every kind of ground, and by such slight signs that an ordinary person would altogether fail to see at all. It is all a matter of practice of experience.[60]

In his chapter on tracking, Baden-Powell outlined a detailed program to teach tracking. The hostile political situation in South Africa and his racial prejudice shaped Baden-Powell's understanding of tracking: "And a white man, although he is seldom so clever at tracking as a native (who has been at it all his life), is generally much better than the native in reading the meaning of the tracks."

Anticipating a relationship between tracking and detective fiction, he recommended that to better understand the meaning of spoors, his readers should consult the "Memoirs of Sherlock Holmes by Conan Doyle, and see how, by noticing a number of small signs, he 'puts this and that together' and gathers important information."[61] Baden-Powell did not know that a future historian would argue that to understand Holmes's adventures one must understand tracking. Ginzburg would show that Holmes was none other but a custodian of trackers' knowledge.

## Conclusion

I end this chapter with a vignette. One morning in July 2015, I was in London at 221B Baker Street, the fabled address—and now museum—of Sherlock Holmes. I stood in a long queue waiting for my turn to get into the museum. Next to me stood some young Indian and Chinese admirers of Holmes. They had read the adventures of Sherlock Holmes and watched movies based on him, and they believed that he was a historical character who had actually lived in the home they were about to enter. I wondered what may have made them believe that Holmes was a real character.

I found one explanation inside the museum, where the curators had transformed fiction into history. The entire house was laid down as described in the works of Conan Doyle. The curators had carefully adhered to the material world of late nineteenth-century England. To me, the boundaries between fiction and history seemed blurred. To any layperson, the materiality of the museum was convincing enough to make him believe that Holmes had existed. The curators' pretension to historicize Holmes had been successful.

How different, I thought, was the representational fate of the tracker, who existed in reality but has been written out of history. Even a renowned historian like Carlo Ginzburg could overlook the history of trackers, who in his classic article exist only among hunters in remote antiquity. For him, trackers are anonymous and ahistorical, confined to folklore. In contrast, Freud, Holmes, and Morelli—the eponymous heroes of his story—possess a definite historical identity. Perhaps Ginzburg's article should also be read, like the numerous objects curated in Sherlock Holmes Museum, as a cultural artifact. In Ginzburg's article, history dances only around Europe. When he writes about European subjects, facts become definite and concrete. When he writes about the rest, his facts become general and imprecise. It appears as though Europe had history, and the rest had only folklore.

It is likely that the trackers' history will remain confined to obscurity. Yet this chapter is an attempt to historicize the tracker, laying down a new layer of history. In the continuous absence of such writings, the visitors to the Sherlock Holmes Museum would never know of trackers' adventures. Had trackers not been deprived of a history, admirers of Holmes might as well have known other enthralling adventures of detection.

### NOTES

1. Daniel R. Headrick, *The Tools of Empire: Technology and European Imperialism in the Nineteenth Century* (New York: Oxford University Press, 1981).

2. David Arnold, *Science, Technology and Medicine in Colonial India* (Cambridge: Cambridge University Press, 2000).

3. On the relationship between canal colonization and the spread of malaria, see David Arnold and Ramachandra Guha, eds., *Nature, Culture, Imperialism: Essays on the Environmental History of South Asia* (Delhi: Oxford University Press, 1994); for a recent perspective on how railway embankments obstructed the natural drains and caused floods, see Ritika Prasad, *Tracks of Change: Railways and Everyday Life in Colonial India* (Cambridge: Cambridge University Press, 2015); on how railway construction led to deforestation, see Madhav

Gadgil and Ramachandra Guha, *The Fissured Land: An Ecological History of India* (New Delhi: Oxford University Press, 1992).

4. See "Introduction," in *Medical Marginality in South Asia: Situating Subaltern Therapeutics*, ed. David Hardiman and Projit Bihari Mukharji (New York: Routledge, 2012).

5. Smritikumar Sarkar, *Technology and Rural Change in Eastern India, 1830–1980* (Oxford: Oxford University Press, 2014).

6. Clive Dewey, *Steam Boats on the Indus: The Limits of Western Technological Superiority in South Asia* (New Delhi: Oxford University Press, 2014).

7. The British in India spelled "Khoji" and "Puggi" in numerous ways; e.g., Khoji is sometimes "Khojee" and even "Khojree," and the like. In this chapter, I have retained uniform spellings except in direct quotations, preferring "Khoji" over "tracker"; however, I have used both the terms interchangeably. I have italicized all the vernacular terms upon first use.

8. See the memoir of sergeant Sardar Bishan Singh, *Tajirbat-i-Hind* [in Urdu] (Amritsar, 1897).

9. The other terms used for "detective" were *guputchar* and *jasoos*, which could also mean "spy." See Markus Daechsel, "*Zalim Daku* and the Mystery of the Rubber Sea Monster: Urdu Detective Fiction in 1930s Punjab and the Experience of Colonial Modernity," *Journal of the Royal Asiatic Society of Great Britain and Ireland* 13 (2003): 21–43.

10. For a history of the vernacularization of the detective genre, see Francesca Orsini, *Print and Pleasure: Popular Literature and Entertaining Fictions in Colonial North India* (New Delhi: Permanent Black, 2009).

11. *Annual Police Administration Report of the Punjab for the Year 1862* (Lahore: Punjab Printing Press, 1863), 82.

12. The report, however, does not provide any comment on the districts of Ambala, Gurgaon, Karnal, and Rohtak. It is also silent on the districts of central Punjab: Amritsar, Ferozepur, Hoshiarpur, Jalandhar, and Ludhiana. This silence was most likely a reflection of official neglect rather than the absence of tracking. As the subsequent records reveal, tracking was well known in many of these areas, including Amritsar, Ferozepur, Gurgaon, and Karnal.

13. Binyamin Blum, "From Bedouin Trackers to Doberman Pinschers: The Rise of Dog Tracking as Forensic Evidence in Palestine," chap. 7, this volume.

14. Vinay Lal, "Everyday Crime, Native Mendacity and the Cultural Psychology of Justice in Colonial India," *Studies in History*, n.s., 15 (1999): 155.

15. David Arnold, *Toxic Histories: Poison and Pollution in Modern India* (Delhi: Cambridge University Press, 2017), 107.

16. G. Hutchinson's *Punjab Police Catechism* (Lodiana: American Presbyterian Mission Press, 1877) and the contemporary police manuals from Britain do not mention it as a detective aid. *The Punjab Police Rules* (Lahore: Civil and Military Gazette Press, 1891) considered the use of plaster of paris only as an aid to tracking. G. W. Gayer, *Foot Prints: An Aid to the Detection of Crime for the Police and Magistrate* (1909) does not mention it. It appears that in colonial India the use of plaster of paris was rare.

17. "Karnal Cattle Lifting Case, 1913," Bryane Papers, IOR Mss Eur F152/2, BL, London; see also David Gilmartin, "Cattle, Crime and Colonialism: Property as Negotiation in North India," *Indian Economic and Social History Review* 40 (2003): 33–56.

18. "Karnal Cattle Lifting Case," 31–32.

19. G. Hutchinson, CSI and CB, began his career in India as second lieutenant of engineers in February 1846. He fought in the First Anglo-Sikh War (1846). During the 1857 revolt, he played an important role in recapturing Lucknow. He served as military secretary in the province of Oude from its reoccupation from the rebels until the end of 1859. During 1860–61, he served as the officiating military secretary in Punjab. In 1861, he undertook the entire recon-

struction of the Punjab Police and the introduction of the organized constabulary. Sir Robert Montgomery, the lieutenant governor of Punjab, praised Hutchinson for his services: "On the new organisation of the police in Punjab his great and varied experience led me to appoint him to organize that force, and his services in that Department have been equally as great and beneficial to the Government as in the other two posts he held." For more information on Hutchinson's life and contribution, see the George Hutchinson Papers, Mss Eur E241: 1826–1899, BL, London, 232–33.

20. George Hutchinson, *Punjab Police Catechism* (Lodiana: American Presbyterian Mission Press, 1877); idem, *Hidayat-namah-i-polis* [in Urdu], trans. Kewal Krishna (Lahore: Matba-i-Ezidi Press, 1873).

21. Hutchinson further wrote, "the benefits arising from this little book have far exceeded what was even hoped for; each Constable has a copy which costs him three annas, and, which I thought it advisable he should pay for himself, as being more likely to be cared for and preserved. All the Police Stations have them. The men are reported as learning it by heart with the greatest zeal." *Annual Police Administration Report of the Punjab for the Year 1861*, 78. Ghulam Muhammad, a police sergeant, mentioned in his memoir that it was must for a new police constable to read Hutchinson's catechism. See his memoir, *Miftah-ul-taftish* [in Urdu] (Lahore: n.p., 1897), 325.

22. Hutchinson, *Punjab Police Catechism*, 104.

23. Hutchinson, *Punjab Police Catechism*, 105.

24. Hutchinson, *Punjab Police Catechism*, 105.

25. William C. Harris, *Police Catechism: Questions and Answers Framed for the Instructions of Constables on Joining the Police* (London: W. Clowes and Sons, 1861). Harris was assistant commissioner of the London Metropolitan Police.

26. C. E. Howard Vincent, *A Police Code* (London: Cassell, 1886), 75; Benjamin Moore Gregg and J. C. McGrath, *A Police Constables Guide to His Daily Work* (London: Effingham Wilson, 1921), xl–xliii.

27. *Annual Police Administration Report of the Punjab for the Year 1866*, 14.

28. *Annual Police Administration Report of the Punjab for the Year 1869*, 26.

29. *Annual Police Administration Report of the Punjab for the Year 1867*, 96.

30. Comment by Major Babbage, DC of Gujranwala, in *Annual Police Administration Report of the Punjab for the Year 1866*, 15.

31. *Annual Police Administration Report of the Punjab for the Year 1872*, 48.

32. *Annual Police Administration Report of the Punjab for the Year 1869*, 26; *Annual Police Administration Report of the Punjab for the Year 1870*, 17.

33. *Annual Police Administration Report of the Punjab for the Year 1871*, 37.

34. *Annual Police Administration Report of the Punjab for the Year 1872*, 48.

35. *Annual Police Administration Report of the Punjab for the Year 1871*, 37.

36. *Annual Police Administration Report of the Punjab for the Year 1871*, 37.

37. See, e.g., *Mitto vs Soojana* (1865); *Crown vs Raja, Dadoo, Boorhanee, and Zeeda* (1867); *Crown vs Aliayar* (1867); and *Crown and Hurdial vs Mujoo, Hyat, and Nugga* (1865).

38. On South Africa and the Middle East, see Blum, "From Bedouin Trackers to Doberman Pinschers," chap. 7, this volume. The story of aboriginal trackers of Australia still awaits a detailed study.

39. See Bryane Papers, Eur Mss F152/2, BL, London, 30–32; Bryane observed that the presence of a tracker in a village deterred thieves from committing crime in that village.

40. *Annual Police Administration Report of the Punjab for the Year 1861*, 38–39.

41. Quoted in *Annual Police Administration Report of the Punjab for the Year 1872*, 51.

42. *Annual Police Administration Report of the Punjab for the Year 1872*, 5.

43. *Annual Police Administration Report of the Punjab for the Year 1872*, 5.

44. Though the police rules mentioned that a Khoji must be issued a uniform, a safa, a belt, and a great coat, it was their discretion to wear them; in most cases, it seems trackers did not wear the police uniform. For more information, see *Punjab Police Rules from 1 March 1861 to 1 July 1900*, vol. 1 (Lahore: Civil and Military Gazette Press, 1900), 520.

45. Eight annas was equivalent to one half of a rupee. In the late nineteenth century, the salary of a constable was around 7 rupees per month. Assuming a rate of 8 annas per day, the monthly salary of a tracker would amount to around 15 rupees per month. *Punjab Police Rules*, 521.

46. *Annual Police Administration Report of the Punjab for the Year 1909*, 20.

47. *Punjab Police Rules, 1934*, vol. 3 (Lahore: Superintendent Government Printing Press, 1935), Rule No. 25. 26 (3).

48. *Punjab Police Rules*, Rule No. 25. 26(3). Emphasis mine.

49. J. A. Scott, "Burglary in an Indian Province," *Police Journal* 1928: 468.

50. Scott, "Burglary in an Indian Province," 464–65. For the working of fingerprinting, see Simon A. Cole, *Suspect Identities: A History of Fingerprinting and Criminal Identification* (Cambridge, MA: Harvard University Press, 2002).

51. Scott, "Burglary in an Indian Province," 468.

52. Scott, "Burglary in an Indian Province," 468.

53. Gayer, *Foot Prints*, i.

54. Gayer, *Foot Prints*, 2–3.

55. Gayer, *Foot Prints*, 1–3.

56. Gayer, *Foot Prints*, 3.

57. Gayer, *Foot Prints*, i.

58. Gayer, *Foot Prints*, 5.

59. R. S. Baden-Powell, *Aids to Scouting for N.C.O.s & Men* (London: Gale and Polden, n.d.), 52.

60. Baden-Powell, *Aids to Scouting*, 55–56.

61. Baden-Powell, *Aids to Scouting*, 62, 65.

# CHAPTER ELEVEN

# Forensic Knowledge and
# Forensic Networks in Britain's Empire

## The Case of Sydney Smith

HEATHER WOLFFRAM

In his memoir, titled *Final Diagnosis*, the Scottish forensic pathologist John Glaister Jr. (1892–1971) reflected on the time he spent in the late 1920s and early 1930s as both the chair of forensic medicine at the University of Cairo and as a medicolegal consultant to the Egyptian government.[1] Beyond describing his efforts to establish a fully equipped forensic laboratory at the university and the scope and variety of medicolegal work carried out in Egypt, Glaister indicated how working and researching in Britain's former colony had persuaded him of the necessity of forging international networks within the medicolegal world. In this regard, he wrote,

> Research had opened to me a subsidiary truth: that [for] medico-legal work to be successful, [it] had to be fully international in its concept. It required a free exchange of ideas and information, an understanding of what other men were working on, a freedom to assist—and, above all, a rejection of the parochial outlook, which, as a Scot, I knew particularly well could hang round the neck of progress. My correspondence with others in the forensic and research fields in countries abroad had shown that those most willing to help, the men who were also acknowledged supreme in their spheres, were without exception outward-looking.[2]

Far removed from the institutional and collegial structures to which he had been habituated during his training and early practice in Scotland and faced with medicolegal questions that were rarely encountered in the British context, it is unsurprising that Glaister, like many medicolegal experts stationed across the empire in the first half of the twentieth century, felt deeply the importance of intellectual networks for the exchange of forensic knowledge and the progress of the discipline.[3] Although Glaister may have envisioned a global network bound only by universal forensic questions, his own networks and those of many of his British colleagues, working in colonies such as Ceylon, Egypt, India, and the Sudan, tended to exist primarily within Britain's empire, the realm in which such men both received their training and shared their knowledge and experience. During the early twentieth century, such networks were frequently built and maintained, as the Scot's reflections demonstrate, through correspondence, but could also be fostered through the movement of personnel to different parts of the empire and through the exchange of specimens, publications, facilities, and expertise between colonies.[4]

Indicative of the existence and importance of forensic networks both within and beyond Britain's empire, Glaister's reminiscences offer a stimulus to the historian to examine the ways in which forensic knowledge circulated during the first half of the twentieth century. In Alison Adam's recent book on the beginnings of British forensic science, colonial criminalistics and laboratories feature as examples of how Britain's empire offered a testing ground for new ideas and technologies as well as a space in which British personnel might "cut their forensic teeth" before returning home to practice.[5] This was undoubtedly the case, but it does suggest that forensic knowledge flowed linearly between Britain and its colonies. The challenges of colonial administration in Britain's empire during the late nineteenth and early twentieth centuries, however, not only led to the adaptation of British or Western institutions and bureaucratic techniques to local conditions, but also inspired a range of creative solutions, often based on indigenous knowledge, that were developed in situ to solve the colony's peculiar problems.[6] Knowledge created at the so-called periphery could be developed, tried, and tested in the "colonial laboratory" before adaptation and implementation at home and abroad. In the realm of criminal identification, the development of fingerprinting in British India is a case in point, as historians, including Simon Cole and Chandak Sengoopta, have shown.[7] But while forensic knowledge created in Britain's colonies undoubtedly found its way back to and influenced the shape of forensic science and forensic medicine in the metropole, it would be wrong to imagine that such knowledge traveled solely in a straight line between center and periphery, periphery and center. Often, colonies, which tended to have more in common with each other culturally and climatically than they had with Britain, applied forensic knowledge and adapted forensic systems developed in one colony for use in another, communicating through correspondence, publication, and the movement of personnel. Medicolegal experts in Egypt, for instance, relied on research conducted in India to confirm their own understandings of the processes of decomposition in hot and humid climates, while toxicologists adapted procedures that had emerged on the subcontinent for cases of poisoning for use in the government analyst's laboratory in Cairo.[8] Such knowledge, shared between colonies as well as between the colonies and the center, is evidence of the existence of multidirectional forensic networks within the British Empire that brought to bear expertise, forged in a wide array of colonial contexts, on both universal and local forensic questions and the conduct and administration of forensic science.

This more complex way of conceptualizing how knowledge was created and circulated in and between Britain's colonies has emerged largely from the work of historians of empire. Since the 1990s, as Tony Ballantyne makes clear, growing calls for transnational histories have seen historians employ new analytical models that seek to reveal "the movement of people, ideas, ideologies, commodities and information across the borders of nation states."[9] Although histories that have employed this approach to empire have often reinscribed London as the imperial center and thereby marginalized those in the colonies or created teleological narratives of national independence, scholarship in certain fields, including the histories of science and medicine, has done much to break down the binaries between center and periphery.[10] Beyond illuminating the importance of networks for the creation and transmission of ideas in colonial contexts and demonstrating how the application of scientific and medical knowledge aided in the consolidation of empire,[11] historians of science and medicine, such as David Arnold, have tended to reject a Western diffusionist model of knowledge transferal between Britain and its colonies and questioned distinctions between center and periphery, metropolitan and colonial science.[12] Instead, these historians have begun to focus on complex networks of knowledge exchange that existed both within colonies, between the indigenous population and the colonizers, and internationally. Moving beyond case studies of specific colonies, however, recent work on scientific and medical networks has considered not just the imperial nature of such connections, but also their trans-imperial dimensions.[13] Deborah Neill's work on networks in tropical medicine, for instance, demonstrates that, in terms of the circulation of scientific knowledge, empires were not hermetically sealed; instead, clinicians and researchers from a variety of European nations worked cooperatively to create professional networks that would establish their collective authority as experts in a new field of scientific inquiry.[14]

Like tropical medicine, early twentieth-century forensics, as Glaister's memoirs suggest, also manifested trans-imperial and international links and were crucial in the attempts of forensic practitioners to establish their authority. Given the gaps in our knowledge about the history of forensics at both the national and imperial levels, however, the exploration of trans-imperial connections would seem premature. A necessary first step in the mapping of forensic networks, then, would appear to be an examination of links forged within one empire. While not addressing the need for more national histories of forensic medicine and science, the study of forensic networks within the British Empire should provide insight into how early twentieth-century British experts communicated with one another across the empire as well as furnish useful clues and stimuli for those working on case studies of specific countries. As Markku Hokkanen has suggested in a work on imperial networks and colonial bioprospecting in Malawi, "colonial knowledge was essentially constructed and constituted 'in travel.'"[15] One of the easier ways of studying networks and their role in producing colonial knowledge, then, is to trace the movements of individuals, what Hokkanen calls "mobile imperial agents," across the network, considering the personal and institutional links that they forged as they traveled as well as the knowledge that they "constituted" in so doing.[16]

There are numerous examples of individuals who moved between Britain and the colonies, given the presence of colonial state chemical laboratories in most colonies and medicolegal facilities in some. Few, however, would provide useful case studies for understanding forensic networks or the impact of what we might call "mobile forensic agents" on the scientific and administrative development of forensics because of a lack of source material relating to them and their careers.[17] One of the exceptions to this rule is the New Zealand–born professor of forensic medicine at the University of Edinburgh, and Glaister's immediate predecessor in Cairo, Sydney Smith (1883–1969). Among a select group of forensic practitioners who engaged in autobiography as a means of shaping both their own self-image and that of the field, and part of a generation of scientists and physicians increasingly committed to representing their field as one reliant on teamwork rather than individual endeavor,[18] Smith not only produced a popular memoir, which provides some indication of the links he forged throughout his career, but also deposited his papers, documenting his colonial work, in an archive.[19] He thus provides a good choice of subject for an exploration of the intellectual and physical links forged by forensic scientists and physicians in Britain and its empire, highlighting not only the ways in which forensic knowledge and resources were shared between colonies, but also the manner in which the movement of personnel helped forge forensic networks and foster the exchange of forensic and administrative knowledge.

Smith's career in forensic medicine, which began with the outbreak of the First World War and continued well into the 1950s, saw him travel between Britain, Ceylon, Egypt, Lebanon, and New Zealand, accumulating and sharing forensic and administrative knowledge, gained in both colonial and metropolitan contexts, as he went. As a "mobile forensic agent," his contributions to constituting forensic knowledge and consolidating connections within the empire are evident in several areas, including the scientific and the administrative. During the period he spent in Egypt, for example, forensic problems resulting from the colony's climate, culture, and political situation saw Smith accumulate a body of both local and universal forensic knowledge, which built upon and confirmed that developed in other colonies. This was then communicated to other parts of the empire through correspondence, publications, and talks. In the two decades that he spent as professor of forensic medicine at Edinburgh on his return from Cairo, Smith's links to and role as an agent of forensic knowledge within the empire continued, not only through correspondence but also through his mentorship of students who came from and returned to the colonies. Experience of Egypt's legal and administrative system, which while it had its commonalities with that found in Scotland differed significantly from that found elsewhere in Britain, provided Smith with an understanding of how forensic medicine and science might be organized and administered differently and more efficiently. No doubt this influenced the way in which he approached medicolegal work on his return to Britain in 1928, and might have played a role in attempts to have Smith participate in the restructuring of British forensic science that was underway during the middle to late 1930s. This administrative knowledge, however, gained in a colonial context, was most fully shared when Smith was sent to Lebanon and Ceylon during the early 1950s in order to address perceived deficiencies in the forensic and medicolegal systems of these former French and British colonies.[20] Here, Smith appears not only to have mobilized knowledge and experience gained in both Egypt and Britain to assist in establishing rigorous scientific and administrative procedures sympathetic to existing structures and financial imperatives, but also to have used his connections with former students to assess the central problems with medicolegal services. While undertaken at the behest of the World Health Organization (WHO) rather than the British state, colonial knowledge and networks forged within the British Empire, prior to the Second World War, remained evident and important to this exercise. Using these examples, it becomes apparent how Smith's experiences in and travel between Britain, Ceylon, Egypt, and Lebanon helped circulate forensic knowledge to different parts of the empire, both establishing and reinvigorating links within the network.

## Smith in Egypt and Britain

In 1917, after a brief period as the Otago Medical Officer of Health, Smith traveled to Egypt, where, for just over ten years, he worked as the head of the medicolegal section of the Parquet (Ministry of Justice) and as a lecturer at Cairo's Medical School.[21] Whereas his Glaswegian colleague Glaister was to roundly condemn parochialism in his memoir, Smith appears to have quickly appreciated, while in Egypt, the role played by legal, cultural, and climatic peculiarities in shaping forensic work and knowledge in the colonies. In this regard, he wrote, "The habits and customs of the people as well as the difference in the laws must alter to a considerable extent the types of crime and ipso facto the type of work done by the medico-legal experts of a country."[22] Faced with a host of medicolegal problems where both climatic knowledge and cultural knowledge were essential in ascertaining identity and causation and in working in a hybrid medicolegal system that combined French law with various British institutions like the government analyst's office, Smith became highly proficient at a range of forensic and deductive techniques that were rarely required in Britain.[23] Because of Egypt's hot and humid climate and the practice of surface burial, Smith was frequently required to identify fragmentary or badly decomposed remains.[24] In so doing, he relied not only on standard forensic techniques, including measurements and x-ray, for discovering the age and sex of unidentified bodies and bones, but also on cultural knowledge that revealed information about religion and occupation.[25] A cross tattooed on the wrist, Smith wrote, would indicate a Coptic Christian, a round callosity in the center of the forehead would suggest a Muslim, while the precise manner of female circumcision might reveal whether a deceased woman was Egyptian or Sudanese.[26] Occupation, Smith maintained, might be ascertained by examining the hands and feet. He wrote,

> the fellah (agricultural labourer) has broad, spade-like hands and feet, with thickened and fissured epidermis ingrained with dirt, and usually with broken nails. The town manual labourer who wears boots will have a thick epidermis on the hands, but comparatively thin skin on the feet. Tradesmen may show certain peculiarities about the hands; thus the bootmaker has a callosity

about the size of a two shilling piece on the palm of the right hand, the carpenter has a marked callosity between the thumb and index finger, the weaver has callosities on his fingers, the dye worker has stains on his hands, the tailor and seamstress have pricks on the fingers of the left hand.[27]

Religious and cultural practices also became central to Smith's thinking about motive in the Egyptian context. The practice of polygamy and Muslim laws on inheritance, divorce, and the maintenance of children, to his mind, provided some of the most frequent motives for murder.[28] This knowledge Smith shared with forensic practitioners in both Britain and abroad by means of publications, including an appendix to his 1925 textbook of forensic medicine, *Forensic Medicine in the East, with Special Reference to Egypt*, and in an article on Egypt's Medico-Legal Institute, which appeared in a Rockefeller Foundation volume on medicolegal laboratories internationally.

Apparent in these two publications were not only Smith's reflections on the issue of the universality of medicolegal questions versus the role of local forensic knowledge, but also his reliance on literature derived from another colonial context, suggesting that imperial links helped facilitate the sharing of forensic knowledge and procedures between colonies. The introduction to Smith's appendix on forensic medicine in the East read, "The principles of legal medicine do not differ in any significant detail in the various parts of the globe. Crime and criminal violence and deaths from misadventure have similar features wherever they occur, and the practitioner who has become familiar with the general principles of the science as laid down in the previous pages is not likely to be baffled by the geographic locus of a case. There are, however, certain differences in the rate of onset and progress of postmortem change under tropical conditions."[29]

In his writing on the processes of putrefaction in hot and humid environments, such as in Egypt, then, Smith used research conducted in India to confirm his own findings. He stated, "In India Dr. Coul MacKenzie made two series of observations during the rainy season and during the month of October . . . He found that external discolourations appeared on an average about twenty-six hours after death. In the rainy season the earliest appearance occurred in seven hours and the latest in forty-one hours . . . He also noted evolution of gas in eighteen hours on the average during the rains and in twenty-nine hours during October. These observations are similar to our experience in Egypt, and may be considered the usual rate of change under tropical conditions."[30]

Smith, of course, was not alone in his use of Indian examples to support his work in Egypt. Alfred Lucas (1867–1945), government analyst in Cairo, also found that India, where the British had been involved in medical jurisprudence much longer than in Egypt, was a useful point of comparison and source of ideas. A reading of Lucas's important book *Forensic Chemistry* (1921) demonstrates that he, too, was conscious of similarities based on criminal and cultural factors between the two colonies and that he believed regulations and procedures developed on the subcontinent would be useful in both the Egyptian context and beyond. In a section of his book on arsenic, for instance, Lucas noted,

> This is one of the commonest forms of poison used. In England oxide of arsenic (white arsenic) is generally employed, but in the East (India and Egypt for example) not only the oxide but the yellow sulphide (orpiment) and occasionally the red sulphide (realgar) are employed. The most important factor conditioning the nature of the particular substance used for poisoning purposes is probably the ease with which the substance in question may be obtained. Thus the extensive use of sulphide of arsenic as poison in Egypt is due to the fact that it can easily be obtained and can be bought clandestinely almost anywhere in the country despite the restrictions imposed on its sale by the Pharmacy Law, which results from its extensive employment for making the depilatory powder used for removing hair from certain parts of the body, a custom which Mohammedans consider essential to personal cleanliness.[31]

Other parts of *Forensic Chemistry* make apparent that Lucas found useful both the Indian government's regulations in suspected cases of poisoning and some of the key texts in Indian medical jurisprudence, including those of Lyons and Hehir and Gribble, in the creation of his own toxicological procedures.[32] Here, he wrote, "Excellent rules for guidance of those forwarding articles for analysis in suspected cases of poisoning are in use in India and may be found described in books dealing with Medical Jurisprudence in India."[33] These examples, derived from the works of Smith and Lucas, demonstrate the practice of looking to longer-established colonies and their medicolegal systems for ideas and suggest that forensic knowledge formed in the colonies often bypassed the metropole entirely, circulating instead between countries on the so-called periphery.

Beyond the climatic and cultural knowledge that he developed while in Egypt, Smith also increased his skills in toxicology and ballistics. Given the wide availability of poison, particularly arsenic, to which Lucas had pointed, homicidal poisoning was a frequent occurrence during Smith's time in Egypt. In 1923, for example, the Cairo laboratory "examined 401 cases of poisoning or attempted poisoning of human beings and 247 animal cases."[34] While the techniques used for testing for poison remained the same as those in Britain, the number of such tests conducted and their frequency allowed Smith and his colleagues to claim some expertise in the area. British visitors to Smith's Cairo laboratory, for example, were astounded by the number of Marsh tests being run simultaneously.[35]

Political violence in Egypt during the early 1920s also afforded interesting forensic opportunities for Smith, who was frequently confronted with both the wounds inflicted by bullets and bombs and the projectiles themselves. Rather unusually for a forensic physician, Smith felt compelled to develop expertise in ballistics as well as in the wounds caused by firearms. He shared the knowledge, techniques, and instruments that he developed as a result of this experience with colleagues in other parts of the empire through publications, correspondence, and talks.[36] Following the publication of his article on the Sirdar assassination in the *British Medical Journal*, for example, Smith sent multiple copies to police forces around Britain, including the Metropolitan Police Special Branch and the West Riding of Yorkshire Constabulary, which the recipients noted would be valuable to those engaged in police instruction.[37] On his return from Egypt, Smith further communicated his colonial and forensic knowledge through a talk about firearms cases, including the killing of the governor general of the Sudan, given at the Medico-Legal Society.[38] His knowledge of firearms and the wounds they inflicted was also put to use in the decades that followed in advising forensic practitioners in other parts of the empire who wrote to him with questions pertaining to wounds and bullets. A 1951 letter from the director of the Medico-Legal Department in Baghdad, for example, asked Smith's opinion on a case where death had putatively resulted from the explosion of ammunition engulfed in fire.[39]

The wealth of scientific and administrative knowledge that Smith gained in Egypt was naturally put into practice once he became professor of forensic medicine in Edinburgh, where it was brought to bear on the numerous medicolegal cases on which he consulted during the 1930s and 1940s. In the decades spent in Scotland, Smith retained his links to Egypt and other parts of the empire in two major ways. The first was through correspondence, which enabled him to exchange forensic knowledge with remote colleagues. A letter from Zomba, Nyasaland, in August 1933, which asked for Smith's help in ascertaining whether a fatal wound to a woman's neck was caused by tripping and falling onto the arrows she was carrying or the result of foul play, is one of numerous examples that demonstrate the manner in which forensic experts across the empire exchanged information and consulted with one another in puzzling cases.[40]

The second means of retaining and extending imperial ties was teaching, which allowed Smith to forge links with students who traveled to Edinburgh to gain forensic training before returning to the colonies to practice. In the mid-1930s, for instance, Smith undertook the training of a doctor from Ceylon, called Dr. Sinnadurai, whom the Ceylon Medical Council intended to employ as a judicial medical officer and lecturer in forensic medicine within the Department of Forensic Medicine.[41] Another Sinhalese doctor, G. S. W. de Saram, trained as a medicolegal expert with Smith during 1946 and 1947 and took a long course in forensic medicine at Smith's former institute in Egypt before also taking up a position in forensic medicine in Colombo.[42] Yet the influence practiced here by Smith was not simply pedagogical, but clearly institutional. When Smith, on his retirement from the University of Edinburgh in the early 1950s, was asked by the WHO to assess the medicolegal systems and facilities in Ceylon, his correspondence revealed how his instruction of Sinnadurai and de Saram had helped shape Sinhalese forensic medicine in the 1930s and 1940s. To his colleague F. S. Fiddes in Edinburgh, Smith wrote in February 1953,

> I have now got more or less settled down in Colombo and have been getting acquainted with some of the conditions pertaining to medico-legal work . . . The trouble is that the plan which I had in mind when I trained Dr. Sinnadurai and afterwards Dr. de Saram, and pressed for a well-equipped department in Colombo, has rather gone astray. At that time the whole of the teaching department came under the control of Sinnadurai and was very satisfactory indeed. Unfortunately, the university, when they decided to have a full time chair, separated the day to day work of the Department from the chair and eventually led to a complete division. We are in the anomalous position now of having a Department of Forensic Medicine at the University with ample accommodation and all the modern equipment, but no work and on the other hand the office of the Judicial Medical Officer with poor accommodation, no equipment whatever and all the work.[43]

Although Smith was disappointed at this turn of events, his connection to his former students, particularly de Saram, proved useful in considering the local situation and suggesting solutions to the former Crown colony's forensic and administrative problems. In a letter to de Saram, dated September 29, 1952, for example, Smith asked for his former student's personal view of "the general conditions in medico-legal work in Ceylon."[44] De Saram's responses ultimately helped Smith conduct his survey and form his recommendations for reform.

## Smith in Lebanon and Ceylon

Between 1953 and 1954, Smith undertook a review of the medicolegal services in Ceylon on behalf of the World Health Organization, also conducting a brief overview of the same in Lebanon during his outward journey.[45] Using his experience in Egypt and in Britain, Smith advised on the conduct and administration of forensic science and medicine in both locations. Although Lebanon was not a former British colony, having won its independence from France in 1943, it is worth considering Smith's work here both because it suggests the growth of trans-imperial and international forensic networks at midcentury and because of the connections Smith attempted to forge between the medicolegal systems of Lebanon and Ceylon, which as a former Crown colony had strong ties with Britain.

Breaking his journey to Colombo in Beirut, Smith undertook a four-day survey of the republic's forensic medicine personnel and facilities in mid-September 1953 to advise the government about the future development of Lebanese medicolegal services. Interviews with the director general of the Ministry of Health, the heads of pathology and bacteriology at the American University, the professor of medical jurisprudence at the French School of Medicine, the technician at the Police Department Laboratory, and the minister of justice, among others, revealed that Lebanon's five districts boasted only nineteen part-time medicolegal officers, who possessed no training in legal medicine.[46] The provision of proper facilities, apparatus, and systems for medicolegal work also appeared inadequate with no central morgue or proper arrangements for safeguarding material under investigation. In addition, Smith expressed concern that specimens sent from the provinces to the capital were sent in unsealed boxes, rather than official containers, and could end up at any laboratory because no government laboratory existed. Furthermore, the Police Department Laboratory lacked a satisfactory ballistics section and appeared to receive little or no help from medicolegal officers in their investigations, leading Smith to state, "It will be seen therefore that a real medicolegal service does not exist."[47]

On the basis of this survey and his knowledge of how the Egyptian Medico-Legal Institute had functioned within a legal system—which, like that of Lebanon, mixed French and Islamic law—Smith made a series of recommendations for the improvement of the republic's medicolegal services. These included the establishment of the Medico-Legal Institute, which, like its Egyptian equivalent, would be under the control of the Ministry of Justice. This institute would be responsible for all the medicolegal autopsies in Beirut and all the medicolegal and forensic science, including ballistics, for the whole country.[48] This was similar to the system over which Smith had presided in Cairo, where the Medico-Legal Institute had also been responsible for all the capital's autopsies; cases of poisoning; examination of blood, semen, hairs, and fibers; and identification of bones for the entire country.[49] In terms of the facilities and equipment immediately required, Smith focused on the issue of mortuary facilities and the secure and safe transport of materials for analysis or pathological examination from the provinces to the capital. In this regard, he specified, "The provision of reasonable post-mortem rooms with cold storage in Beirut and the main provincial towns. The provision of proper containers such as vacuum jars for the conveyance of pathological material from the provinces to Beirut."[50]

Although the Ministry of Justice had passed a law in March 1953 to set up a Department of Legal Medicine and his own recommendations had suggested procedures, personnel, and facilities that would improve Lebanon's medicolegal services, Smith warned against overconfidence in such measures if they were not backed up by a rigorous system of medicolegal training.[51] In terms of appointments to the new department, Smith wrote, "I am of the opinion that no appointments should be made until some arrangement for training be devised, and a nucleus of trained personnel is available. Even as an interim measure, it is doubtful if there can be any prospect of success in appointing untrained personnel to these important posts."[52] The necessity, to Smith's mind, of experience and thoroughgoing training systems was also evident in his recommendation of "The provision of training in the Institute for selected medical officers in pathology and forensic science prior to their appointment to the provinces . . . The provision in due course of postgraduate training for medical men and special training for police officers attached to the criminal Investigation Department."[53] It was here also, in the area of forensic education, that Smith appeared most clearly to assume the role of mobile imperial or perhaps postcolonial agent, not only advocating but also personally facilitating the extension of forensic networks between former

French and British colonies. His report recommended the creation of forensic ties between Lebanon and Ceylon, asking for "The appointment of a trainee to be trained in Colombo for approximately a year."[54] That Smith would personally help forge such a link, by means of the networks to which he was already connected in Ceylon, through his former students and the WHO, was evident in his offer to help personally train a Lebanese doctor while stationed in Colombo. Referring to the French School of Medicine's Dr. Simon Khalaf, who had impressed him with his good clinical training and enthusiasm for a career in forensic medicine, Smith wrote: "I would suggest that he be sent to Colombo on a WHO fellowship for training for six months or more under my direct supervision . . . If this were done it would mean there would be at least one well-trained officer available after some months. He would act as a second in command to the Director of the Institute and he could assist in training further personnel . . . from time to time one could be sent abroad for additional training."[55]

In this way, not only might a link with Ceylon be forged in the person of Dr. Khalaf, but also an enduring system of training and knowledge exchange might be established between Lebanon and other nations as further trainees gained experience abroad.[56] But even though it was true that Ceylon offered excellent opportunities for medicolegal training, given both the frequency of serious crime in that country and the excellent facilities available within the Forensic Medicine Department of the University, at the time of Smith's report to the Lebanese government, Sinhalese legal medicine was itself in disarray, having lost the connection it had once possessed between practice and teaching.

During the 1930s and 1940s, Smith's protégés, Sinnadurai and de Saram, were instrumental in transporting forensic knowledge acquired in Britain back to Colombo, where it helped shape medicolegal practice in their combined roles as judicial medical officer and lecturer in forensic medicine. But the flow of forensic knowledge between Britain and Ceylon during this period had not been unidirectional. With the appointment of the retiring government analyst in Ceylon, C. T. Symons, as adviser to the home office on the application of scientific aids to police work in February 1935, for example, it became apparent that the experience afforded by forensic work in colonies such as Ceylon—which went beyond chemical analysis to include examination of wounds, bloodstains, dust, mud, paint, documents, and counterfeit currency—was regarded as a valuable addition to forensic knowledge developed in the British context.[57] When Smith arrived in Ceylon, he contrasted his criticism of the island's medicolegal services with praise for Symons and the Government Analyst's Department, which he said had "for many years, been of considerable assistance to the Authorities in criminal investigation."[58] Smith went on to say that much of the work one would expect to be carried out by medical experts, including the examination of blood, hairs, fibers, and weapons causing wounds, had in fact been carried out by the government analyst because of the lack of trained medical officers.[59] This was long a problem in Ceylon's provinces, but it was of more recent vintage in Colombo, dating from the split in 1951 of the roles of judicial medical officer and professor of forensic medicine.

In 1933, dissatisfied with the lack of medicolegal training among the island's medical officers, the head of the Ministry of Health sent Dr. Sinnadurai to Edinburgh to train with Smith.[60] Becoming a judicial medical officer and part-time lecturer in forensic medicine on his return to Ceylon in 1935, it was hoped that Sinnadurai would be in a position not only to improve the facilities available for medicolegal work, of which there was much in Colombo, but also to train many more judicial medical officers, some of whom could be sent to the provinces. Too reliant on one individual, this plan was compromised when Sinnadurai died in 1945 and was further undermined when his successor, de Saram, was obliged to relinquish the position of judicial medical officer in order to take up the university's chair of forensic medicine in 1951.[61]

Thus on Smith's arrival in Ceylon he found, as in Lebanon, medical officers, particularly in the provinces, without specific medicolegal training, as well as a lack of faith on the part of the legal profession in forensic medical services, cases of bribery and corruption among medical officers, and an illogical division of medicolegal work and facilities between the judicial medical officer, the university's forensic department, and the government analyst's office. In his report to the WHO, therefore, Smith stated, "I am of the opinion that an efficient Medico-Legal Service can be assured only if THE VARIOUS DEPARTMENTS DEALING WITH THESE MATTERS IN COLOMBO ARE MERGED; THE PROVINCIAL JUDICIAL MEDICAL OFFICERS ABSORBED INTO THE SERVICE; THE WHOLE OF THE SERVICES PLACED UNDER THE DIRECTION OF ONE HEAD AND THAT ALL BE RESPONSIBLE TO THE MINISTRY."[62]

Discussion of Smith's assessment of the state of forensic medicine in Ceylon at two conferences of the Sinhalese ministries, officers, and personnel involved in the provision of the island's medicolegal services and revealed discontent with parts of the report and some of its recommendations. The police, for instance, were unhappy with Smith's comments about Ceylon's crime rate, which they felt ignored the work they had done to suppress crime.[63] Others disliked the idea of a centralized medicolegal institute, pointing to the lack of a similar institute in England and even suggesting that the postgraduate teaching of legal medicine was largely unnecessary. The strongest resistance, however, was reserved for the recommendation that the role of the head of the new institute be combined with that of the professor of forensic medicine.[64] There was some suggestion by those who opposed such a move that the whole idea of the new institute was an attempt by Smith to create a position for de Saram.[65] By the time Smith left Ceylon, sending his report to the WHO, it was clear that while an institute and new morgue might eventuate in the not-too-distant future, the reunification of practice and teaching would remain an unfulfilled desire as far as medicolegal services in Ceylon were concerned.

## Egypt Again

On a return trip from the antipodes in 1955, Smith spent three months in Egypt, during which time he gave lectures and attended meetings with staff at the university.[66] Smith's role as a mobile agent of forensic knowledge came full circle, with his experiences in Britain, Ceylon, Egypt, and Lebanon informing his critique of the current state of Egyptian legal medicine and his suggestions for its reform. The department that he had done much to develop and make into one of the world's best medicolegal institutes was apparently suffering from much the same ailment that had afflicted their Sinhalese counterparts, having separated teaching from actual medicolegal practice around the time of Smith's departure from Cairo for Edinburgh in 1928. That this separation was understood to be problematic among both practitioners and administrators both locally and internationally had been evident for some time. A 1937 Home Office analysis of the work of the Cairo laboratory, based on an Egyptian government report for the period 1929 to 1934, for example, noted, "This whole Egyptian scheme makes no provision for teaching. The whole time is taken up with actual cases."[67] Similarly, the pathologist Keith Simpson, when describing his own trip to Cairo in March 1950, stated that the medical school and Government Public Mortuary, where medicolegal autopsies were carried out, were distant from one another, and the practice of allowing forensic autopsies to be carried out at teaching hospitals by university staff, a practice cultivated to great advantage by London coroners, unfortunately did not exist in the Egyptian capital.[68] Smith, having just attempted rectification of this problem in Ceylon, could thus not help but express his disappointment in the state of legal medicine in Cairo during his stay, encouraging his former colleagues to address the unnatural separation of teaching and practice.[69]

In a report that he supplied to the university in 1955, which borrowed heavily from his 1954 report to the WHO on Ceylon, Smith argued that the separation of the teaching of legal medicine from its practice in 1928, at which time the Medico-Legal Department of the Ministry of Justice was cut off from the university, had severely handicapped the promotion and interest of research into medicolegal problems.[70] To Smith's mind, there was no better way of keeping practitioners up-to-date than by allowing them to come into contact with students in their daily work. Students, similarly, required practical experience in medicolegal work in order to become competent practitioners. On the consequences of this lack of practical experience, Smith wrote, "This must inevitably lead to a lowering of the standard of training of medical students in Legal Medicine, which would seriously handicap them in their future practice, as it is almost certain that all of them will be involved in some medicolegal case at one time or the other. The position would become even more serious if such an inefficiently trained practitioner gets a post in the Medico-Legal Department, where he will be responsible for this special type of work, and where upon his unsupported opinion, the liberty or the life of an accused person may depend."[71]

Shortly after his return to Edinburgh, Smith received a letter from M. A. Soliman, who assured his correspondent that he intended to follow up on his recommendations. Soliman wrote, "Meanwhile, we shall be trying our best to change things which made you a bit disappointed about medicolegal work in Egypt and with the stimulus you gave us, we shall never yield to anything beyond total and complete union of medicolegal practice and teaching in the universities. If we succeed and I hope we will, the credit will always be yours and everybody will know that Sir Sydney Smith who started the modern scientific medicolegal work in Egypt, came in due time to lift it up from a deep pitfall."[72]

While the impact of Smith's attempts to shape medicolegal practice and institutions in Britain's colonies and beyond was perhaps ultimately limited by political and financial interests over which he had no control, particularly after these colonies won their independence, it is nonetheless clear that as a "mobile forensic agent" he had some success not only in

sharing both local and universal knowledge between colonies via publication, correspondence, teaching, and advising, but also in molding medicolegal practice from afar. The networks of which Smith was a part and
that he helped forge within the British Empire to a large extent survived
the immediate impact of decolonization, expanding beyond their former
confines to link new nations to forensic knowledge. Although Smith's story,
despite his mobility and his intellectual links to other place and practitioners, provides only a limited perspective on the nature of medicolegal
and forensic networks in the twentieth century, it does provide a starting
point for further exploration. This chapter has made apparent, I hope, what
a fruitful avenue for research the tracing of such networks has the potential to be.

NOTES

1. John Glaister, *Final Diagnosis* (London: Hutchinson, 1964), 57.

2. Glaister, *Final Diagnosis*, 57–58.

3. On the institutions and structures of Scottish forensic medicine as well as Glaister's
training and early career, see M. Anne Crowther and Brenda White, *On Soul and Conscience:
The Medical Expert and Crime* (Aberdeen: Aberdeen University Press.)

4. Scholarship on how scientific knowledge is circulated considers a variety of ways in
which information travels, highlighting in particular the movement of people, texts, artifacts, and specimens. On the circulation of scientific knowledge in general, see, e.g.,
James A. Secord, "Knowledge in Transit," *Isis* 95, no. 4 (2004): 654–72. For an example of the
circulation of scientific knowledge in and between colonies, see Jan Golinski, "From Calcutta to London: James Dinwiddie's Galvanic Circuits," in *The Circulation of Knowledge
between Britain, India and China: The Early Modern World to the Twentieth Century*, ed.
Bernard Lightman, Gordon McOuat, and Larry Steward (Leiden: Brill, 2013), 75–94.

5. Alison Adam, *A History of Forensic Science: British Beginnings in the Twentieth Century*
(Florence, KY: Taylor and Francis, 2015), 67–72, 159–61.

6. A good example of the adaptation of Western criminological and criminalistics knowledge to a colonial context is that of the translation and adaptation of Hans Gross's *Handbuch
für Untersuchungsrichter als System der Kriminalistik* [Handbook for examining magistrates
as a system of criminalistics] (Munich: Schweitzer Verlag, 1893) by two Madras-based lawyers,
John and J. Collyer Adam, for use in India as *Criminal Investigation: A Practical Handbook
for Magistrates, Police Officers and Lawyers* (Madras: A. Krishnamachari, 1906). Alison Adam
has explored this example in some depth in *History of Forensic Science*, 67–72.

7. Simon A. Cole, *Suspect Identities: A History of Fingerprinting and Criminal Identification* (Cambridge, MA: Harvard University Press, 2002); Chandak Sengoopta, *Imprint of the
Raj: How Fingerprinting Was Born in Colonial India* (London: Pan, 2004).

8. Sydney Smith, *Forensic Medicine: A Textbook for Students and Practitioners* (London:
J & A Churchill, 1925), 475; Alfred Lucas, *Forensic Chemistry* (London: Edward Arnold, 1921),
179.

9. Tony Ballantyne, *Webs of Empire: Locating New Zealand's Colonial Past* (Vancouver:
University of British Columbia Press, 2012), 27.

10. Ballantyne, *Webs of Empire*, 27–28.

11. On the way in which scientific networks could contribute to imperial consolidation and expansion, see, e.g., John McAleer, "'A Young Slip of Botany': Botanical Networks, the South Atlantic and Britain's Maritime Worlds, c. 1790–1810," *Journal of Global History* 11 (2016): 24–43.

12. David Arnold, *Science, Technology and Medicine in Colonial India* (Cambridge: Cambridge University Press), 9–13. For recent scholarship that challenges the centrality of the metropole within scientific and medical networks, see Lightman et al., *Circulation of Knowledge between Britain, India and China*; McAleer, "'A Young Slip of Botany,'" 25–27.

13. David Arnold, "Introduction," in *Warm Climates and Western Medicine: The Emergence of Tropical Medicine, 1500—1900*, 2nd ed. (Amsterdam: Rodopi, 2003), 11; Deborah Neill, *Networks in Tropical Medicine: Internationalism, Colonialism, and the Rise of a Medical Speciality, 1890–1930* (Palo Alto, CA: Stanford University Press, 2012), 3; Maureen Malowany, "Unfinished Agendas: Writing the History of Medicine of Sub-Saharan Africa," *African Affairs* 99 (2000): 326–27.

14. Neill, *Networks in Tropical Medicine*, 3.

15. Markku Hokkanen, "Imperial Networks, Colonial Bioprospecting and Burroughs Wellcome & Co.: The Case of Strophanthus Kombe from Malawi (1859–1915)," *Social History of Medicine* 25, no. 3 (2012): 593.

16. Hokkanen, "Imperial Networks, Colonial Bioprospecting," 603.

17. Alison Adam also notes the limited amount of biographical material, largely limited to obituaries, available on early twentieth-century forensic scientists and physicians. See her *History of Forensic Science*, 169.

18. Other examples of autobiography and memoir include works by Frank Camps, John Glaister Jr., and Keith Simpson. On the role of autobiography in shaping forensic self-images, see, in particular, Nicholas Duvall, "Forensic Medicine in Scotland, 1914–39" (PhD thesis, University of Manchester, 2013), 210–28; Adam, *History of Forensic Science*, 168–76.

19. Smith's papers are located at the Royal College of Physicians in Edinburgh; hereafter RCPE.

20. On Smith's time in Lebanon and Ceylon, see RCPE, SMS/7/76 and SMS/7/77–83.

21. Sydney Smith, *Mostly Murder* (London: Companion Book Club, 1959), 53–54. On Cairo's medical school under British colonial rule, see Hibba Abugideiri, *Gender and the Making of Modern Medicine in Colonial Egypt* (Farnham: Ashgate, 2013), chap. 4.

22. Glaister, *Final Diagnosis*, 57; Sydney Smith, "Medicolegal Institute, Ministry of Justice of the Egyptian Government," in *Methods and Problems of Medical Education*, 9th ser. (New York: Rockefeller Foundation, 1928), 51.

23. On Egypt's Medico-Legal Institute and the legal system of which it was a part, see Smith, "Medicolegal Institute," 49–51.

24. Smith, *Mostly Murder*, 66.

25. Smith, "Medicolegal Institute," 51–52;

26. Smith, *Forensic Medicine*, 476.

27. Smith, *Forensic Medicine*, 479.

28. Smith, "Medicolegal Institute," 52.

29. Smith, *Forensic Medicine*, 471.

30. Smith, *Forensic Medicine*, 475.

31. Lucas, *Forensic Chemistry*, 203.

32. Lucas, *Forensic Chemistry*, 177.

33. Lucas, *Forensic Chemistry*, 179.

34. Smith, *Forensic Medicine*, 473.

35. Smith, "Medicolegal Institute," 53–54; Smith, *Mostly Murder*, 63.

36. On Smith's novel approach to the question of firearms, see Duvall, "Forensic Medicine in Scotland," 133–36.

37. RCPE, SMS/1/1/1–20.

38. Sydney Smith, "The Identification of Firearms and Projectiles: As Illustrated by the Case of the Murder of Sir Lee Stack Pasha," *British Medical Journal* 1, no. 3392 (January 2, 1926): 8–10. The paper given at the Medico-Legal Society was published as: Sydney Smith, "The Investigation of Firearm Injuries," *Transactions of the Medico-Legal Society* (1928–29): 81–106;

39. See, e.g., Dr. A. I. Kayssi, Director of Medico-Legal Department, Baghdad to Sydney Smith, May 17, 1951, RCPE, SMS/1/1/61–80.

40. Medical Laboratory, Zomba, Nyasaland to Sydney Smith, August 29, 1933, SMS/1/1/21–40.

41. Office of the Director of Medical and Sanitary Services, Colombo to Sydney Smith, February 21, 1935, RCPE, SMS/7/77; Sydney Smith to Dr. F. S. Fiddes, Edinburgh, November 17, 1953, RCPE, SMS/1/1/88.

42. "Gerald Samuel William de Saram: 17 April 1896–18 June 1963," *Journal of Pathology and Bacteriology* 89, no. 1 (1965): 411–14.

43. Sydney Smith to Dr. F. S. Fiddes.

44. Sydney Smith to G. S. W. de Saram, September 29, 1952, RCPE, SMS/7/79.

45. Documentation relating to his position as a WHO medicolegal adviser can be found at the RCPE, SMS/7/77.

46. "Medico-Legal Services in Lebanon, November 1953," RCPE, SMS/7/76, 1–2.

47. "Medico-Legal Services in Lebanon, November 1953," 2.

48. "Medico-Legal Services in Lebanon, November 1953," 4.

49. Smith, "Medicolegal Institute," 49.

50. "Medico-Legal Services in Lebanon, November 1953," 5.

51. "Medico-Legal Services in Lebanon, November 1953," 3.

52. "Medico-Legal Services in Lebanon, November 1953," 3.

53. "Medico-Legal Services in Lebanon, November 1953," 6.

54. "Medico-Legal Services in Lebanon, November 1953," 6.

55. "Medico-Legal Services in Lebanon, November 1953," 5.

56. Shortly after sending his report, Smith received word that his recommendations were being considered seriously with a decision not to launch the Ministry of Justice's proposed Department of Legal Medicine during 1953 or 1954, as well as a decree, while the government investigated the possibility of implementing his proposed plans, which would ensure that medicolegal cases were only examined at the laboratories of the two universities. N. N. Beyhum, Ministry of Public Health, Beirut, Lebanon to Sydney Smith, November 3, 1953, RCPE, SMS/7/76.

57. "Memorandum, 24 August 1934," National Archives, Sub-File: 679973/1, HO45/17085. On the scientific aids movement and Symons's role, see Adam, *History of Forensic Science*, 118–48; Ian Burney and Neil Pemberton, *Murder and the Making of English CSI* (Baltimore: Johns Hopkins University Press, 2016), 100–152.

58. Sydney Smith, "World Health Organisation, Project Ceylon—Report, March 1954," RCPE, SMS/7/76, 7.

59. Smith, "World Health Organisation, Project Ceylon," 7–8.

60. Smith, "World Health Organisation, Project Ceylon," 3.

61. While this initially caused few problems because close cooperation continued between the judicial medical officer and the university, the head of the Department of Health intervened later in 1951 to stop such collaboration, deeming it inappropriate. Smith, "World Health Organisation, Project Ceylon," 3–5.

62. Smith, "World Health Organisation, Project Ceylon," 22.

63. "Notes of a Conference Held on Friday, 13 May, 1954, at 9am. Re. the Creation of a Medico-Legal Service," RCPE, SMS/7/77.

64. "Notes of a Conference Held on Friday, 13 May, 1954."

65. Nasa to Smith, August 19, 1954, RCPE, SMS/7/79.

66. Smith, *Mostly Murder*, 318.

67. This analysis was undertaken as part of an assessment of how feasible a London Medico-Legal Institute would be. "Egyptian Medico-Legal Institute, Report on Work, 1929–1934," Sub-file: 694370/10, HO45/17568, National Archives, Kew.

68. Keith Simpson, "Forensic Medicine in Egypt," *Guy's Hospital Gazette* 64, no. 1620 (1950): 334–35.

69. M. A. Soliman, Forensic Medicine Department, Cairo to Sydney Smith, May 7, 1956, RCPE, SMS/1/1/130.

70. "Report by Prof. Sir Sydney Smith to the Ain-Shams University, 1955," cited in Mohamed Emara and Mohamad Ahmad Soliman, *Forensic Medicine and Toxicology*, 5th ed. (Cairo: Dar El-. Kitab El-Arabi Press, 1961), 3–4.

71. "Report by Prof. Sir Sydney Smith to the Ain-Shams University, 1955," 4.

72. M. A. Soliman, Forensic Medicine Department, Cairo to Sydney Smith, May 7, 1956, RCPE, SMS/1/1/130.

# A Tale of Two Cities?

## Locating the History of Forensic Science and Medicine in Contemporary Forensic Reform Discourse

SIMON A. COLE

In July 2015, I attended the Locating Forensic Science and Medicine Conference in London, from which the volume that is now in your hands (or on your screen) was developed. I arrived at the conference having flown overnight from another conference on forensic science with a quite different location in Washington, DC.

The conference in Washington was convened by the National Institute of Standards and Technology (NIST)—they are the people in change of weights and measures in the United States. The conference was called "Forensic Science Error Management."[1] That I was the only person to make the transit between these two conferences provoked me to reflect upon the relationship between forensic history and what we might call "contemporary forensic reform," as well as on my own role as a historian of forensic science who has fortuitously become an actor in contemporary forensic reform.

I am interested in the way that the past is represented in artifacts from contemporary forensic reform discourse, like the NIST conference. I would suggest that the conference was typical of such discourse in two principal ways: (1) in portraying the present historical moment—which I date as the period from 2009, when the landmark US National Research Council (NRC) report *Strengthening Forensic Science in the United States*[2] was published, to the present—as a moment of new beginning, a fresh start for forensic science; (2) in portraying the history of forensic science, even until very recently, as a sort of "dark age" that either will, should, must, or might be swept away by this new beginning. The sense of history-in-the-making was palpable in Washington: the conference was billed as "the first-ever international symposium devoted exclusively to the topic of forensic science error management." Is it true that no one thought much about forensic error until now? My own work has shown that there is some truth to that—that fingerprint examiners at least, and some other disciplines as well, were claiming to have a "zero error rate" well into the twenty-first century. The conference website's characterization of forensic error as a "taboo topic" suggests that the NIST conference was in some sense meant to commemorate the liberation of the notion of error in forensic science. Somewhat extraordinarily, this 2015 conference was meant to mark the historical moment when interested parties could finally talk about error in forensic science without immediately getting bogged down in argument about whether such a thing even exists.

This way of locating forensic history is not confined to the United States. Consider a suite of papers published around the same time in the *Philosophical Transactions of the Royal Society* as the output of "the adventure that culminated in the Royal Society's twin events designed to enable and encourage the forensic science community to come together to discuss and plan a vision for the future, not only for the UK, but as partners in a global ecosystem." The editors' introduction is titled "Time to Think Differently: Catalysing a Paradigm Shift in Forensic Science," signaling both the initiation of a new historical period and the obligatory reference to pop-Kuhnianism. The sense of a pivotal historical moment is palpable when the editorial declares: "There is no doubt that the forensic science ecosystem stands at a critical crossroads and there must be a common responsibility taken for the changes that need to be enforced. Of one thing we are absolutely certain, our current path is destined for disaster if we choose to carry on simply doing more of the same."[3] In a rather extraordinary outburst of millenarianism, the editors' introduction then announces:

> The time has come to reject the inadequacies of the past and embrace a healthy new culture that can steer our ecosystem into calmer and more productive waters, a culture of confidence and professionalism which supports openness and trust, where research, technology, leadership and workforce development are all valued as a collegiate part of the holistic community. If this is the bright future that we wish to achieve and the legacy we wish to leave behind, then we must genuinely work together and not let the current woes of the discipline and the egos of an intransigent old guard smother the green shoots of a paradigm shift that broke new ground in the octocentennial year of the Magna Carta.[4]

In another article, the same authors invoke the widespread sense of crisis that pervades contemporary forensic science:[5] "There is no doubt that forensic science is in crisis, and it currently faces its most uncertain future.

However, our future is in our own hands and what we, as a criminal justice community, choose to do next will be our legacy."[6]

Of course, the notion that today is the period of forensic reform is itself ahistorical. Forensic science has been arguably reforming itself throughout its history. But I do think the present moment can be characterized as one in which a forensic reform effort of a certain kind at a certain scale is occurring. Although the seeds of the present forensic reform movement are decades old, it is still useful to treat 2009 as a sort of watershed moment when the impetus to reform was taken seriously, by players that mattered—and in this case, that appears to be primarily governments, with a somewhat smaller role played by scientific institutions like the US National Academies, the American Association for the Advancement of Science (AAAS), the Royal Society, the American Statistical Association, and so on. Just to review: in the United States, in response to the NRC report, the White House Office of Science and Technology Policy created a Subcommittee on Forensic Science with a suite of working groups; a National Commission on Forensic Science was created; NIST created an elaborate Organization of Science Area Committees (OSACs), consisting of committees and subcommittees populated by more than four hundred individuals; the AAAS is carrying out a "gap analysis" of scientific research in forensic science; and more.

I don't want to overstate the erasure of history in contemporary forensic reform discourse. Many contemporary forensic thinkers write with a profound sense of history.[7] Jeffrey Jentzen's contribution to this volume is an excellent example. David Stoney's recent review paper, which attempts to lay out future directions for trace evidence, begins with an extensive historical review of debatable necessity for his overall argument.[8] Alex Biedermann and James Curran recently wrote a devastating critique of another article. When I wrote Biedermann to congratulate him on the critique, he responded, "My main motivation behind this was to keep some historical sources alive." And indeed that is what his critique does.[9]

Nonetheless, I do want to argue that, in general, contemporary forensic reform discourse *locates* the history of forensic science in a past that is somehow clearly distinct from the reformist present. There are perhaps several presumed bases for the disjuncture between past and present, but chief among them is the rise of statistical thinking in forensic science—the notion that we have finally come around to conceiving of all forensic evidence in a probabilistic manner. And, obviously, this has something to do with the NIST conference's notion that it is now—just now—acceptable to talk about uncertainty in forensic science. But other presumed bases may be the involvement of important actors in reforming forensic science. Governments come to mind, but also scientific institutions like the National Academies and the Royal Society.

Having myself transited between the two worlds, I want to suggest that these two groups thinking about forensic science might be able to do more for one another than either has entirely realized. The historical work presented in this volume can offer a useful perspective on the discussions that are occurring in Washington. The voluminous contemporary discussion of the 1993 US Supreme Court decision *Daubert v. Merrell Dow*, for example, might be informed by Marcus Carrier's account of the development of implicit set of "standards" for the use of German forensic toxicology in court. The very topic of the Washington conference—forensic error—is fraught by its fundamental unknowability: we know about forensic error because of a few exposed cases, but we don't know about the unexposed cases.[10] Contemporary discussions of this issue might be informed by Mitra Sharafi's comments about the "limits of the archive" in detecting instances of the fabrication of blood evidence in colonial India. Contemporary discussions of the urgent need to apply algorithms and statistics for forensic science might be informed by Projit Mukharji's discussion of the "steady push toward mathematization and mechanization of forensic work" among a dynasty[11] of document examiners in colonial India. Lawyers and scholars who are agitated about the inability of criminal defendants to inspect the proprietary algorithms that produce an increasing proposition of forensic evidence might be informed by Binyamin Blum's description of the inability to access the reasoning process behind the conclusions of tracking Doberman Pinschers in early twentieth-century Palestine. For a different case study, one might consider Gagan Singh's account of Punjabi trackers who as humans were considered partially accountable for their knowledge, but as natives were also considered partially inscrutable. The use of tracking evidence without validation presages contemporary concerns about the use of forensic techniques absent validation. Contemporary debates over the increasing ubiquitous use of biometric identifiers might be informed by José Ramón Bertomeu-Sanchez's discussion of Federico Olóriz's efforts to develop a national fingerprint register in early twentieth-century Spain. And the combatants in the battles over the abusive head trauma diagnosis for infant death (known as "shaken baby syndrome") would be informed by Bruno Bertherat's discussion of Ambroise Tardieu's theories about infanticide in late nineteenth-century France. Those

concerned with the effect of American politics on the politics of forensic science (e.g., the failure to renew the National Commission on Forensic Science;
see below) may be informed by Trais Pearson's account of forensic science as
"civic epistemology" in Thailand.[12]

In a somewhat different vein, as a new member of the team of scholars who
edit the National Registry of Exonerations, currently the definitive source on
wrongful convictions in the United States, I read with growing unease Ian
Burney's cultural history interpretation of our distant ancestor, Erle Stanley
Gardner's Court of Last Resort. If, as Burney argues, Gardner's efforts to expose wrongful convictions took on the cultural form of the "frontier posse,"
how will future cultural historians interpret the efforts of my co-editors and
me, who naively imagine ourselves as rather less flamboyant and culturally
neutral historical actors?

Here I want to emphasize two ways that I think location is important in
contemporary forensic reform discourse, ways that will be important in
shaping the role that forensic science plays in future society. The first
locational issue concerns what we might call the institutional location of
forensic science. One of the recommendations—perhaps the principal
recommendation—of the NRC report was to *relocate* forensic science, that
forensic laboratories should be removed from law enforcement. Second,
the report recommended that the new regulatory agency that it posited
should be created in order to reform American forensic science must not be
located in the US Department of Justice (DOJ) or any other law enforcement
or prosecutorial agency. The NRC report explicitly considered a variety of
locational possibilities for this hypothetical new agency—called the "National
Institute of Forensic Science," or NIFS—rejected them all, and pronounced
that the NIFS must be an independent agency.

These recommendations are interesting because the location of forensic
science in law enforcement seems to me to be precisely one of those institutional formations that exist only because a certain contingent set of historical circumstances has made it so. To me, there is reasonably broad assent—
I discuss an exception below—regarding the notion that, in a perfect world
being designed de novo today, forensic science would *not* be located in law
enforcement. It would have some sort of institutional location that more
strongly conveyed its allegiance to science, rather than to law enforcement.
It is harder to reach agreement that it is possible, feasible, or desirable to
change the institutional location of forensic science *now* that it has been
historically embedded in law enforcement. At least one recent work in

contemporary forensic reform claims that the embedding of forensic science in the "prosecutorial entities dates back as far as 13th century China."[13] The chapters of this volume trace many of the small parts of that history by which forensic science got embedded in the state, and more specifically law enforcement.

"Locating Forensics" took place in the United Kingdom. The centrality of the British Empire as a network for the dissemination of forensic knowledge and techniques is explored in Heather Wolffram's "Forensic Knowledge and Forensic Networks in Britain's Empire: The Case of Sydney Smith" (chap. 11, this volume). In the United States, the United Kingdom represents the epitome of locating forensic science outside law enforcement, in a sense: the privatization of the Forensic Science Service (FSS) into a sort of quasi-state, for-profit entity; the subsequent disastrous closing of the FSS; and the resulting devolution of forensic service to either (1) truly private, for-profit corporations or (2) true law enforcement locations inside local police forces. Interestingly, another model of an independent forensic laboratory recently imploded, with the firing of the leadership of the District of Columbia crime laboratory.

In the end, the NCFS was located at the intersection of the DOJ and NIST; it was jointly administered by the two agencies. The NCFS was officially chaired by the deputy attorney general of the United States. A recent change in leadership of the DOJ, then, brought about the precipitous demise of the NCFS, and its replacement by a Forensic Science Working Group controlled entirely by the DOJ and consisting entirely of people who work for the DOJ.[14] Thus the creation of the NCFS, which was in some sense the realization of the forensic reformist dream expressed most prominently by the NRC report, had another side to it. Contrary to what the NRC report envisioned, the NCFS has more firmly embedded forensic science, first in the law-enforcement-prosecutorial complex and, more broadly, in the state in general. Current events may cause us to wonder whether we will see the increasing concentration of forensic science in the state, even if we can also discern a concomitant trend toward private, for-profit enterprise as well.

The second locational issue is not unrelated. The NRC report specifically— and, I would argue, contemporary forensic science reform discourse generally—has been taken to task from many different angles for presuming that the location of forensic science is in the laboratory. And I think that is certainly a fair characterization of both the report and much of contemporary discourse. These critiques argue that the true location of forensic science is at least equally, if not more so, the crime scene. These critiques argue that contemporary forensic reform discourse has taken on the relatively easy task of trying to reform laboratory procedures—and make them more like familiar, mainstream scientific laboratory procedures—while evading the far more difficult task of reforming the handling of crime scenes. The investigation of crime scenes, after all, is a peculiar activity that is not easily categorized as either science or policing. As such, it does not have ready analogues in mainstream scientific activity the way that laboratory analyses do. And yet, as these critics points out, most forensic analyses start with recovery from the crime scene.

At least three streams of such criticisms are apparent. The first is represented by Jennifer Laurin, a law professor who, from a legal perspective, argues that contemporary forensic reform discourse gives short shrift to crime scenes, with legal implications that I don't have time to discuss here.[15] The second was represented at the NIST conference by Peter DeForest, who I think would not be insulted if I described him as an "old-school" generalist criminalist and who was trained by Paul Kirk (a character in Ian Burney's "Spatters and Lies: Contrasting Forensic Cultures in the Trials of Sam Sheppard, 1954–66," chap. 4, this volume) in the legendary, now-defunct criminalistics program at the University of California, Berkeley. As DeForest put it, "laboratory scientists have become increasingly disconnected from the scene investigation." He argued that "forensic scientists must be in control of their own investigations."[16] In short, he seeks to physically relocate forensic scientists from the laboratory to the crime scene.

The third stream is the forensic school of thought that goes under the title "intelligence-led policing" (ILP), the most prominent advocates of which are Swiss and Australian forensic scientists such as Olivier Ribaux, James Robertson, and Claude Roux. These thinkers contend that historically forensic science has become too oriented around the law and the trial, and not enough around other sorts of state actions that can be responsive to information about criminal activity, such as policing or the rather vaguely defined suite of activities that today we call "security."[17]

As an aside, the ILP literature falls into the category of forensic literature that I described above with a strong sense of history. Intelligence-led policing sees itself as the true fulfillment of the future envisioned by figures like Hans Gross and Edmond Locard. This vision, they argue, got derailed by a number of factors during the twentieth century, but the principal one seems to have been the differentiation, rather than integration, of the roles of police officer and scientist. I have described ILP as "unabashedly nostalgic" in an article. I was sad to see that that was interpreted as an insult. I am confident that historians understand that it was intended as, at worst, a neutral observation and perhaps even as a compliment.

Without going into too much detail, the vision of ILP is one that is centered on exploiting all traces at the crime scene that might have intelligence value and deploying that information for the state security apparatus, while focusing much less on criminal prosecutions and trials. One fascinating thing about ILP is that it pushes forensic science in the opposite direction as that of human factors discourse, which is today probably the more dominant discourse, at least in the United States. Rather than arguing for isolating forensic scientists from the criminal investigation—and thus firmly locating them in the laboratory—ILP calls for more closely integrating them into the investigation. ILP does not deny the existence of bias; it simply believes that the benefits of integration outweigh the costs in bias. I think this is a philosophically defensible position—unlike the position that bias does not exist—even if I don't agree with it. But what really worries me about ILP is that its vision taken to its logical conclusion, in seeking to turn police into scientists, will probably result in turning all forensic scientists into police. ILP's orientation toward the state security apparatus looks a lot like a state police force.[18]

I predict we will see in the next few years more of this showdown between the ILP and the human factors visions, a showdown that has already begun. In part, this battle will concern the location of forensic science: at the crime scene or in the laboratory, in the police or "in science," whatever that means. However this battle goes, it illustrates just one way in which "locating forensic science" matters not just to the history of forensic science, but to contemporary forensic science as well.

## NOTES

1. See the 2015 International Forensics Symposium website, accessed July 27, 2018, https://www.nist.gov/director/2015-international-forensics-symposium.

2. National Research Council (NRC), *Strengthening Forensic Science in the United States: A Path Forward* (Washington, DC: NRC, 2009).

3. Sue Black and Niamh Nic Daeid, "Time to Think Differently: Catalysing a Paradigm Shift in Forensic Science," *Philosophical Transactions of the Royal Society B* 370 (2015): 20140251, 3.

4. Black and Daeid, "Time to Think Differently," 3.

5. Kelly M. Pyrek, *Forensic Science under Siege: The Challenges of Forensic Laboratories and the Medico-Legal Investigation System* (Amsterdam: Academic Press, 2007).

6. Éadoin O'Brien, Niamh Nic Daeid, and Sue Black, "Science in the Court: Pitfalls, Challenges and Solutions," *Philosophical Transactions of the Royal Society B* 370 (2015).

7. E.g., Christophe Champod, Keith Inman, Pierre Margot, Norah Rudin, David Stoney, and Franco Taroni and colleagues come to mind.

8. David A. Stoney and Paul L. Stoney, "Critical Review of Forensic Trace Evidence Analysis and the Need for a New Approach," *Forensic Science International* 251 (2015): 159–70.

9. Alex Biedermann and James Curran, "Drawbacks in the Scientification of Forensic Science," *Forensic Science International* 245 (2014): e38–e40.

10. Chief among these exposed cases, and prominently featured at the NIST conference, is the Mayfield fingerprint misidentification. Sharia Mayfield and Brandon Mayfield, *Improbable Cause: The War on Terror's Assault on the Bill of Rights* (Salem, NH: Divertir, 2015).

11. Mukharji's discussion of a family dynasty of document examiners evokes Sulner's discussion of his own family dynasty. Andrew Sulner, "Critical Issues Affecting the Reliability and Admissibility of Handwriting Opinion Evidence—How They Have Been Addressed (or Not) since the 2009 NAS Report, and How They Should Be Addressed Going Forward: A Document Examiner Tells All," *Seton Hall Law Review* 48, no. 3 (2018): 631–717.

12. See also Samson Lim, *Siam's New Detectives: Visualizing Crime and Conspiracy in Modern Thailand* (Honolulu: University of Hawai'i Press, 2016).

13. Suzanne Bell et al., "A Call for More Science in Forensic Science," *Proceedings of the National Academy of Sciences* 115, no. 18 (2018): 4541–44.

14. Spencer S. Hsu, "Sessions Orders Justice Dept. to End Forensic Science Commission, Suspend Review Policy," *Washington Post*, April 10, 2017.

15. Jennifer E. Laurin, "Remapping the Path Forward: Toward a Systemic View of Forensic Science Reform and Oversight," *Texas Law Review* 91 (2013): 105–18.

16. Peter DeForest, "Forensic Science—The Quality Assurer?," in *Proceedings of the 2015 International Symposium on Forensic Science Error Management*, ed. John M. Butler (Arlington, VA: National Institute of Standards and Technology, 2015).

17. E.g., Olivier Delémont, Sonja Bitzer, Manon Jendly, and Olivier Ribaux, "The Practice of Crime Scene Examination in an Intelligence-Based Perpective," in *The Routledge International Handbook of Forensic Intelligence and Criminology*, ed. Quentin Rossy, David Décary-Hétu, Olivier Delémont, and Massimiliano Mulone (London: Routledge, 2018), 86–101; Olivier Ribaux and Stefano Caneppele, "Forensic Intelligence," in *Routledge International Handbook of Forensic Intelligence and Criminology*, 136–48.

18. Simon A. Cole, "Forensic Science Reform: Out of the Laboratory and into the Crime Scene," *Texas Law Review* 91 (2013): 123–36.

BRUNO BERTHERAT is an assistant professor of contemporary history at the University of Avignon and a member of the Centre Norbert Elias. His research focuses on the history of identification, forensic medicine, and crime, as well as on the history of corpses and their representations, including funerary practices, in France. He is currently writing a book on the Paris Morgue in the nineteenth century.

JOSÉ RAMÓN BERTOMEU-SÁNCHEZ is the director of the Institute for the History of Medicine and Science at the University of Valencia. His research is focused on the history of forensic science in France and Spain (Research Grant HAR2012-36204-C02-02). His latest book, in Spanish, is *The Truth about the Lafarge Affair: Science, Justice and Law in the Nineteenth Century* (2016). He is now finishing a biography of Mateu Orfila, *Between the Prosecutor and the Executioner* (in press).

BINYAMIN BLUM is an associate professor at the Hastings College of the Law of the University of California. He teaches and researches in the fields of criminal law, criminal procedure, and evidence. His current book project, *Forensic Culture in the Age of Empire*, explores the colonial origins of forensic science.

IAN BURNEY is a professor of the history of science, technology, and medicine and director of the University of Manchester's Centre for the History of Science, Technology and Medicine. He is the author of several books on the history of forensic medicine and science, including *Bodies of Evidence: Medicine and the Politics of the English Inquest* (2000) and *Murder and the Making of English CSI* (2016), both published by Johns Hopkins University Press. He is now working on two book projects, one based on the Sam Sheppard case and the other on a history of "innocence."

MARCUS B. CARRIER holds a BA in history and chemistry (2013) and an MA in history, philosophy, and sociology of science (2016) from Bielefeld University. During his master's studies, he spent a semester in the History and Philosophy of Science Program at the University of Notre Dame. He is currently working on his PhD on the comparative history of nineteenth-century forensic toxicology in France and Germany, also at Bielefeld University.

SIMON A. COLE is a professor of criminology, law, and society and the director of the Newkirk Center for Science and Society at the University of California, Irvine. He is the author of *Suspect Identities: A History of Fingerprinting and Criminal Identification* (2001) and, with Michael Lynch, Ruth McNally, and Kathleen Jordan, *Truth Machine: The Contentious History of DNA Fingerprinting* (2008).

CHRISTOPHER HAMLIN is a professor in the Department of History and the Program in the History and Philosophy of Science at the University of Notre Dame and an honorary professor at the London School of Hygiene and Tropical Medicine. His research focuses on the application of knowledge to public needs, mainly in areas relating to health. In nearly six dozen articles and several books, he has examined concepts of disease and disease causation; forensic science and expert disagreement; the assessment of water and air; the regulation of environmental nuisances; social epidemiology, focusing on issues of hunger and exposure; and cultural and religious concepts of nature. His recent conceptual overview of the history of forensic science, "Forensic Cultures in Historical Perspective: Technologies of Witness, Testimony, Judgment (and Justice?)" was published in *Studies in History and Philosophy of the Biological Sciences* 44 (2013): 4–15. Recent books include *Cholera: The Biography* (2009) and *More Than HOT: A Short History of Fever* (2014).

JEFFREY JENTZEN is a professor of pathology at the University of Michigan, where he is the director of autopsy and forensic services. He has practiced forensic pathology for over thirty years and was the chief medical examiner in Milwaukee, Wisconsin. He holds a doctorate in the history of science from the University of Wisconsin–Madison. He has authored numerous publications on the practice and history of death investigation. He is the author of *Death Investigation in America: Coroners, Medical Examiners and the Quest for Medical Certainty* (2009). His forthcoming book for Harvard University Press is titled *Instruments of Empire: A Global History of Death Investigation in the Colonial Perspective.*

PROJIT BIHARI MUKHARJI is an associate professor in the history and sociology of science at the University of Pennsylvania. He was educated in India and the United Kingdom and works on the history of science in early modern and modern south Asia. His articles have appeared in several journals, including *Comparative Studies in Society and History, Journal of Asian Studies, Indian Economic and Social History Review,* and *Bulletin of the History of Medicine.* Mukharji is also the author of two monographs, *Nationalizing the Body* (2009) and *Doctoring Traditions* (2016).

QUENTIN (TRAIS) PEARSON received his PhD in Southeast Asian history from Cornell University. He is currently a visiting assistant professor of history at Boston College. His work has appeared in journals including *Modern Asian Studies* and *Bulletin of the History of Medicine.* His first book, *Sovereign Necropolis: The Politics of Death in Semi-Colonial Siam,* is forthcoming with Cornell University Press.

MITRA SHARAFI is a professor of law and legal studies at the University of Wisconsin–Madison, with history affiliation. She is the author of *Law and Identity in Colonial South Asia: Parsi Legal Culture, 1772–1947* (2013), which won the Law and Society Association's Hurst Prize in 2015. Sharafi is currently working on a book on the history of forensic science in India during the nineteenth and twentieth centuries.

GAGAN PREET SINGH, PhD, teaches at Indraprastha College at the University of Delhi. In his doctoral thesis, "Policing in Colonial Punjab," he studied the interaction between Western and local forms of policing practices. During his research, he was a visiting student to the King's College London and a recipient of the Charles Wallace India Trust Fellowship. He has published several research articles and reviews.

HEATHER WOLFFRAM is a senior lecturer in history at the University of Canterbury in New Zealand. She is the author of *The Stepchildren of Science: Psychical Research and Parapsychology in Germany, c. 1870–1939* (2009) and *Forensic Psychology in Germany: Witnessing Crime, 1880–1939* (2018). Her current project is a study of forensic medicine in interwar Egypt.